Supervision and Safety of Complex Systems

Supervision and Safety of Complex Systems

Edited by
Nada Matta
Yves Vandenboomgaerde
Jean Arlat

iSTE

WILEY

First published 2012 in Great Britain and the United States by ISTE Ltd and John Wiley & Sons, Inc.

ISTE Ltd
27-37 St George's Road
London SW19 4EU
UK

www.iste.co.uk

John Wiley & Sons, Inc.
111 River Street
Hoboken, NJ 07030
USA

www.wiley.com

Library of Congress Cataloging-in-Publication Data

Supervision and safety of complex systems / edited by Nada Matta, Yves Vandenboomgaerde, Jean Arlat.
pages cm
Includes bibliographical references and index.
ISBN 978-1-84821-413-2
1.System safety. I. Matta, Nada. II. Vandenboomgaerde, Yves. III. Arlat, Jean.
TA169.7.S87 2012
658.20028'9--dc23

2012024378

British Library Cataloguing-in-Publication Data
A CIP record for this book is available from the British Library
ISBN: 978-1-84821-413-2

Printed and bound in Great Britain by CPI Group (UK) Ltd., Croydon, Surrey CR0 4YY

MIX
Paper from responsible sources
FSC® C013604

Table of Contents

Foreword

With the development of science and technology, our societies ask for increasingly sophisticated systems that are optimized to better meet our needs while controlling costs. At the same time, our societies should beware of the drawbacks these systems may present, particularly of the risk they can pose to both people and the environment. Society demands that the safety of these complex systems be guaranteed in a normal situation and in the case of an incident or accident.

These issues involve several areas of industry. They obviously concern nuclear electricity generation, which society watches very carefully. France has created legislation on the continuing improvement of nuclear safety, thanks to specially periodic safety controls that take into account the evolution of knowledge. The accident in Fukushima reminded us that the simultaneous occurrence of events that are considered individually very unlikely is possible. Through the additional safety evaluations undertaken by the nuclear safety authority, the lessons are learned from this accident.

Research in safety is therefore essential, and I am delighted by the quality of the results gathered in the present book. I am particularly happy about the active participation of all the nuclear operators, including those who are in charge of the management of radioactive waste, the innovative nature of which is too often underestimated.

Eric BESSON
Former French Minister of Industry, Energy and Digital Economy

Foreword

Small is beautiful…

The world's population growth and its increasing needs, technological progress, the evolution of the economy, the "shortening" of space and time, etc. – i.e. the "globalization" of life and activities – have led to the development of large cities, to advanced optimizations of manufacturing processes implemented in increasingly large factories and to the daily use of equipment (public transport, nuclear power plants, thermal power plants, etc.) to its full extent. This can lead to the increased risks of large-scale accidents caused by a design issue, through the use of a facility outside its operating range, through human error, or through an act of sabotage.

Our society is currently very "fragile" and we have to do everything in our power, from the facility design-stage onwards to detect a slight drift from the operating parameters, to analyze its causes and the potential danger, thus enabling solutions to be implemented at the earliest possible time. Similar operational strategies must be carried out. The supervision of systems and the processing of collected information are not always straightforward. Finally, for all of the main problems that might occur, we must create an operational plan to be put in place in case of a major accident. French and European authorities, in particular, are working towards that goal.

Several university laboratories are leading research projects in the area of supervision and several aspects linked to safety and security. They also consider the role of human error in these projects. "GIS Surveillance, Sûreté et Sécurité", founded in 2007, gathers together several university laboratories and other research and industrial organizations to analyze and study approaches to the supervision, safety and security of complex systems by using complementary approaches. This book gathers together most of the studies they have led in the first four years of GIS.

There is no such thing without risk…

Christian LERMINIAUX
President of GIS 3SGS

Introduction

The "supervision, safety and security of large systems" theme is currently present at all levels, particularly at the European and national levels. It is relevant to every large facility, infrastructure or organization, whether public or private. It appears, however, that public and private research facilities currently tackle this theme from different angles and in a "scattered" way. This scattering can be explained by the strongly multidisciplinary nature of this area, which does not come under the umbrella of an academic subject. The gathering together of these structures and the industrialists concerned is the best response that can be taken for France to play a leading role in the worldwide development of this major and rapidly emerging field. It is within this framework that at the end of April 2007, MESR (*Ministère de l'Enseignement supérieur et de la Recherche* – French Ministry of Higher education and Research) founded the scientific interest group: "surveillance, sûreté et sécurité des grands systèmes" (supervision, safety and security of large systems) – GIS 3SGS.

This foundation was prompted by several university laboratories that had noticed that research on models of safety and security assessment is usually contextual, i.e. associated with an industrial object. Scientific initiatives are therefore not really structured around these themes: there is a true scientific challenge to create coherence between global models representing a system and models developed at the scale of a component or subsystem, the latter being necessary because of the source of industrial specifications and maintenance policies. These two scales are necessary, but strongly dependant.

The scientific problems in the area of surveillance, safety and security of large systems, which pose a high risk in the long term, are in constant evolution because of:

Introduction written by Yves VANDENBOOMGAERDE, Christian LERMINIAUX and Nada MATTA.

– the evolution and replacement of the systems currently in operation;

– the constant complexity of systems;

– the increasing demands in terms of industrial safety;

– the availability constraints; and

– the evolution of information and communication technologies.

All this requires the development of new methodologies, the design of new models, the implementation of new simulation methods, the diffusion of new knowledge, etc.

The supervision approach

Supervision plays a leading role in the operation of large systems. From the real-time analysis of data collected online, it requires quick decision-making and thereby implies the consideration of the time variable. Faced with the complexity of large systems, surveillance has to be robust in relation to the uncertainties and errors that are associated with both the models and the data. The surveillance methods need to be integrated to the systems of control and command (remote machine monitoring) and need to be improved by using information on the reliability of components as well as on the maintenance operations.

The increasing use of systems including software and digital data requires total control of all aspects of safety and security, often already required by current regulations. These aspects play a leading role in the adoption of digital systems in the economic, legal and the societal context, and are an essential factor for innovation, and hence for economic development.

The merging of the currently independent approaches towards surveillance issues (automatics, signal and computer science) requires the cooperation of actors from these different communities.

The safety approach

The control of risks requires a systemic approach when large complex systems are involved. Methods relating to operating safety are developed with this goal in mind and allow for this control from the point of view of the reliability, maintainability and security of large systems.

The development of models and methods of assessment and the optimization of the safety of large systems is therefore organized around several structuring themes:

– "Human factor" (HF) models that are adapted to large systems and/or supporting studies upstream from their design: the operator is a potential source of an undesirable event, but he or she is also a fundamental element in the control or recovery chain following such an event. It is therefore necessary to improve the existing HF models, in particular to help designers at the time of the upstream studies.

– The analysis of positive feedback (APF), or precursors, needs to be developed to make the system effective and efficient. In systems where several components interact that are based at different geographical localizations (several separate rooms, physical separation, etc.), the coexistence of varied human activities requires the use of models enabling us to ensure a common representation of the state of the facility or system. Such tools have yet to be created from an operational point of view. Similarly, the development of models and industrial tools enabling the *a priori* simulation of risky operations should be encouraged (for instance, studies such as decentralized multifactor operations).

– Modeling of the damage and ageing of systems and optimization of maintenance: the surveillance/maintenance interaction (conditional or predictive maintenance, *health monitoring*) where maintenance decisions are based on a diagnostic or prognosis of the state of the system, established from surveillance data.

– Safety of instrumented/programmed systems of security and command and control: safety, assessment and certification of software and systems, particularly in the case of embedded systems, in order to obtain fault-tolerant systems, amongst other things.

– Study of fault-tolerant systems with the intention of suggesting new methodologies for the reconfiguration of system control laws subjected to failures: the modeling and development of methods for reliable systems by using (semi)formal methods and demonstration tools in order to define the proven development cycle of a system with a predominant software from a specification, to design a system integrating software elements whose safety or security properties are guaranteed by design, forward-looking maintenance and collaborative maintenance (multiagent systems) for e-maintenance strategies.

The security approach

The issue of the security of large systems integrating software plays a leading role in the adoption of digital systems. The systems that are of interest to us can be "software", "hardware" or "hybrid" based, as in the case of process control systems.

The security of systems is of increasing importance in the prevention of and protection against hacking by the incorporation of confidentiality, access control or anonymity, and by controlling the information flow and its coherence with respect to individual freedom and national constraints.

To address issues regarding security, it is essential to define security policies and their mechanisms of implementation. The definition of a security policy is important since it determines an acceptable level of security. The actions will concern the proven security of services, secured protocols, cryptography, computer virology, the validation of services, the management of certifications and revocations. Particular attention should be paid to the control strategies of resources dependent on economic models, without forgetting the methods of identification and authentication and the control of information management.

The concepts of surveillance, safety and security are complementary and strongly interact. There are tight links between aspects of security (resistance to acts of sabotage) and safety, whose operation is validated against unintentional faults. These links must be developed. The problems of the two systems are often interwoven. As an example, coding errors are taken advantage of to create security breaches: it is therefore also necessary to formally prove the correction and robustness of security. On the other hand, the plans for the continuity of activity are essential. Their design emphasizes a necessary link between safety and security.

The contributions of this area to global security are characterized by two aspects. The first concerns the adaptation of risk analysis methods for the identification and assessment of risks associated with threats of human origin (sabotage). The second aspect is the use of probabilistic and scenario approaches, in order to evaluate the "security" performances of systems (integrated security system).

The scope and organization of GIS 3SGS

The following university laboratories are working with GIS 3SGS in order to cover all of the scientific and technical areas to be implemented within the GIS 3SGS framework:

– CRAN, Nancy research center for automatics (Nancy University, CNRS);

– CReSTIC, Research center for information and communication sciences and technologies (URCA, Reims Champagne-Ardenne University);

– Heudiasyc Laboratory for the heuristics and diagnostics of complex systems (UTC, University of Technology of Compiègne, CNRS);

– ICD, Charles Delaunay Institute (University of technology of Troyes, CNRS);

– LAGIS, laboratory of automatics, computer science engineering and signaling (University of sciences and technologies of Lille, Ecole Centrale de Lille, CNRS);

– LAMIH, Laboratory for automatics, mechanics, and industrial and human computer science (University of Valenciennes and Hainaut-Cambrésis, CNRS);

– LORIA, Laboratory of Lorraine for research in computer science and its applications (Henri Poincaré University, Institut National Polytechnique de Lorraine, INRIA, Research institute in computer science and automatics, CNRS).

The industrial problems have, mainly, been suggested by EDF (Electricity of France), the CEA (French Atomic Commission) and ANDRA (the National Radioactive Waste Management Agency), who are founding members of GIS.

Let us mention that, besides MESR and CNRS, INRIA and the General Council of Aube have also contributed to the running of GIS 3SGS. DGA (the General Army Agency) and SGDSN (the Agency for Defense and National Safety) have also provided constant support.

From a practical point of view, GIS 3SGS has worked under the supervision of a gathering council, led by Christian Lerminiaux, director of the University of Technology of Troyes, and with the help of a very active scientific council, initially led by Sylviane Gentil (INPG), then by Jean Arlat, director of the Laboratory of Systems Analysis and Architecture at Toulouse University (CNRS).

GIS 3SGS aims to make the approaches relative to the research on safety, surveillance and security of large systems transversal and complementary. Within this framework, it has been a breeding ground for collaborative projects between industry and research laboratories in the following application areas: energy, transport, information and digital systems, networks and critical infrastructures. The actions and projects led by GIS 3SGS include:

– call on different methods and complementary disciplines;

– the use of a generic methodology applicable to different areas;

– the favoring of flexibility: small projects have led to more ambitious projects, dealing with complex problems; and

– the implementation of collaborations between laboratories and favoring laboratory/industry networks.

This book presents the research projects carried out within the framework of this scientific interest group, particularly those on surveillance and operating safety. All of the projects supported by GIS 3SGS started as a problem suggested by an industrialist and at least two different university teams had to be assembled. They concerned several novel aspects of supervision, the predictive assessment of maintenance operations, diagnostic and prognosis methods, operating in faulty mode, reliability, performance, command and control, reconfiguration, and uncertainties. This was achieved by using dynamic or probabilistic modeling and by taking into account human factors, environmental factors and feedback.

Organization of the book

Part 1: the presentation of three industrial problems relating to nuclear energy:

– the first being related to aspects of the maintenance of a nuclear power plant in operation;

– the second to the surveillance of the operation of the steam generators of fourth-generation nuclear reactors that are currently being studied (sodium-cooled fast reactors);

– the third to the optimization of the distribution of the instrumentation of an underground nuclear waste storage in space and time.

Part 2: a presentation of research projects carried out within the framework of six projects on the supervision and modeling of complex systems in the areas of transport and energy. The results obtained (fault indicators, tolerance to faults, reliability model for complex, hybrid and dynamic systems) are applicable in many other industrial areas.

Part 3: the presentation of prospective studies of surveillance and analysis relating to the means of operation of a steam generator within the framework of studies of a fourth-generation nuclear reactor. The research re-analyzes the acoustic signals recorded in 1994 in the steam generators of the PFR Scottish reactor during the deliberate injections of gas into the liquid sodium.

Part 4: the presentation of tools and methods enabling us to simultaneously analyze organizational, human, technical and environmental factors and their interdependence; and to identify the factors whose conjunction can weaken the defense system (accidents/incidents on large systems). The research carried out applies to two industrial problems – a methodology enabling us to apprehend the systems according to the organizational levels: “action” (the human operator) and technical.

PART 1

Industrial Issues

Chapter 1

Safety and Performance of Electricity Production Facilities

In an increasingly demanding regulatory, legal and social context, power utilities always look for the improvement of the safety and performance of their plants. In the long term they do this by preserving, or even extending the lifetime of their industrial assets. For instance from EDF's (Electricité de France) point of view, this leads to three challenges.

The first challenge is complexity, the complexity of a nuclear, hydraulic or thermal power plant, of an electricity transport network, etc. Risk assessment and management requires understanding a sociotechnical system at every level and in every dimension: the component (itself constituted of elementary components), the equipment (made up of components), the technical system, (combination of pieces of equipment), the individuals and teams organized in complex structures who design or run the system and, last but not least, the environment of the system (natural, technological, organizational, regulation, etc.)

For this challenge, our research projects deal with the methods and tools for the assessment of the risks of a system run by humans, for the design of work situations adapted to humans, and for the organizational diagnostic of the safety and resilience of an organization.

The second challenge is uncertainty. It is everywhere: in physical phenomena and associated hazards, in their measure, their modeling, and in any human activity.

Chapter written by Gilles DELEUZE, Jean PRIMET, Philippe KLEIN, Carole DUVAL and Antoine DESPUJOLS.

To manage uncertainty, we have to decrease its sources (with more observations, and data) or its consequences (by taking conservative margins), and to use reference methodologies to evaluate all the sources and consequences of uncertainties.

The third challenge is decision-making and action. Here we have the difficulty of managing potential antagonisms inherent to industrial goals. The most obvious, for instance, involves safety and performance, in order to increase at once the short-term performance and durability of the facility. The decision must be made based on a problem that is properly defined and associated with an exhaustive analysis of the stakes carried by the different parties involved. It must also be based on a shared process that associates the parties involved in the decision (experts, managers, external parties etc.), and which accounts for the diversity of risk appraisal.

The GIS 3SGS has enabled developments that will contribute to these challenges being taken up. It is about developing methods, models and tools enabling the probabilistic evaluation of safety and reliability before making a decision, or guiding and justifying the choices of maintenance and investments during operation. It also involves the multicriteria modeling or the cost–benefit approach, bringing together strong and opposable decision elements for the management of industrial risks.

Among the projects presented in this book, we will mention for instance APPRODYN, DEPRADEM and MARATHON.

APPRODYN contributes to the challenge of complexity. It is about experimenting and comparing approaches of dynamic reliability to model the availability of critical systems in a probabilistic way. It focuses in particular on the complex interactions between the physical processes (modeled by continuous variables) and command and control (modeled by discrete variables), in the presence of (stochastic) faults or rare contexts.

MARATHON contributes to the challenge of uncertainty. It focuses on a methodology of risk analysis that enables us to prioritize the risks of sociotechnical systems interacting strongly with their environment. The goal is to identify the different types of uncertainties linked to the knowledge created in this type of analysis (coming from operating experience and experts' claims) and to choose the best method adapted to this integration approach from between the methods of propagation of uncertainties.

The DEPRADEM project focuses on modeling the damage and prognostics in a decision-aid related to maintenance. Its goal is to integrate quantitative data in the maintenance decisions by implementing methods that are able to consider damage mechanisms, their effects and the efficiency of the actions undertaken. It is about introducing rationality to the choices made through a better prediction of the results of maintenance strategies on pieces of equipment in operation. The purpose is to fulfill the safety, security, availability, sustainable development and economic efficiency requirements. The study lies at the interfaces between damage/prognostic and prognostic/decision of maintenance. It is about identifying the methodological adaptations and the tools to be developed so that the prognostics can take into account the results of damage models, and maintenance can take into account the results from a prognostic process.

Chapter 2

Monitoring of Radioactive Waste Disposal Cells in Deep Geological Formation

2.1. Context

The French agency for radioactive waste management (Andra) is in charge of the long-term management of radioactive waste generated in France. Within the framework of this mission, it uses its expertise and its know-how to enable the state to implement safe management solutions for all French radioactive waste in order to protect the current and future generations from the risk this waste represents. Andra manages two disposal facilities in operation in the Aube district accepting very low-level radioactive waste to low- and intermediate-level short-lived radioactive waste, as well as a closed center in the monitoring phase in the Manche district.

In 2005 Andra concluded that deep geological disposal for high-level long-lived waste and intermediate-level long-lived waste was feasible. It was put in charge of designing and implanting a disposal center for these wastes (called Cigéo, the Industrial Geological repository Center) by the program law of June 28, 2006. An underground research laboratory was built in the early 2000s in the town of Bure in Meuse district. This laboratory implements experiments of a scientific and technological nature. It is envisaged that by 2015 all the elements necessary for an authorization request for the creation of Cigéo will be ready. Subject to the necessary authorizations, it is envisaged that Cigéo will be operational in 2025, with the necessary construction work starting in 2017. The underground facilities of the repository center will be progressively built upon and will run for a century. The

---

Chapter written by Stéphane BUSCHAERT and Sylvie LESOILLE.

authorization will determine the minimal duration during which, for the sake of caution, the disposal reversibility must be ensured; the law from June 28, 2006 states that this duration of reversibility cannot be fewer than 100 years.

Within the framework of this geological repository project, the monitoring (called by Andra "observation and surveillance") of the environment and structures must incorporate the knowledge required to run the disposal and its reversible management. It must also contribute to the safety analyses while Cigéo is in operation and after closure.

The complete monitoring system must fulfill a set of regulations and societal expectations; expressed:

– in the law of June 26, 2006 on the management of radioactive waste and the law of June 13, 2006 on transparency and security;

– in the "safety guide relative to the permanence of nuclear waste in deep geological formation", coming from a revision process initiated by the authority on radioactive safety of the fundamental safety rule n° III.2.f;

– in the environment code, requesting an environmental reference state in agreement with the extent of the industrial project; and

– within the framework of public debate, exchanges on reversibility, etc.

To meet these requirements, the monitoring system considered by Andra is based on a global strategy including the monitoring of structures and the environment.

2.2. Monitoring of the environment

Andra has set up a perennial observatory of the environment. This observatory, recently labeled SOERE (systems of long term observation and experimentation for the environmental research) by AllEnvi (French National Alliance of Environmental Research) allows Andra to collect the data necessary to carry out an impact assessment before construction. Cigéo is going to be run over a century, beside the evolutions generated by industrial activities in the area. Several changes will take place during this period, be it at the local scale with socioeconomic evolutions (agriculture and forestry) or at a more global scale with climate change. It will be important for Andra to be able to discriminate against the relative contributions of these different evolutions. On top of a fine mesh of sampling and analysis, Andra is putting a network of instrumented stations in place to monitor water, the air and biogeochemical cycles in the soils of the region of interest. An environmental specimen bank is also under construction.

2.3. Monitoring of geological repository structures

The second key element in the monitoring strategy of the geological repository is structural monitoring. Besides the equipment relatively classically implemented to ensure the operational safety of the centre's exploitation, a complementary and parallel monitoring system will be developed. The time range will be increased: the system will be implemented from construction onwards and maintained beyond the first partial closures of the disposal cells.

In contrast, the specifications on the reactivity time or the availability of the information from monitoring are less strict. The goal is to check the expected behavior and to bring additional elements of knowledge to those coming from the modeling and studies currently being carried out in surface laboratories and in the underground laboratory at Meuse Haute-Marne, to supply the decision-making along the repository process. The focus here is placed on the verification of phenomenological evolutions by aiming in particular to:

– Confirm the knowledge involved in the evaluation of the long-term safety and accuracy of the models, on the basis of data obtained *in situ* (scale, construction, operation), within the framework of periodical re-evaluations of the structures. It is about the confirmation of phenomenological models and parameters involved in the evaluation of the long-term safety of Cigéo. By comparing the measurements carried out *in situ* on the first decades of the repository's life with the results of the calculations, the confidence in the evaluation of the long-term safety is strengthened.

– Observe the evolutions of the structures and environmental conditions necessary for reversible management, or even the possible withdrawal of waste packages (re-evaluation of the life spans, etc.). The dimensioning of the disposal cells are otherwise often based on envelope hypotheses and fault scenarios. The monitoring must enable the margins thus involved to be defined.

– Monitor certain evolutions that are likely to influence the safety of the operation, complementing and linked to the operational safety regulations foreseen by "command and control".

The monitoring system considered must supply a thermo-hydro-mechanical-chemical-radiological characterization of phenomena around the disposal of radioactive waste cells. The constraints to fulfill are rigorous:

– The duration of operation needs to be in multiples of 10 years. Certain important evolutions for long-term safety, however, have a characteristic time that is almost inaccessible at the century-scale, so that the variation in the time range can be decreased and the accuracy required is very high;

– The monitoring equipment placed in the waste disposal cells, as soon as the first radioactive waste package is placed inside, is inaccessible. This restricts the maintenance of the instruments (apart from a few automated devices).

– The instruments must not be invasive, since they should not affect the conditions favorable to long-term safety.

– The locally aggressive environmental conditions (temperature, chemistry, irradiation, etc.) must be taken into account.

In order to suggest solutions adapted to these specificities and constraints, many studies and reflections have been carried out. The developments deal with: (i) the monitoring strategy, in particular on the distribution of instrumented structures within the repository; (ii) the design of monitoring units according to a thorough qualification procedure; and (iii) research and development to adapt, complete and qualify the sensing devices.

The response to monitoring requirements lies in the classical and proven approaches, which have been adapted to the context of the geological repository by taking into account the available feedback in the monitoring of some tunnels and confinement areas of nuclear power plants and dams. Andra is also leading an ambitious research and development program into instrumentation and aims to develop specific monitoring equipment. At each significant step in technological progress in a relevant monitoring method for the geological repository, a "qualification" process is started. It involves testing and qualifying the complete measurement chain in progressive steps, knowing these steps and being able to anticipate them, determining the failure rates and mastering the possible long-term drifts.

To design a perennial monitoring system (to last multiple 10-year durations), the strategy focuses first on the implementation of complementary monitoring approaches: embedded sensors completed by the classical approaches of visual inspection and non-destructive evaluation methods. This redundant and complementary procedure is also applied to the embedded instrumentation. Excess sensors would have to be placed among the monitoring units and be linked up according to their complementarities: proven technologies next to innovative sensors, and localized measurements associated with devices providing distributed measurements. This combination of approaches and means enables us to strengthen our confidence in the endurance of the global system but it creates difficulty during the merging of data from these multiple measurement systems that are disparate from the point of view of technologies used and the parameters measured (direct and indirect measurements, localized or distributed, etc.).

It is one of the first expectations Andra has of GIS: the implementation of decision aid systems. The merging of data coming from the monitoring system of the repository aims to supply indicators that are of direct use to the operator. This could be progressive in order to benefit from what is learned from the spatial and temporal redundancy of the global monitoring system. The focus of Andra towards more classical industrial applications means that the acceptable reactivity times for monitoring long-term safety are very slow – at the day-scale which lifts technological constraints that are often strong (calculation times). The duration of monitoring and the risks of associated drifts, the inaccessibility and the associated impossibility of maintenance, however, are relatively original optimization constraints.

To find the equilibrium between the monitoring requirements already expressed, the implementation constraints and the associated costs of monitoring equipment, Andra looks to maintain a reasonable density of embedded sensors. For this purpose, Andra will take advantage of the complementarities of different technical approaches to monitoring that are available by putting progressively more emphasis on visual inspections and on non-destructive tests, equipment that can be mutualized, and decreasing the number of embedded sensors. For this purpose, it optimizes the arrangement of sensors to spread the instrumentation in a largely inhomogeneous way by taking advantage of the similarity between structures, particularly the kilometers of access tunnels and the thousands of disposal cells for long-life high-activity waste (5,000 cells are currently foreseen). By taking advantage of the similarity of some expected phenomenological evolutions (linked to the large homogeneity of the host rock and reserving the selection of disposal cells for similar radioactive package waste), the monitoring strategy suggests following a sequence of structures, referred to as dummy, current and non-instrumented levels, whose density of embedded instrumentation is progressively decreased. Each structure thus labeled has a precise function in monitoring. The "dummy" structure is chosen from among the first structures built. It must exhaustively fulfill the technical monitoring goals. Beside the first constructions, dummy structures will be chosen in specific locations enabling a relevant monitoring. The "current" structure is less well instrumented, monitored by comparison with a "dummy" structure. The non-instrumented structures will only contain essential equipment for operating safety and will be the target of occasional inspection and control. Andra has integrated GIS in order to benefit from optimization methods enabling the optimal number of dummy structures to be chosen by taking into account the variability of phenomena and limits of reproducibility of measurements.

It is for this reason that Andra got involved in the PARDI project, whose results are detailed in this book. This project will enable it to apply the optimization of the instrumentation at the level of disposal. One of the specificities involves spreading the dummy structures both in space and time:

– Almost 5,000 foreseen high-level long-lived cells are spread in zones or modules, each containing a few hundred cells. Each module will be built in one step and then put into operation. The question considered by the PARDI project is how we can statistically define how many cells to use, knowing that the first qualitative evaluations led by Andra are in the range of a few percent, at least for the first modules built.

– The construction and operation of the disposal facility will be very progressive and will last almost a century. The need for information can be progressively decreased as soon as the operation has shown that the evolutions agree with the initial expectations and are sufficiently similar between structures of the same type. In other words, the monitoring strategy must be integrated into the temporal phasing of the exploitation: progressive construction of the typical disposal cells (mineshafts, wells, tunnels, cells, plugs, sealing and embankments) and parallel operation of new phases of the repository with possible evolutions in design. The PARDI project and the concept of learning that is developed therein must enable us to deepen this reasoning by supplying quantitative elements to take advantage of feedback from the first structures monitored.

2.4. Conclusion and perspectives

The inclusion of Andra in GIS is relatively recent and happened after the launch of the organization. Andra is integrated in this community and has an interest in operating safety. It wants to transpose the knowledge and models available in this area to the system of monitoring of the geological repository. In order to optimize the instrumentation plan and the development of measurement interpretation methods, it will establish the diagnostics involved in operating the disposal. Afterwards, the area surrounding the security and safety operation of facilities that Andra manages, and the transport of radioactive materials, will also be able to benefit from the results of GIS.

Chapter 3

Towards Fourth-generation Nuclear Reactors

3.1. Context

In the context of increasing energy needs, an increasing number of countries want to integrate nuclear energy into their energy mix (which will result in a decrease in green house gases emission and avoid the reliance on fossil resources that are progressively being exhausted).

This "nuclear rebirth" should lead to a greater pressure on uranium ore. Thus it must be accompanied by a policy of resource preservation. This is one of the big stakes in fourth-generation nuclear reactors, which appears to be the keystone on which tomorrow's nuclear industry will be able to be built.

This fourth generation will be fast neutron reactors that are able to transform a large portion of uranium 238 into plutonium 239, and can thus be used for electricity generation. Through this transformation it becomes possible to use not just 0.5–1.0% of natural uranium, but >90% for the purposes of electricity generation. It also enables us to consume important stocks of depleted and reprocessed uranium that alone could supply the current level of electricity production in France for a few thousand years. The world "availability" in primary fissile resources can thus be

---

Chapter written by Jean-Philippe NABOT, Olivier GASTALDI, François BAQUÉ, Kévin PAUMEL and Jean-Philippe JEANNOT.

multiplied by 100. These reactors also enable us to recycle plutonium by taking advantage of its energy potential.

The other big stake in fourth-generation reactors is the facilitation of the management of nuclear waste by decreasing the volume and intrinsic long-term radio-toxicity of ultimate waste. These reactors would, in fact, be able to burn a significant part of the long-life radioactive elements that make up the waste – the minor actinides (americium, neptunium, curium, etc.).

The Generation IV International Forum, launched in 2000 by the American Department Of Energy, was born from this desire to create an international research and development framework able to catalyze the research efforts led by different countries to enable emergence of the highest performing technologies at a greater speed.

Four main objectives have been defined to characterize the systems of the future, which should be: sustainable (economical in terms of natural resources with the minimal production of waste); economical (in terms of investment cost and production cost by kWh, which must be competitive with respect to that of other energy sources); safe and reliable; and resistant towards proliferation and easy to protect against external attacks.

In France, program law n° 2006-739 of June 28, 2006 relating to the sustainable management of nuclear materials and radioactive wastes demands that a fourth-generation nuclear reactor be in operation by 2020. The choice has been made to use ASTRID (the *Advanced Sodium Technological Reactor for Industrial Demonstration*), a sodium fast reactor (SFR)-only path that is able to fulfill this requirement by the imposed deadline because it benefits from maturity and sizeable feedback in France and abroad.

To fulfill the aforementioned goals, the fourth-generation SFR should make significant progress in the area that is of particular interest to us by decreasing the probability of an accident leading to a core meltdown. It will achieve this through increased prevention, improving the inspectability of sodium structures, particularly those that have a safety function, and decreasing the risks linked to the affinity of sodium for oxygen: sodium fire and sodium water reaction.

ASTRID therefore wants to show significant advances in the qualification of innovative options in the areas identified above. Its features should be able to be extrapolated to future high-power industrial SFR, particularly with regards to safety (which should be of a level at least equal to that of third generation power plants) and operability. Its features will allow it to be representative of the core and main components and it will offer experimental capabilities. Its design should therefore have some flexibility so that more innovative options can be tested in a second step that would not have been chosen in the initial design. Novel inspection and instrumentation technologies should be tested, as should new fuels and even new components on the circuits.

3.2. Surveillance and acoustic detection

In this context, the first project, relative to acoustic detection and what it could bring to the early detection of some evolutions intervening in steam generators, has been suggested within the framework of GIS-3SGS.

The second phase consisted of competent laboratories processing acoustic signals recorded in 1994 in the steam generators of the Scottish PFR reactor, during the voluntary injections of water vapor and argon gas in the liquid sodium. It also included processing of signals that were collected during the "end of life" scientific program carried out by Phenix, a French SFR prototype of 250 MWe. This phase was completed in 2010. We want to characterize and classify the different sequences recorded, which will enable us to update the analyses and treatment already carried out in 1996 and 1997 by CEA (the French Atomic Commission) on the recorded PFR signals. The different tools used by the universities involved can then be compared.

The results were conclusive (providing an interesting enlightenment on the applicability of the detection technique of loss-of-coolant in the steam generators by passive acoustic detection). A wider program was then established in order to define the techniques to be used to obtain a gain considered useful by the reactor designers from the point of view of safety.

Acoustic detection is also a promising method under development in the area of instrumentation and surveillance. It could enable us to detect the boiling of sodium within the core and has particularly interesting potential in the earlier detection of accidents linked to this phenomenon, thus constituting a new line of defense. The first line of defense is the detection of delayed neutrons.

It is very useful for the future of sodium reactors for us to ensure that among the stakes linked to this issue that the knowledge gained by the current operators of the Phenix reactor in Marcoule (stopped in 2009) has capitalized on, and that the associated simulator, in its current version, has been properly taken into account. Generally, it would be useful if we could be sure that there is no bias between the physical reality and the hypotheses of the software of the simulator. An ergonomic study would also be interesting.

The decade that will pass between the stopping of Phenix and the starting of Astrid leads us to favor transmission feedback from the operators of Phenix to that from the future operators of Astrid. This action would involve interviewing operators and appraising the Phenix simulator in order to better define the specifications of the future Astrid simulator. For control and surveillance sensors, we could study how the physical parameters measured can/must be completed by parameters from models and simulations (for example, the material balance or thermal energy balance), as well as the possible correlations between different sensors (the data fusion method).

The crosscheck between measurements and models also has a stake. The coherence between the different information layers must be evaluated, for instance, the sensors that enable us to follow the load operation of the reactor and the automatic regulation loops. There are a number of questions that must be answered here. How this information can be taken into account at the "surveillance" level? Is there communication between the diverse layers? Are there methodological locks that must be investigated?

The current reflections on the surveillance conditions of sodium, the coolant for heat transfer in SFR reactors, aim to enable the emergence of new processes to complete, improve and diversify the current systems (towards the rich feedback) and those whose study is ongoing. These new techniques should be adapted to the environment of the coolant sodium by taking into account its specific properties, for instance the risk of inflammability at temperatures close to the fusion point and its exothermal reaction with water generating hydrogen and sodium hydroxide. It is also necessary to define the relationships between the surveillance level required, the potential level of damage and the safety requirements.

The surveillance of SFR (sodium-cooled fast reactor) and the associated controls lies initially on the fulfillment of three safety functions:

– control of the nuclear reactivity:

- insertion of rods absorbing the neutrons at the level of the core (rods referred to as command or control); and

- neutron measurements (*in core* or deported);

– evacuation of the thermal power by the coolant fluid:

- thermal measurements; and

- flow measurements (or rotational speed of mechanical pumps);

– confinement of products that are dangerous to the environment (radioactive products and sodium):

- detection of clad failure (if any);

- detection of sodium leakages (within the energy conversion system exchangers, and in air); and

- pressure measurements.

Furthermore, it is necessary to have continuous surveillance for inaccessible zones at our disposal. For instance, zones in the main vessel situated under the core are far from the slab through which the sodium instrumentation can enter.

3.3. Inspection during operation

The opacity of sodium forbids the use of optical processes for inspection when a reactor is in operation. This is why ultrasonic techniques are used: it enables us to test the sodium zones (telemetry, metrology of immersed surfaces, defectometry and imaging).

The associated accuracy, however, remains linked to that obtained by the eye of an operator (a specification required by the regulation).

Research and development in the use of ultrasound is enabling us to face the drawback due to the opacity of sodium – the distance measurements, imaging, thermometry and flowmetry, the detection of phenomena leading to the presence of gas bubbles in sodium and volume inspections –. Ultrasounds are poorly attenuated in sodium (less than in water), except when it is engased: the gas rate of the liquid must be monitored because it has very important effects on its acoustic properties.

The performances of the inspection techniques carried out while the reactor is in operation depend on the performances of the ultrasound techniques and on the constructive arrangements instigated during the design. Iterative work enables us to adapt the design towards the implementation of inspection and to develop inspection tools to meet the required controls.

These include:

– putting of sodium immersed targets in place that "grasp" the waves (or any other intrusive vector) well and reflect them towards the signal receivers;

– design favoring accessibility to include the sensors up until the zones to be controlled;

– recourse to inspectable witness zones;

– recourse, if necessary, to structures that can be taken apart (components such as heat exchangers, pumps, sodium purifiers);

– the development of repair techniques *in situ* (under an argon gas bell);

– the design of telemanipulation techniques (rigid rod, articulated arms, etc., whose trajectory must be defined and controlled); and

– the development of robots that embed the control and repair tools.

3.3.1. *The case of acoustic measurements*

Local or global acoustic measurements seem to be an interesting path to consider because of their non-intrusive nature, despite the need to locate acoustic sensors above the reactor slab (which avoids or removes the introduction through the upper slab that is already obstructed, or the risk of generating additional impurities in the liquid sodium) and because of the speed of information transmission.

The complexity of such a measurement will come from the multiplicity of noise sources (moving pumps, flows internal to the main tank, etc.) and their variation (according to the regime) and hence from the difficulty to measure and discriminate against the origin of an acoustic variation.

Knowledge of the source noise of the defect to be detected, gained through experiments carried out in experimental devices, for instance, will not be sufficient to establish the feasibility of the detection principle. For this, data about the background noise is also necessary: its intensity, spectral distribution, attenuation, etc. In the absence of a French reactor in operation that could be representative of the future SFR, these data will come from simulations produced by *ad-hoc* tools (possibly validated by data coming from the operation of Superphenix, Phenix, or even models if need be).

3.4. Conclusion

As you will have noticed, the acoustic processes are a major part of the means of surveillance and inspection of future SFR reactors: the complexity of the propagation modes of ultrasound in structures immersed in liquid sodium has led to the development of simulation tools and signal processing methods, an activity that fulfills the goals of the actions promoted by GIS-3SGS.

PART 2

Supervision and Modeling of Complex Systems

This part describes work carried out within the framework of six projects on the modeling and supervision of complex systems. Work presented in these chapters puts forward the industrial issues expressed by the needs of the industrial partners of the projects, the theories and approaches implemented and the expected results. These six projects (FD2S, merging of data for the supervision of a complex system application to guided transport systems; FDADI, merging of multisensor data in diagnosis for the monitoring of the trajectory of a vehicle convoy; DEPRADEM I and II, modeling of the process for maintenance decision support; COSMOS I and II, design and observation of systems with multiple modes of safe operating; APPRODYN, approaches of the dynamic reliability to model critical systems; and PARDI, contribution of the retailer effect for the operation diagnosis of systems) mainly deal with:

– the processing of faults in the data sensors embedded in public transport or FRET;

– the application of a forecast type approach to the maintenance and decision support in a dynamic context;

– the development of supervision and multitask learning systems by integrating the concept of retailer effect of machines; and

– the experimentation of approaches of reliability for critical command and control systems.

These approaches have been tested in the areas of transport and energy. They can be applied in all other areas where the complexity and the dynamics of the systems generate problems relating to supervision and maintenance.

Chapter 4

Fault-tolerant Data-fusion Method: Application on Platoon Vehicle Localization

4.1. Introduction

For several years, active security methods as well as intelligent technical devices have mostly been developed in public or private transport vehicles. Sophisticated methods from robotics, such as localization, obstacle detection and trajectory planning, are now available in intelligent transport systems. The implementation of these new functions strongly depends on the quality of data coming from different sensors embedded in vehicles as well as their merging, particularly when it comes to critical information that must be supplied for autonomous navigation systems.

The goal of our research within the framework of the FD2S project carried out by GIS 3SGS was to integrate the concept of tolerance to faults, failures or inaccuracy of sensors. The goal was to integrate the detection, diagnostics and processing of faults in the sensor data for data fusion, which requires the validation/invalidation of data collected, and the decision-making concerning the reconfiguration of the data fusion algorithm. These functionalities play an essential role in providing autonomy in order to give a vehicle (or a robot) the capability to adapt and react to unforeseen events, thus allowing it to fulfill the goals of its mission. This is a decisive element in improving the operational efficiency of intelligent transport systems and decreasing the cognitive load of their piloting.

Chapter written by Maan EL BADAOUI EL NAJJAR, Cherif SMAILI, François CHARPILLET, Denis POMORSKI and MIREILLE BAYART.

Our research has enabled us to use Bayesian networks for multisensor data fusion and a decision algorithm [SMA 10] via the use of probabilistic graph theory. These Bayesian networks enable us to establish a probabilistic graph model from the joint probability distribution on a set of random variables. This formalism enables an intuitive representation of the knowledge on a given application area and facilitates the implementation of high-performance and clear models. The representation of knowledge is based on the graphical description of the relationships existing between the variables describing the system. Each variable is associated with a local probability distribution quantifying this relationship. The Bayesian network is shown as a formalism that is potentially useful because it can integrate data fusion, diagnostics, reconfiguration and learning in a same formalism. In this chapter, we also suggest a formulation for the problem of localizing a vehicle convoy using a Bayesian network. The approaches developed were tested on benchmarks of real data obtained by the "TRAVEL[1]" platform used by LAGIS. It was extended using LORIA's CyCab vehicle. The results were used to suggest appropriate validation procedures.

This chapter is organized in the following way. After a review of the theoretical tools of multisensor data fusion, a description of Bayesian networks is suggested. Then, in section 4.4, we describe the method proposed for fault-tolerant data fusion for the localization of a single vehicle. In section 4.5, we present an approach for the modeling and localization of a vehicle convoy that is an extension of the approach suggested in section 4.4. Finally, we present the testing and results of approaches using real data. We end this chapter with a conclusion and perspectives.

4.2. Review

In filtering, we look for an algorithm that is able to carry out the calculation of the optimal filter that also fulfills real-time constraints. In concrete terms, the calculation time of an estimated variable must be smaller than the observation period. To fulfill this constraint and to spare the memory of the embedded computer, we preferably want a recursive algorithm. Thus, the calculation of the estimated variable at the current time must depend only on the current observation and on the previous estimated variable; the past observations are not reused. These constraints are paramount for the good efficiency of the algorithm.

In 1960, Kalman and Bucy [KAL 60] revolutionized estimation theory by providing the first recursive filtering algorithm. Kalman-Bucy's filter enables us to precisely and quickly calculate an optimal filter when the state and observation

1 The TRAVEL platform is a set of three electric overequiped RobuCar type vehicles supplied by Robosoft.

models only include linear functions and additive Gaussian noise. It has been shown, in fact, that in this case the probability distribution is assumed to be a Gaussian one. It is therefore sufficient to propagate its mean and its variance: the problem is of finite dimensions. Generally, however, the nonlinear filtering problem does not have a solution of finite dimensions. In practice, we often keep using versions derived from Kalman-Bucy's filter, such as the extended Kalman filter (EKF) that linearizes the model around the current estimate. It owes its success to:

– its very good robustness due to its "least square" nature;

– its numerical stability, which is well controlled because the implementations are often factorized; and

– its very good aptitude for real-time calculations (EKF being a particular version of a predictor/estimator observer).

Even though for the past 40 years this algorithm has been the most widely used nonlinear filter, it exhibits some weak points when the system is too nonlinear or when the filter is poorly initialized.

Another variant of EKF is the "unscented" Kalman filter (UKF) that was presented by Julier and Uhlmann [JUL 97] to address the following weaknesses of EKF:

– the difficulty of tuning an EKF (it must be done by trial and error);

– the difficulty of implementation (the filter must be rewritten in quadratic form); and

– the behavior of the system must be sufficiently linear between two registrations.

A UKF filter is a predictor/estimator that does not linearize the transition function f and observation function h in the way the EKF does. This method directly evaluates the first- and second-order moments of the images by the f and h of the variables of interest by using an unscented transformation. The numerical implementation of a UKF can be high performance if we use Cholesky's factorization.

The UKF filter has met with huge success since the end of the past century, particularly in robotics. It has replaced the EKF in several applications. It must be noticed that UKF has several variants, such as numerical derivative filters (first-order DD1 and second-order DD2) [NOR 00] and the *central difference Kalman filter*. These filters all use a strategy of deterministic sampling. They have been used by several researchers for localization [BON 05, DIS 01].

More recently, the combination between Kalman variants (EKF, UKF, etc.) and *interacting multiple model* (IMM) algorithms has achieved some success. The IMM algorithm is the most frequently used in the literature [LI 05, BLO 88] for the classification of the movement of a target according to several types of movement models. With this combination, the estimation of a state vector is based on the combination of a finite number of sub-models, to which one of the variants of a Kalman filter is simultaneously applied [BAR 82].

The particle filter has also gained some recognition since the beginning of this century. The method of particle resolution lies in the exploration of the state space by particles randomly drawn (Monte Carlo drawing). The principle resembles "grid" (or mesh) methods. It has the particularity of generating a progressive mesh as a function of the stochastic flow of the system studied and of its observations. The essential tools used in this method are Bayes' rule and a dynamic adaptation of the law of large numbers [ARU 02, GUS 02, BER 99].

The key idea behind the success of particle filtering, other than the evolution of numerical calculators, is the fact that the estimation and its quality depend little on the dimension of the state model. Under some conditions, we can hope to have the same estimation quality with the same number of particles for a system of a higher order [CRIS 00].

Any user of particle filtering worries about the number of particles and wants to minimize their number to reduce the calculation load. The minimization amounts to selecting the particles that are representative of the probability density function. Since the quality of the estimation of the density depends on the number of particles, the decrease in the number of particles is tricky. In the literature there are several methods that can be used to decrease the number of particles. By introducing an additive noise to the particles, we can decrease their numbers. This method is known as *jittering* [FEA 98]. Another similar approach was introduced in [GOR 93] under the term *roughening*. In [GOR 93], the *prior editing* method is also discussed. In this method, a step is added to the filtering algorithm. In this step, a particle likelihood test is performed. The idea is to eliminate the particles that have an insufficient probability; they will not be selected during the following step.

Finally, our choice of theoretical tool for the *f* multi-sensor data fusion has been based on criteria that appear to be essential to solve the problem of the localization of a vehicle convoy:

– the aptitude to accept various information sources, both redundant and complementary, on top of the ease of integrating them into the method;

– the capability to provide a consistent estimation of the localization error;

– the tolerance to sensor failures such as faults, inaccuracies, inconsistencies, incoherences or abnormal data (the possibility of managing the multi-hypotheses aspect and the multimodality of measurements could be a way to be tolerant to sensor inaccuracies);

– the possibility to integrate a diagnostic layer for the sensors;

– the possibility to reconfigure the fusion method after the detection of a sensor failure in order to restrict the perturbating effect of a faulty sensor; and

– the aptitude for real time.

4.3. Bayesian network for data fusion

Among the probabilistic approaches, Bayesian networks (BN), also referred to as probabilistic graphical models, are sometimes defined as being a marriage between probability theory and graph theory (Figure 4.1). Probabilities allow these models to take into account the aspect of uncertainty when modeling a given event. The graphical part offers an intuitive tool in several areas of application where the users need to "understand" the model they are using [JEN 01, MUR 99, MUR 02, SMA 10]. The joint use of probabilities and graphs gives us a family of knowledge models that are very rich with dynamic Bayesian networks, hidden Markov models and their derivatives, Kalman filter, etc., for instance.

A	0	1
P(A)	0.9	0.1

B	0	1
P(B)	0.3	0.7

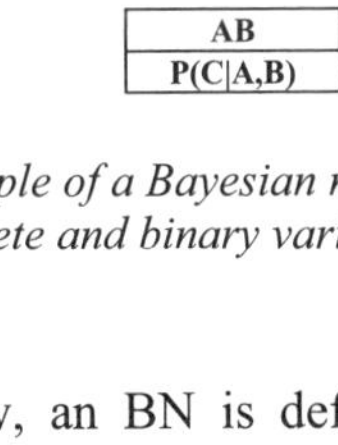

AB	00	01	10	11
P(C|A,B)	0.3	0.7	0.9	0.1

Figure 4.1. *Example of a Bayesian network with three discrete and binary variables*

From a formal point of view, an BN is defined by a set of discrete and/or continuous random variables, by a structure expressed in the form of a directed acyclic graph that indicates the dependence between these variables and a set of conditional probabilities that links the value of a node to that of its parents. Figure 4.2 gives an example of a Bayesian network with several discrete and binary variables.

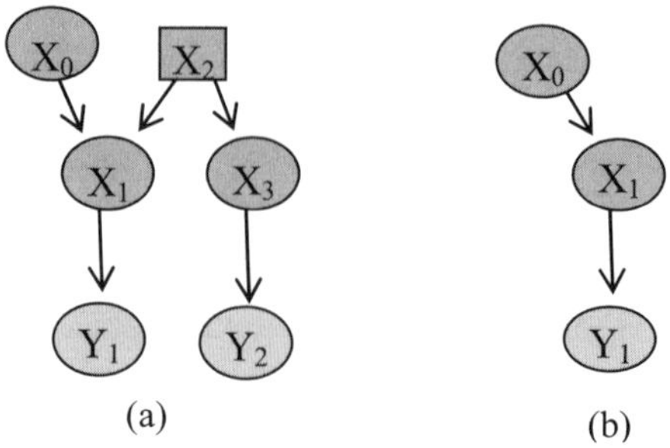

Figure 4.2. *a) Example of a Bayesian network where the circles represent the continuous variables and the rectangles represent the discrete variables. b) Example of a Kalman filter. Variable Y_i represents the observations from which we can update the hidden variables, X_i*

In this project, we were interested in dynamic Bayesian networks (also referred to as temporal Bayesian networks), which are directed graphical models of stochastic processes. They generalize hidden Markov models and the linear dynamic systems by representing the hidden (and observed) states as state variables with complex interdependences. The graphical structure provides an easy way by which to detail these conditional independences, and as a consequence also provides a reduced parametrization of the model.

The description of this theory is omitted in this chapter. More details can be found in [MUR 02, SAM 10].

4.3.1. *Bayesian network and Kalman filter*

As we saw previously, a Kalman filter is a tool to estimate the state of a dynamic system from noisy observations. More formally, to estimate state X_k at time k, we have information $Y_{0:k}=(Y_0,\ldots,Y_k)$ and of the previous state X_{k-1} at our disposal, and the goal is to obtain the most information possible relating to X_k. In terms of probability, it involves calculating the conditional law of random vector X_k knowing that $Y_{0:k}$: $P(X_k|Y_0,\ldots,Y_k)$.

Let us assume that we know the conditional law of random vector X_{k-1} knowing $Y_{0:k-1}$. The Kalman filter proceeds in two phases:

– prediction: we calculate the conditional law: $P(X_k|Y_0,\ldots,Y_{k-1})$, with the help of the evolution model; and

– correction: we use new observation Y_k to correct the state predicted in order to obtain a more accurate estimation.

A classical Kalman filter such as described previously can be represented by a Bayesian network with continuous variables. In [MUR 02], the author shows that a Kalman filter is a special case of a Bayesian network, which can be represented by the network shown in Figure 4.3.

A Kalman filter assumes that state $x_t \in \Re^{N_x}$ observation function $y_t \in \Re^{N_y}$, filter input $u_t \in \Re^{N_u}$ and the observation and transition functions are linearly Gaussian, i.e.:

$$P(X_t = x_t | X_{t-1} = x_{t-1}, U_t = u) = N(x_t; Ax_{t-1} + Bu + \mu_X, Q) \qquad [4.1]$$

and:

$$P(Y_t = y | X_t = x, U_t = u) = N(y; Cx + \mu_Y, R) \qquad [4.2]$$

In other words:

$$X_t = AX_{t-1} + BU_t + V_t \qquad [4.3]$$

where $V_t \sim N(\mu_X, Q)$ is a Gaussian noise. Similarly:

$$Y_t = CX_t + W_t \qquad [4.4]$$

where $W_t \sim N(\mu_X, Q)$ is another Gaussian noise independent of V_t.

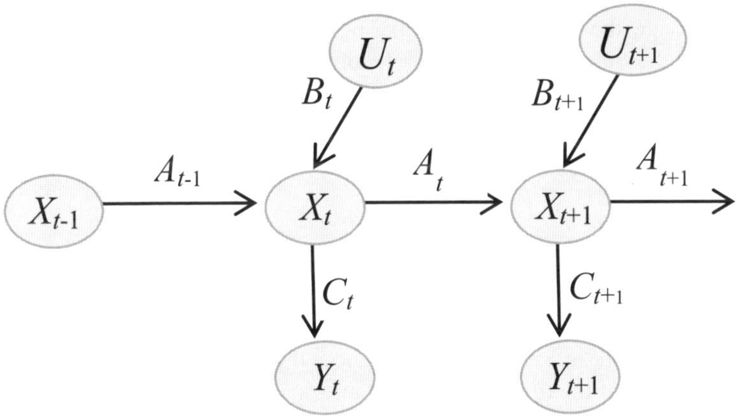

Figure 4.3. *Representation of a Kalman filter by a Bayesian network*

In the pieces of work carried out by [ELB 03, BON 05, QUD 06], Kalman filtering is used to match an estimate on a road segment. If we want to manage or manipulate several segments at the same time (multi-hypotheses), however, the Kalman filter as it has previously been described is not adapted, even though we could use N Kalman filters corresponding to the number of candidate segments.

4.3.1.1. *The choice of use of Bayesian networks*

In order to suggest a solution to the issue of dynamic localization of a vehicle, we use the Bayesian networks for the following grounds:

– The Bayesian network generalizes Kalman filters: the Kalman filter is only one particular case of a Bayesian network. Furthermore, this tool gives us the opportunity to handle the discrete and continuous variables in the same network, which is not allowed by a Kalman filter. The possibility to pair discrete and continuous variables in the same network is very interesting because it will allow us to generate Bayesian networks in the configuration of a *switching Kalman filter*, *switching state–space model* or *jump-Markov model* [MUR 02]. These are configurations that we have attempted to take advantage of and to manage the multi-hypotheses aspect (with more than one estimation) and the multi-modality aspect of our observations. These two aspects are necessary for our localization application on a 2D map in order to manage the inaccuracies in the data and, in our case, to manage the ambiguity by manipulating the multi-modality of the observation. This enables the management of several segments at the same time.

– To provide the possibility of integrating a diagnostic layer for the sensors in order to detect the possible failures.

– To provide the possibility to carry out a reconfiguration of the data fusion method after the detection of a sensor failure by using, for instance, the Bayesian network with a configuration of the *switching state–space model.*

– To provide the possibility of learning with the same theoretical tool that opens other research horizons.

4.4. Localization of a single vehicle: multisensor data fusion with a dynamic Bayesian network

The road network in a cartographic database is represented by polyline modeling of the central axis of the roads. Attributary information is attached to these polylines that we try to extract, for instance, to perform driving assistance functions.

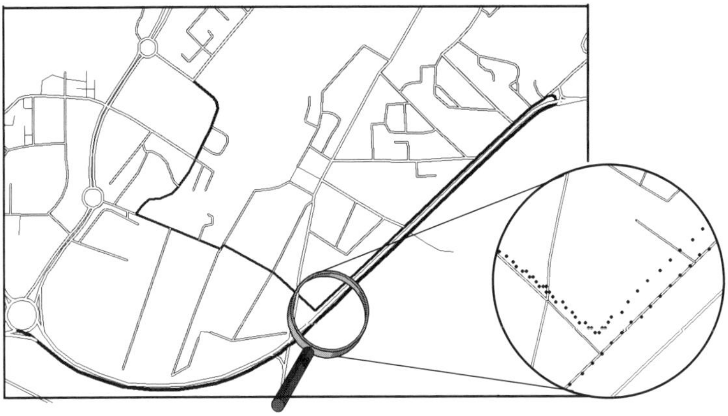

Figure 4.4. *Illustration of the localization problem on a map (trial carried out with a differential GPS and an IGN Géoroute map)*

There are two ways in which the localization of a vehicle on a map can be found:

– a first way involves the selection of the basis segment representing the road on which the vehicle is moving; and

– the second way consists in determing the curvilinear abscissa of the vehicle on the segment with respect to one of its extremities (this method is often used in navigation systems that are currently on the market).

In the rest of the chapter, the first definition will be considered.

In [ELB 03], there is a review covering the different methods of map localization existing in the literature. The reader can also refer to [QUD 06] for more details.

For an automated vehicle equipped with a GPS receiver and proprioceptive sensors such as odometry, the dynamic approach to the issue of map localization consists of searching for a recurrent method taking advantage of the map in order to register the drifts of dead reckoning.

For this, we can distinguish two approaches:

– a localization forced to remain attached to the road network: the solution is on the segment representing the road; and

– an absolute localization in the map referential, then a projection on the closest segment when we look for an attribute.

It is the second approach that we have considered in this project: it enables us to consider GPS and cartographic measurements equally. It therefore better lends itself to a formulation of the problem from the angle of data fusion.

4.4.1. *Presentation of the approach developed*

Let us focus on the multi-hypotheses approach for the rest of the section, managing multimodal observations that were studied in Cherif Smaili's thesis [SMA 10]. This approach belongs to the class of Bayesian methods [ZHA 97].

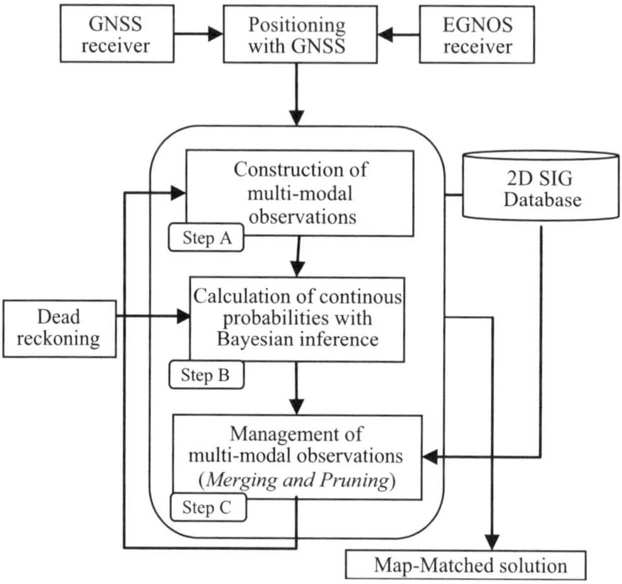

Figure 4.5. *Synopsis of a multi-hypotheses cartographic localization algorithm*

Figure 4.5 describes the approach based on a *switching Kalman filter* (SKF). Initially, in step A, a discrete variable is formed by using the segments supplied by Système d'Information Géographique (SIG) in order to build the multimodal observation. These observations are updated by using rules based on the topology of road networks. This discrete variable aims to generate as many states as there are modes in the multimodal observation. Afterwards, in step B, inferences are performed by using Bayesian networks to provide estimates that result from the merging of different observations and each state from the preceding time. These calculated inferences can be simulated using Kalman filters generated in parallel for each state. Besides the state estimation, the Bayesian inference gives a variance–

covariance matrix associated with each state vector that represents an estimation of the error made by the inference. In step C, a policy of the creation and elimination of hypotheses is worked out to supply a ranking of hypotheses at a given time. This policy is based on heuristic rules such as the connectivity between segments and the movement history of the vehicle, as well as other criteria detailed later in this chapter. This is an essential step because it avoids the combinatorial explosion in the case where a large number of segments are selected and processed by the method.

The Bayesian network model developed for the issue of *map-matching* is shown in Figure 4.6. In this model, we use the discrete variable S_k, which represents the segments on which the vehicle can be situated: $seg_k^1, seg_k^2, \ldots, seg_k^n$. This variable is initialized *a priori* at each step by using two criteria:

– the course of the segment that is the closest to the course of the vehicle; and

– the proximity of the GPS point with respect to the segments.

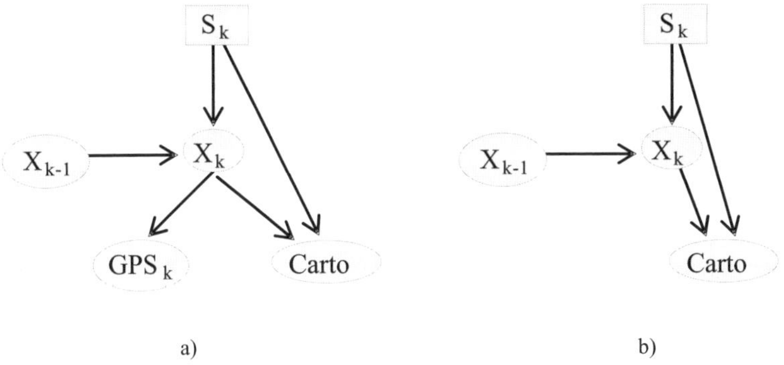

Figure 4.6. *a) Model of a Bayesian network for the localization of a vehicle on a map in the presence of GPS signals. b) Model of a Bayesian network for the localization of a vehicle on a map in the absence of GPS signals*

The segment whose GPS point and whose course is the closest is given a probability greater than the other candidates. Unlike the selection approaches of the most likely segment (using proximity criteria, course, etc.), we favor the segment that fulfills the criteria cited above. Under no circumstances, however, do we give this segment a probability equal to one. This strategy strengthens the uncertainty concept on which we are working.

The second variable used in the network in Figure 4.6 is continuous. Let us note that this variable is hidden: $X_k = (x_k^i, y_k^i, \theta_k^i)$, representing the position and course of the vehicle for each candidate segment of variable S_k.

The graph given in Figure 4.6 allows us to properly represent the causal links between variables. Thus, continuous variable X_k is updated by the observations (obviousness or likelihood of measurements):

$$Carto_k = \left\{(x_k^{seg_1}, y_k^{seg_1}, \theta_k^{seg_1}),...,(x_k^{seg_n}, y_k^{seg_n}, \theta_k^{seg_n})\right\}$$

and:

$$GPS_k = (x_k^{gps}, y_k^{gps})$$

if GPS signals are available. Otherwise (hidden GPS), we only use the cartographic observations when the odometric estimations are being updated. These two variables ($Carto_k$ and GPS_k) are continuous.

The Bayesian network calculates joint probability $P(S_k, X_k | GPS_k, Carto_k)$, which represents the probability that the vehicle lies in each candidate segment $S_k = seg_k^1, seg_k^2,..., seg_k^n$ as well as the estimation of the placement of the vehicle (position, course): $X_{seg_k^1}, X_{seg_k^2},..., X_{seg_k^n}$ on each segment.

Let us note that the relationship between discrete variable S_k and continuous variable X_k plays the role of a generator containing N Kalman filters (where N represents the number of candidate segments).

In order to take into account evolution over time, we use a dynamic Bayesian network. This network is created by duplicating the static Bayesian network shown in Figure 4.6a or 4.6b and linking these networks from one step to next (every time we have a new observation).

Figure 4.7 represents the dynamic Bayesian network for the localization of a vehicle on a map in the presence of GPS signals. According to this figure, we have two variables that are linked from one step to the bext. The variables at stake are S and X.

Let us underline the resemblance between the model suggested above and what we refer to in the literature as the SKF represented by Figure 4.8. There are a number of different names for the SKF: *switching linear dynamical system*, *switching state–space model*, *jump-Markov model*, *jump-linear system*, *conditional dynamic linear model*, etc. This type of model is often used to approximate the nonlinear models (we assume that these models are piecewise linear) [MUR 02]. We can underline the fact that the methods known as IMM algorithms are particular cases of this representation.

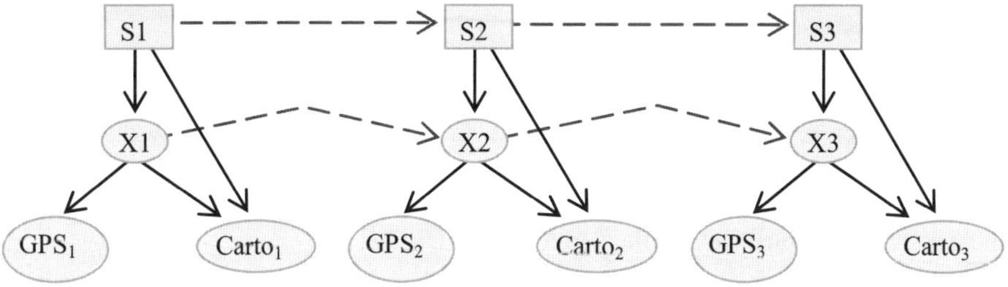

Figure 4.7. *Dynamic Bayesian network with three time steps (T=3) for the localization of a vehicle on a map in the presence of GPS signals*

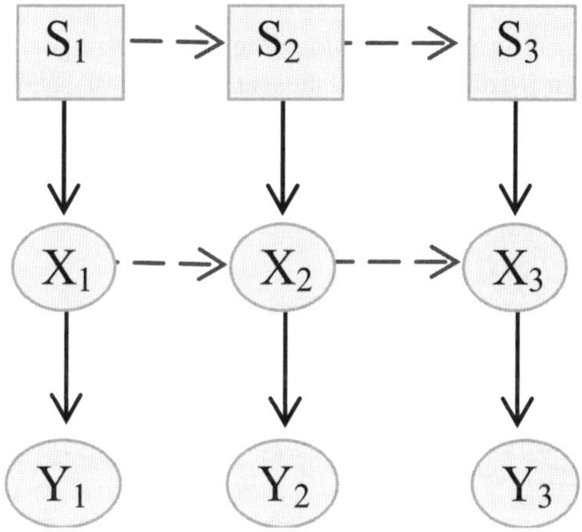

Figure 4.8. *Model of a Bayesian network representing the switching Kalman filter*

4.4.2. *Inference in switching Kalman filter*

The fundamental problem with the SKF type of network is that the inference is almost impossible. The number of Gaussians in the distribution of the state increases exponentially with time. To illustrate this, let us assume that the initial distribution $p(X_1)$ is a mixture of K Gaussians, one for each value of S_1. During the marginalization stage in S_1, each Gaussian must be propagated through K equations (one for each value of S_2) so that $p(X_2)$ is a mixture of K^2 Gaussians. In general, at time t, the probability distribution $p(X_t \mid y_{1:t})$ is a mixture of K^t Gaussians, one for each possible combination of $S_1,...,S_t$.

To cope with the number of Gaussians, which increases exponentially, particle filters and approximative inference algorithms have successfully been used in this type of model [URI 02].

Even though inference is almost impossible in the case of models of the SKF type, there are particular cases in which this inference is possible. The trivial case is when the discrete variable can be observed. In this case, the probability distribution of the continuous variable is unimodal (a single Gaussian). The second case where inference is possible is when the discrete variables are not linked from one step to the next. An illustrative example is given in [MUR 02].

Algorithms referred to as *pruning algorithms* decrease the number of Gaussians by eliminating some of them. These algorithms keep N Gaussians with strong probabilities, discard the others and normalize the probabilities so that the sum is equal to one [BAR 88, URI 02].

Another approach referred to as *collapsing*, or *general pseudo-Bayesian* (GPB) *algorithms*, consists in partitioning the Gaussians into N subsets. The GPB algorithm restricts the number of Gaussians of the state to N Gaussians that correspond to the number of subsets. For more details on these approaches, we refer the reader to [LAU 89] and [URI 02].

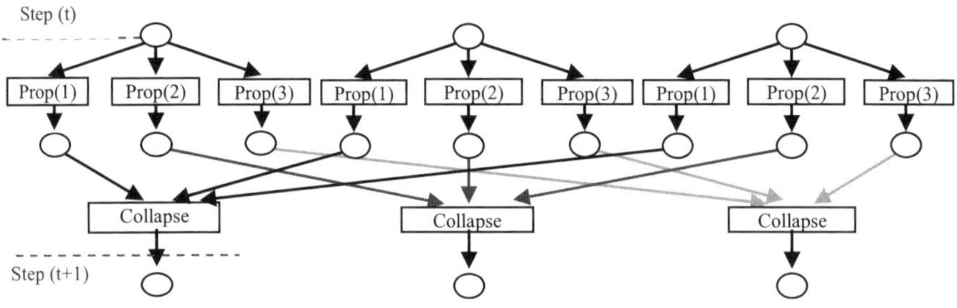

Figure 4.9. *An example of running the GPB algorithm for M=3. Each circle represents a Gaussian (the rectangles with the label prop(), represent the propagation of each Gaussian)*

To better understand the principles of this approach, let us give an example that illustrates it. Let us assume, at time $t = 1$, that we have M Gaussians to represent the state (M is the number of combinations of the discrete variable that are possible). After the propagation phase, we obtain M_2 Gaussians. These M_2 Gaussians are collapsed into M Gaussians. Figure 4.9 illustrates this approach for a number of Gaussians equal to three ($M = 3$).

In the case of the Bayesian network represented in Figure 4.7, the number of Gaussians in the distribution of state X_i does not increase (exponentially) with time. A single Gaussian per candidate segment is considered at each time, t. The distribution along a segment at time t is assumed to depend only on the distribution along this same segment at time $t-1$. This can be translated into the introduction of a conditional probability $P(S_t|S_{t-1})$ that is equal to the identity matrix. The intuition behind this conditional relationship is that the segments are "disconnected" from each another. The hypotheses are propagated along each segment but not between segments. Let us note that this approach resembles the reduction approach given by *pruning algorithms*.

4.4.3. *Detailed synopsis of the method based on Bayesian networks*

We use the fusion of GPS measurements, the measurements of proprioceptive sensors and data on the map in order to generate an estimation of the posture of the vehicle. We are particularly interested in what follows in the search for the path (road segment) on which the vehicle drives, with a quantification of the degree of certainty as a function of the quality of the information handled and the geometric configuration of the road network around the vehicle. The synopsis of this approach based on Bayesian networks is given in Figure 4.10.

Initially, let us assume that we have a GPS or odometric estimate (when the GPS is hidden) of the position of the vehicle. This estimation is given to SIG in order to select a set of segments around this estimate. The result of this step is a set of segments that are numbered as follows: $seg_k^1, seg_k^2, ..., seg_k^n$, where k represents the selection time.

In order to build cartographic observations (evidence), we project the estimation onto each candidate segment. If there is no orthogonal projection, then we take the point that is closest to the segment of interest. The result of this step is the set:

$$Carto_k = \left\{(x_k^{seg_1}, y_k^{seg_1}, \theta_k^{seg_1}), ..., (x_k^{seg_n}, y_k^{seg_n}, \theta_k^{seg_n})\right\}$$

The inclusion of $Carto_k$ and GPS_k observations in the Bayesian network enables us to update the probability distributions of hidden variables, X_k. After inference, we obtain the probabilities of each candidate segment in which the vehicle is likely to be (represented by variable S_k) as well as the position and course of the vehicle in each segment (represented by variable X_k).

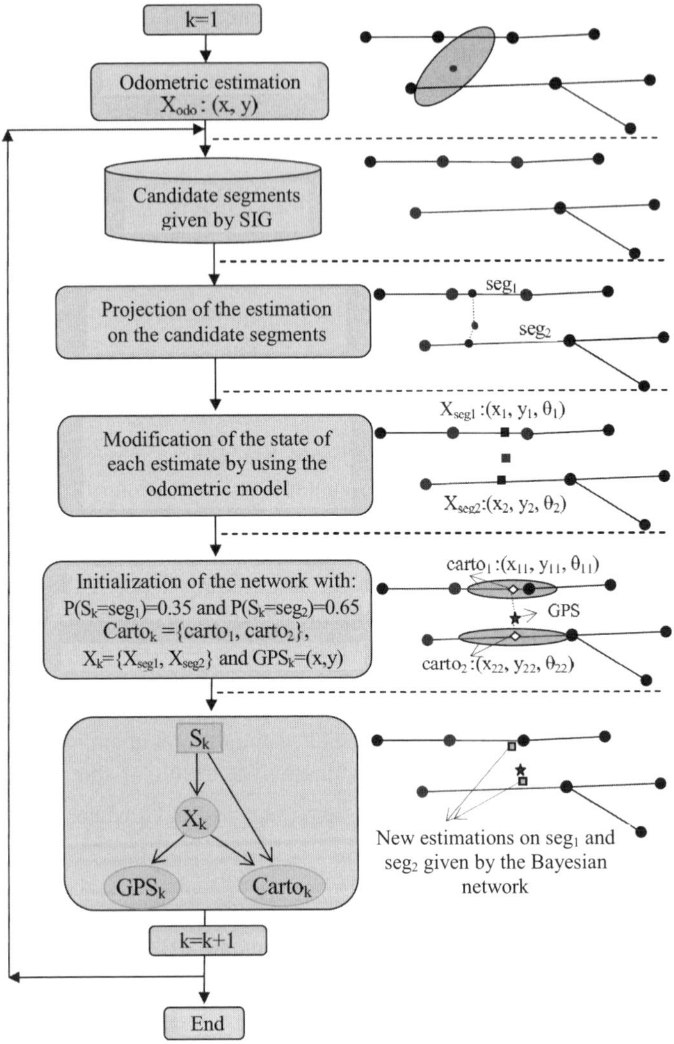

Figure 4.10. *Synopsis of the approach based on Bayesian networks*

4.4.4. *Example of management of multi-hypotheses by a Bayesian network*

In order to clarify ideas, let us give an example that illustrates how our method works. In Figure 4.11, the vehicle moves from road segment number 1 towards road segment number 8. Let us assume at time $t = 1$ that an estimation is given by the GPS (or odometer). From this position we select all of the possible segments (this step is given by SIG). Here there is only one single candidate segment. We project the GPS estimation onto this candidate segment (segment number 1) to build a cartographic observation:

$$Carto_1 = \left\{(x_1^{seg_1}, y_1^{seg_1}, \theta_1^{seg_1})\right\}$$

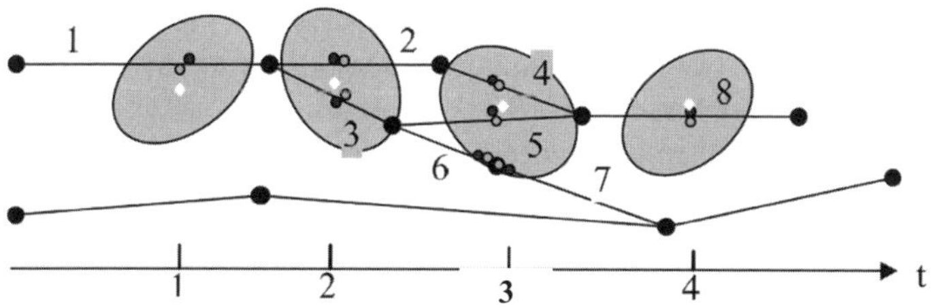

Figure 4.11. *Example of the management of several segments using a Bayesian network. The white points represent the estimations given by the GPS as well as the uncertainty around this estimate. The cartographic observations are represented by dark gray points and the estimations given by the Bayesian network are given by light gray points*

Observations $Carto_1$ and GPS_1 are used to update the probability distributions of hidden variable X_1. In this trivial case, we are certain that the vehicle is driving on road segment number 1 ($P(S_1{=}seg_1|Carto_1, GPS_1){=}1$). The position of the vehicle on segment seg_1 is given by the hidden variable

$$P(X_1|Carto_1, GPS_1) \sim N(\mu_1, \Sigma_1),$$

where $\mu_1 = (x_1, y_1, \theta_1)$ and Σ_1 represents the covariance matrix (the uncertainty of this estimation).

t	$Carto_t$	$P(S_t \mid Carto_t, GPS_t)$	$P(X_t \mid Carto_t, GPS_t)$
1	$\{seg_1\}$	$P(S_1 = seg_1 \mid Carto_1, GPS_1) = 1$	$\mu_1^{seg_1}, \Sigma_1^{seg_1}$
2	$\{seg_2, seg_3\}$	$P(S_2 = seg_2 \mid Carto_2, GPS_2) = 0.75$	$\mu_2^{seg_2}, \Sigma_2^{seg_2}$
		$P(S_2 = seg_3 \mid Carto_2, GPS_2) = 0.25$	$\mu_2^{seg_3}, \Sigma_2^{seg_3}$
3	$\{seg_4, seg_5, seg_6, seg_7\}$	$P(S_3 = seg_4 \mid Carto_3, GPS_3) = 0.55$	$\mu_3^{seg_4}, \Sigma_3^{seg_4}$
		$P(S_3 = seg_5 \mid Carto_3, GPS_3) = 0.25$	$\mu_3^{seg_5}, \Sigma_3^{seg_5}$
		$P(S_3 = seg_6 \mid Carto_3, GPS_3) = 0.15$	$\mu_3^{seg_6}, \Sigma_3^{seg_6}$
		$P(S_3 = seg_7 \mid Carto_3, GPS_3) = 0.05$	$\mu_3^{seg_7}, \Sigma_3^{seg_7}$
4	$\{seg_8\}$	$P(S_4 = seg_8 \mid Carto_4, GPS_4) = 1$	$\mu_4^{seg_8}, \Sigma_4^{seg_8}$

Table 4.1. *Details of the inference process given by SKF for each road segment*

At time $t = 2$, the uncertainty of the sensors (odometer, map, GPS) means that segments 2 and 3 are two candidates. These segments are used to generate two cartographic observations:

$$Carto_2 = \left\{ (x_2^{seg_2}, y_2^{seg_2}, \theta_2^{seg_2}), (x_2^{seg_3}, y_2^{seg_3}, \theta_2^{seg_3}) \right\}.$$

The update of the Bayesian network is done using new observations, $Carto_2$ and GPS_2, and we obtain two new estimations corresponding to the two segments $P(S_2{=}seg_2|Carto_2,GPS_2){=}0.75$ and $P(S_2{=}seg_3|Carto_2, GPS_2){=}0.25$. In this case, we have a 75% chance that the vehicle is driving on road segment number 2 and a 25% chance that the vehicle is on road segment number 3. The position of the vehicle on each segment is given by the joint probability: $P(X_2, S_2|Carto_2, GPS_2)$. Table 4.1 summarizes the inference process for the example where $1 \leq t \leq 4$.

4.4.5. *Illustration of the map localization method using SKF*

The trajectory chosen to test the validity of our approach is presented in Figure 4.12a. This trajectory is chosen because it presents two parallel roads and two crossings, which are the sources of ambiguity. This journey was carried out in Compiègne with the experimental vehicle Strada from the Heudiasyc laboratory at UTC.

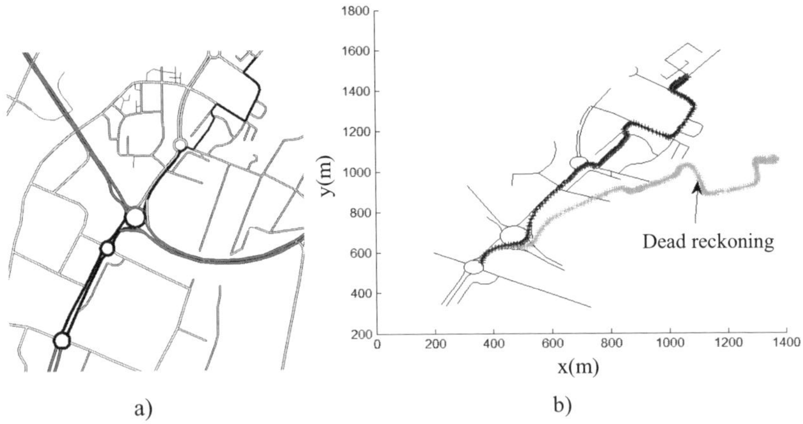

Figure 4.12. *Global view of the test journey*

We drew the estimations given by the odometric model on Figure 4.12b. The major drawback with odometry is its drift over time due to slipping of the wheels and the imperfection of the roads (holes, bumps, etc.). This drift is well represented in this figure.

In the first experiment (see Figure 4.12a), we used the GPS data at the beginning of the test; then a hidden GPS was simulated for the rest of the experiment. Figure 4.13 showed the speed of the vehicle throughout the test. In the absence of the information given by the GPS, the localization algorithm becomes an algorithm using only odometry and cartography.

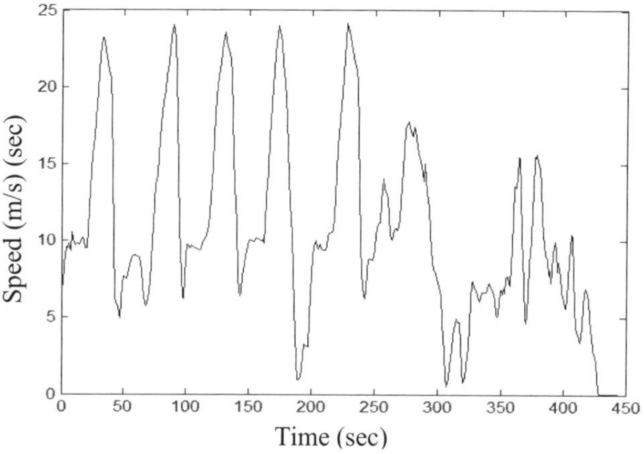

Figure 4.13. *Speed of the vehicle during the test*

4.4.5.1. *First ambiguous situation: the case of parallel roads*

In the situation of parallel roads, we want to show how the Bayesian network deals with this type of ambiguous situation, even in the case of GPS hiding.

In the case shown in Figure 4.14, there are two parallel roads. This is a typical case of ambiguity. With the suggested approach, the Bayesian network manages several segments during several time steps until the presence of more accurate GPS data enable it to minimize the uncertainty of the estimate. Let us note that we have not taken the direction of the traffic into account in any of our experiments.

Figure 4.15 provides more details about the ambiguity zone in Figure 4.14. The most likely segment in each step is represented by a dotted line. The probability distribution of each segment is given in Table 4.2.

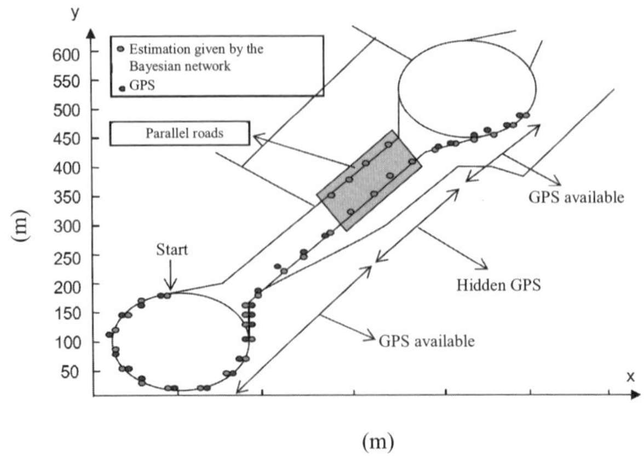

Figure 4.14. *Management of ambiguity (due to parallel roads) when using SKF*

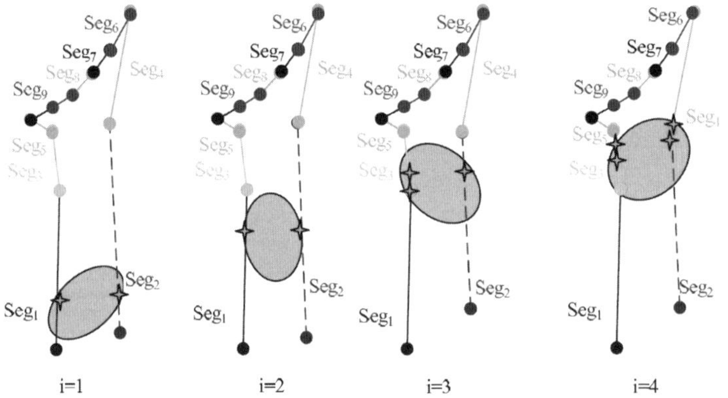

Figure 4.15. *Details on the ambiguity zone.*
The most likely segment is shown by the dotted line

i	Segments	$P(Seg_i)$
1	Seg_1, Seg_2	$P(Seg_1) = 0.20, P(Seg_2) = 0.79$
2	Seg_1, Seg_2	$P(Seg_1) = 0.20, P(Seg_2) = 0.79$
3	Seg_1, Seg_2, Seg_3	$P(Seg_1) = 0.10, P(Seg_2) = 0.71, P(Seg_3) = 0.18$
4	$Seg_1, Seg_2, Seg_3, Seg_4$	$P(Seg_1) = 0.06, P(Seg_2) = 0.64, P(Seg_3) = 0.15, P(Seg_4) = 0.14$

Table 4.2. *Details of the inference process given by SKF*
for each road segment

4.4.5.2. *Second ambiguous situation: the case of a road junction*

With this part of the trajectory, we can see how SKF deals with ambiguous situations in the case of a road junction, which is often found at the entrance or the exit to a crossing.

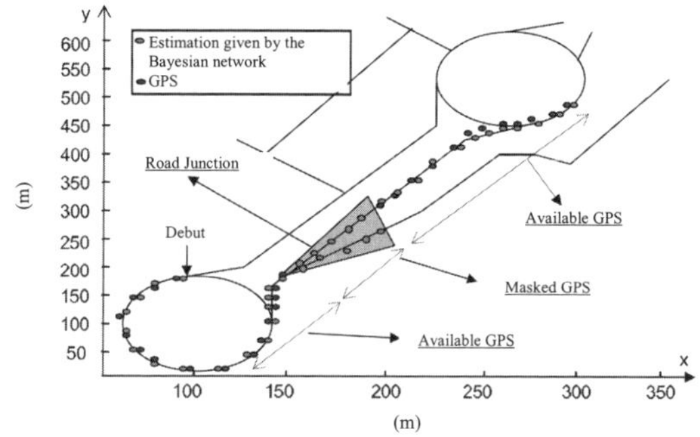

Figure 4.16. *Management of the ambiguity (a road junction) when using SKF*

To simulate the hiding of the GPS data, we stop using these data just before the road junction (see Figure 4.16). In the absence of accurate information, the localization algorithm only uses odometry and cartography. The uncertainty surrounding the estimate increases, which forces the SKF to deal with several segments over several time steps until the presence of GPS data that minimize the uncertainty around the estimate, and as a consequence the number of candidate segments decreases.

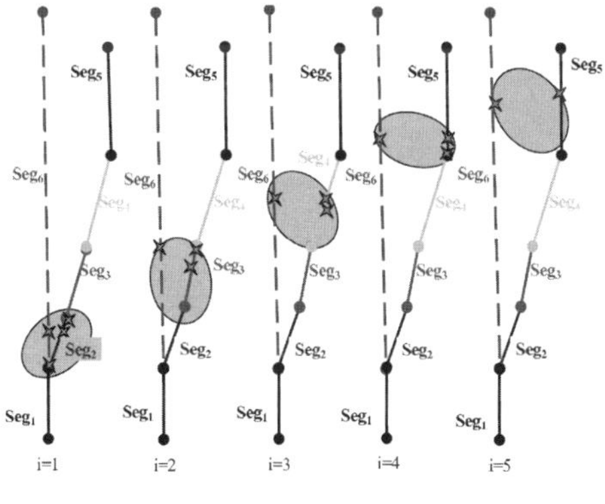

Figure 4.17. *Details on the ambiguity zone: most likely segment as a dotted line*

Figure 4.17 gives more details on the ambiguous zone in Figure 4.16. According to Table 4.3, we can see from among the different hypotheses (candidate segments) that SKF gives a greater probability to the right-hand segment. These results were published in [SMA 08a, SMA 08b].

i	Segments	$P(Seg_i)$
1	$Seg_1, Seg_2, Seg_3, Seg_6$	$P(Seg_1) = 0.08, P(Seg_2) = 0.13, P(Seg_3) = 0.14, P(Seg_6) = 0.65$
2	Seg_2, Seg_3, Seg_6	$P(Seg_2) = 0.07, P(Seg_3) = 0.18, P(Seg_6) = 0.74$
3	Seg_3, Seg_4, Seg_6	$P(Seg_3) = 0.13, P(Seg_4) = 0.15, P(Seg_6) = 0.72$
4	Seg_4, Seg_5, Seg_6	$P(Seg_4) = 0.2, P(Seg_5) = 0.11, P(Seg_6) = 0.69$
5	Seg_5, Seg_6	$P(Seg_5) = 0.28, P(Seg_6) = 0.71$

Table 4.3. *Details of the inference process given by SKF for each road segment*

4.5. Multi-vehicle localization

Over the past decade a large number of vehicles referred to as "smart" cars have appeared. These cars provide the user with a mode of short-range transport adapted to the city, decreasing travel expenses, reducing stress generated by driving, reducing sound and atmospheric pollution, etc. Among these smart cars, there are instances when cars move in a convoy.

4.5.1. *The problem studied*

The task of navigating a convoy involves moving a set of vehicles in a queue from one point to another (following a reference path or the head car) by keeping a predefined gap between the vehicles (see Figure 4.18). In order to carry out this task, each vehicle in the convoy must be able to place itself with respect to the reference path and to the preceding vehicles.

In this context, most studies and experiments are based on control. The authors of these pieces of work assume that each vehicle is able to locate itself with respect to the reference path via a very accurate GPS [BOM 06, THU 02]. The use of the GPS for localization is not always possible, however, because of the irregularity of the signals given by the satellites: the absence of GPS in a tunnel, an erroneous estimation if the number of satellites is insufficient, etc.

We are interested in the issue of the localization of a convoy (a chain of vehicles). The merging of data will allow us to overcome the aforementioned GPS problems. Let us note for the sake of simplicity that during the simulation we will use a simple command (a proportional command).

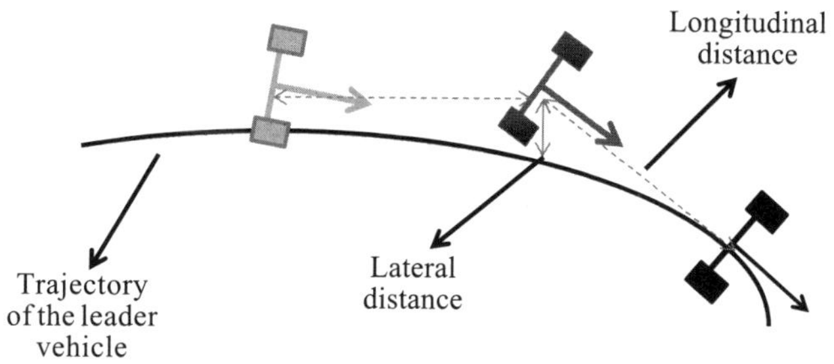

Figure 4.18. *Representation of a convoy consisting of a lead vehicle and two followers*

4.5.2. *Communication within the convoy*

For a set of vehicles in a queue to follow a reference path while keeping a predefined gap between the vehicles, we assume that they can communicate with each other. The question that is then raised is:

– what do we want to transmit?

– and to whom?

In fact, if we assume that the vehicles follow the reference trajectory given by the lead vehicle, this trajectory must be transmitted to at least the first vehicle or to all of the vehicles in the convoy. In order to avoid the issue of cumulative errors, we transmit the trajectory of the lead vehicle to all of the cars making up the convoy.

The second thing to be taken into account when we want to control a set of vehicles is the maintainance of a safe distance between each pair of vehicles. Thus, each vehicle must know the speed (or position) of the preceding vehicle.

From a practical point of view, we only need these two pieces of information (the trajectory of the lead vehicle and the speed of the preceding vehicle) for the management of the convoy.

4.5.3. *Sensors used on each vehicle in the convoy*

The flexibility of the use of Bayesian networks allows us to add or remove sensors quite easily. Each sensor is modeled and represented in the Bayesian network by a variable referred to as an observation that is used to update the estimation of the position and course (x, y, θ).

The trajectory of the lead vehicle (known by all of the vehicles in the convoy) must be highly accurate. Thus, the leader vehicle is equipped with a centimetric localization system such as GPS-LRK. This GPS provides measurements with a sampling frequency of 10 Hz, with a precision of 2 cm [BOM 06]. In order to provide the lead vehicle with a good estimation of its course, we provide this vehicle with a second "gyroscopic" sensor.

In several pieces of work, the follower vehicles are equipped with a GPS-RTK centimetric positioning system [BOM 06, MAR 05] and a telemeter.

In order to emphasize the concept of merging using Bayesian networks, the follower vehicles are assumed to be equipped with a GPS of metric accuracy. A telemeter is placed at the front of each car to give an estimation of the distance between a vehicle and the one preceding it and the angle with respect to the axis of the abscissa.

We believe that merging the distance given by the telemeter and the centimetric position given by the GPS-LRK allows us to decrease the uncertainty surrounding the estimation given by the GPS of the first follower vehicle. Thus, each vehicle can correct its position with respect to its predecessor.

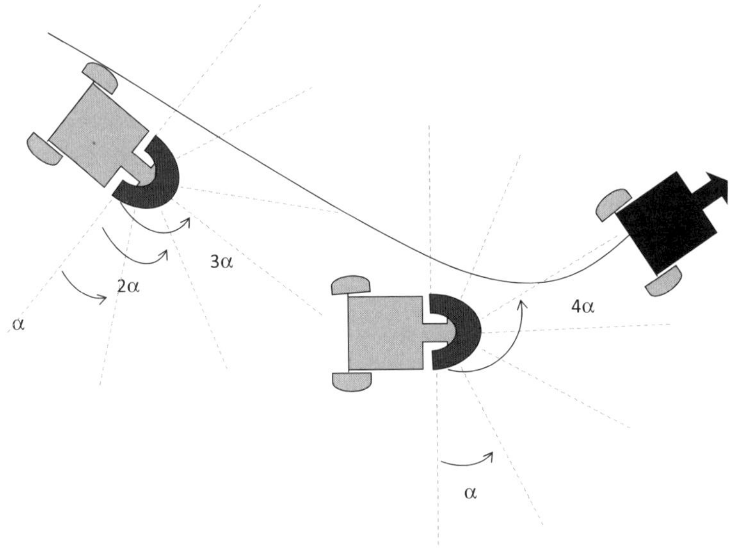

Figure 4.19. *The telemeter gives the distance and the angle with respect to the axis of the abscissa between two vehicles*

4.5.4. *Bayesian network for the localization of a chain of vehicles*

To ensure that a chain of vehicles follows a reference trajectory while keeping a predefined gap between the vehicles, we consider the following hypotheses:

– the lead vehicle is equiped with a centimetric localization system (GPS RTK) and a gyroscope to measure the angle of the bend;

– the lead vehicle is driven by a human being;

– the reference path is transmitted to all of the vehicles in the convoy;

– the follower vehicles are equiped with a GPS receiver and a telemeter to give the distance between the vehicles;

– each vehicle in the convoy must maintain a safe distance (DistS = longitudinal distance) from the preceding vehicle and control the direction actuator to decrease the lateral gap (DistL = lateral distance) with the reference path (see Figure 4.18); and

– the evolution model of each vehicle is an odometric model [ELB 03], with integration of the elementary displacement and rotation. The elementary displacement is used to maintain a safe distance (DistS) between a vehicle and its predecessor. The elementary rotation is used to actuate the direction of the wheels to decrease the lateral gap (DistL) with the reference path.

4.5.5. *Extension of the approach: modeling and localization of a chain of vehicles*

The model of a Bayesian network for the localization of a convoy of vehicles is given in Figure 4.20. The lead vehicle X_T is equipped with a centimetric localization system modeled by variable GPS = (x_{GPS}, y_{GPS}). A second sensor (gyroscope) is used by the lead vehicle in order to determine the bend angle (represented by variable Gyro). The position and course of the lead vehicle $X_T=(x_T, y_T, \theta_T)$ are updated by the observations made by the GPS and Gyro.

Since the distance given by the telemeter requires two positions – (x_1,y_1) and (x_2,y_2) – a relationship linking the two points is written in the following form:

$$\text{Telemeter} = \sqrt{(x_2 - x_1)^2 + (y_2 - y_1)^2} \quad [4.5]$$

The telemeter variable is thus represented in the graph (Figure 4.20) between X_{si} and X_{si+1} for the distance between the position of follower i and follower $i + 1$.

The idea is that, knowing the centimetric position of the lead vehicle and the distance separating it from the first follower (which is given by the telemeter), we can decrease the uncertainty around the estimation of the first follower. Given that the mean and covariance of the first follower have changed (there is more accuracy of the position), this new position and the distance linking the two vehicles give a new estimation for the second vehicle (see Figure 4.21).

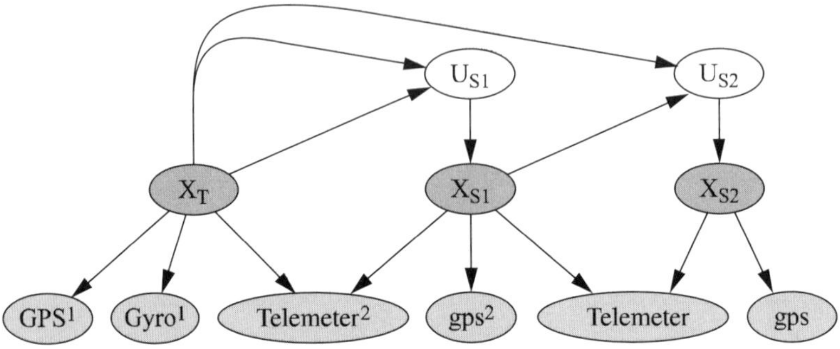

Figure 4.20. *Model of a Bayesian network for the localization of a convoy made up of a lead vehicle and two follower vehicles*

Each follower vehicle is modeled in Figure 4.20 by its state X_{Si}. Each vehicle is equipped with a GPS localization system (differential or natural) and a telemeter. Merging these two sensors provides a new estimation of the position of each vehicle. The trajectory of the lead vehicle must be transmitted to all of the follower vehicles and as a consequence links (X_T,U_{S1}) and (X_T,U_{S2}) will be used to control the course of the vehicle and decrease the lateral gap (DistL) with the reference path (these links only represent the transmission of the trajectory of the leader vehicle to the follower vehicles and not the probability $P(U_{S1}|X_T)$, $P(U_{S2}|X_T)$). Finally, to maintain a safe distance (DistS), each vehicle must know the speed (or position) of the vehicle in front, which is represented in Figure 4.20 by the links (X_T,U_{S1}) and (X_{S1},U_{S2}). (These links represent only the transmission of the speed or the position of the follower vehicle and not the probability $P(U_{S1}|X_T)$, $P(U_{S2}|X_{S1})$.)

Let us note that the lead vehicle (driven by a human) does not possess control variable U, contrary to the follower vehicles (U_{S1}, U_{S2}). Let us also note that we have removed the cartographic observation. The flexibility of Bayesian networks allows us to add other observations at any time.

Estimated position X_{Si} is the result of the fusion of observations provided by the GPS, the distance given by the telemeter and input U_{Si}. Let us remind ourselves that input $U_{Si} = (d_k, w_k)$ represents the command given to the vehicle to follow the trajectory of the lead vehicle and keep a predefined gap with the preceding vehicle.

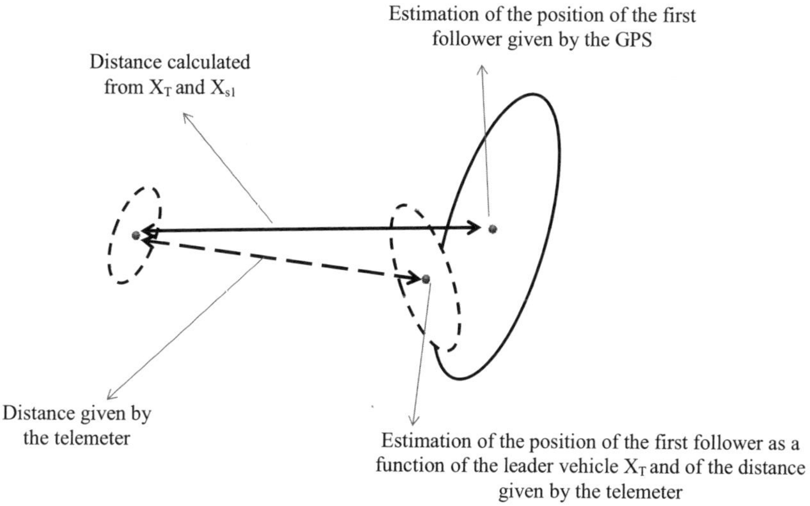

Figure 4.21. *This diagram shows how to decrease the uncertainty produced by the GPS receiver of the follower vehicles by using the distance given by their telemeter and the position of the leader vehicle. This correction is made step-by-step*

Let us note that the new observation equation (GPS, Gyro) for the lead vehicle is given by equation [4.6]. The observation equation (GPS) for the follower vehicles is the same as for the case of a single vehicle:

$$Y_k^T = \begin{bmatrix} x_k^{GPS} \\ y_k^{GPS} \\ \theta_k^{GPS} \end{bmatrix} = \begin{pmatrix} 1 & 0 & 0 \\ 0 & 1 & 0 \\ 0 & 0 & 1 \end{pmatrix} \cdot \begin{pmatrix} x_k \\ y_k \\ \theta_k \end{pmatrix} + \alpha_k \qquad [4.6]$$

where $\alpha_k \sim N(\mu,Q)$. Covariance matrix Q is given by:

$$Q = \begin{pmatrix} \sigma_x^2 & \sigma_{xy} & 0 \\ \sigma_{xy} & \sigma_y^2 & 0 \\ 0 & 0 & \sigma_\theta^2 \end{pmatrix} \qquad [4.7]$$

where σ_x^2, σ_y^2 and σ_θ^2 respectively represent the deviations of the measurements in longitude, latitude and course. σ_{xy} represents the covariance between x and y.

4.5.6. *The issue with this model*

The observation equations handled so far (GPS, cartography and gyroscope) have been linear equations. The only observation that cannot be written in a linear form, however, is the distance given by the telemeter (see equation [4.5]). In order to keep the observation equations linear, a transformation is carried out on the Bayesian network indicated in Figure 4.20.

4.5.7. *New model for the localization of a chain of vehicles*

The variable representing the telemeter in the network in Figure 4.20 is a nonlinear function of Gaussian variables (X_T, X_{S1}, X_{S2}).

The issues we faced were the following:

– we needed to rewrite the distance given by the telemeter in a linear form; and

– take advantage of the accuracy of the estimation given by the GPS of the lead vehicle and the telemeter in order to give a more accurate estimation of each follower vehicle.

In order to take into account these remarks, we have suggested the following idea. As long as each follower vehicle has a telemeter to estimate the distance between it and the preceding vehicle and the trajectory of the lead vehicle, we proceed as follows (see Figure 4.22).

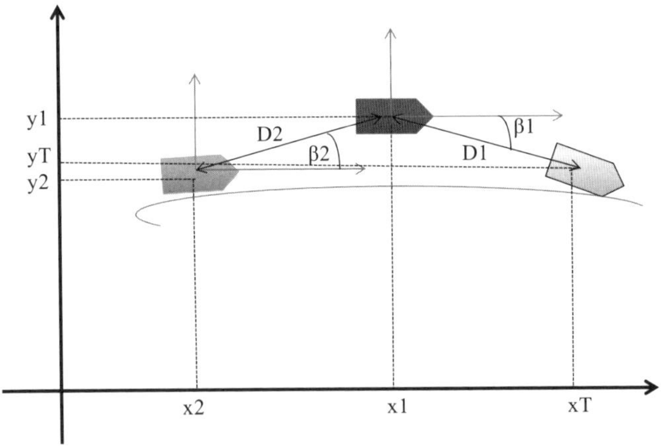

Figure 4.22. *New estimation built from the position of the preceding vehicle and from the distance measured by the telemeter*

The first follower vehicle S_1 estimates its location from the position and the distance separating it from the lead vehicle. Knowing distance D_1 (given by its telemeter) and the position of the leader vehicle (x_T, y_T), which is transmitted by the latter, we build a new observation as follows:

$$\begin{cases} x_{S_1}^{new} = x_T - D_1 \cos(\beta_1) \\ y_{S_1}^{new} = y_T - D_1 \sin(\beta_1) \end{cases} \quad [4.8]$$

Similarly, the second follower vehicle estimates its position from that of the preceding vehicle (S_1) and from the distance D_2 given by its telemeter. Thus, each follower vehicle S_i where $2 \leq i \leq n$ estimates its position is as follows:

$$\begin{cases} x_{S_i}^{new} = x_{i-1} - D_i \cos(\beta_i) \\ y_{S_i}^{new} = y_{i-1} - D_i \sin(\beta_i) \end{cases} \quad [4.9]$$

The new observation built from the distance given by the telemeter and the position of the predecessor is therefore written:

$$Y_k^{new} = \begin{bmatrix} x_k^{new} \\ y_k^{new} \end{bmatrix} = \begin{pmatrix} 1 & 0 & 0 \\ 0 & 1 & 0 \end{pmatrix} \begin{pmatrix} x_k \\ y_k \end{pmatrix} + \alpha_k \quad [4.10]$$

where α_k is the Gaussian noise generated by the telemeter. x_k and y_k are the position of the vehicle.

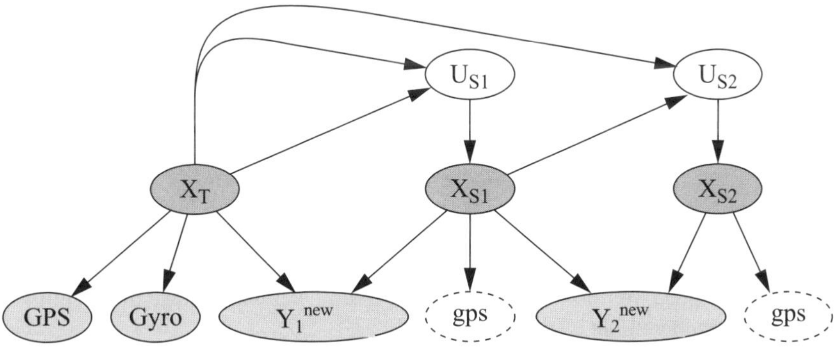

Figure 4.23. *New model of a Bayesian network for the localization of a convoy consisting of a lead vehicle and two follower vehicles*

The new model for the localization of a convoy of vehicles is given in Figure 4.23. This model is equivalent to that given in Figure 4.20. The position estimated for each follower vehicle (X_{Si}) is the result of merging the newly-generated observation Y_i^{new} and GPS measurements.

4.5.8. *Proportional commands*

In order for a set of vehicles aligned in a queue to follow a reference trajectory given by the lead vehicle while keeping a predefined gap between each other, each vehicle in the convoy must know its position and the position (or the speed) of the vehicle that precedes it. Let us remind ourselves that in this piece of work, we were interested in localization and not command. To validate our localization experiments, we used simple commands referred to as proportional commands.

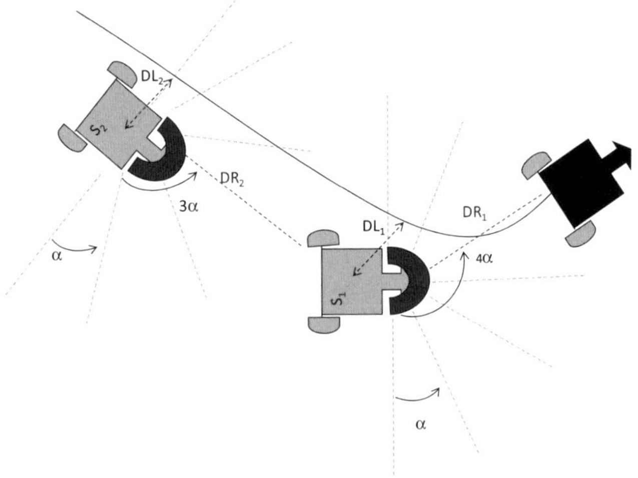

Figure 4.24. *A model of a convoy consisting of a lead vehicle and two follower vehicles. The true distance given by the telemeter is represented by DR_1 and DR_2. The lateral distance between each vehicle with respect to the trajectory of the lead vehicle is given by DL_1 and DL_2*

By referring to Figure 4.24, each vehicle S_i in the convoy can adjust its speed V_{Si} knowing the speed (or position) of the preceding vehicle. Thus, the longitudinal command of each vehicle is given by:

$$VS_i = VS_{i-1} + k_1(DR_i - DS) \qquad [4.11]$$

where DR_i represents the true distance given by the telemeter, and k_1 is a proportionality factor enabling us to specify the rate of convergence towards a safe distance, *DS*.

By still referring to Figure 4.24, each vehicle in the convoy must adjust its course so that it matches the trajectory of the lead vehicle. Thus, knowing the course of the lead vehicle θ^T at time *k* (the tangent to the curve at the perpendicular point of the position of each vehicle) and its own course θ^{Si}, we calculate the difference:

$$\Delta cap = \left|\left(\theta^T - \theta^{S_i}\right)\right| \quad [4.12]$$

The lateral command of each vehicle is given:

$$\theta_{nouv}^{S_i} = \theta^{S_i} \pm \left(\Delta cap + k_2.DL_i\right) \quad [4.13]$$

where DL_i represents the lateral distance of each vehicle with respect to the reference trajectory and k_2 is a proportionality factor enabling us to converge towards the reference trajectory indicated by the lead vehicle.

Let us note that equation [4.13] does not take into account the lock angle of the wheels. In this case, the vehicles are considered (represented) to be two-wheel robots.

Figure 4.25. *Experimental vehicle (CyCab) created by the INRIA team in Nancy*

Figures 4.25 and 4.26 show the experimental vehicles used within the framework of the FD2S and FDADI projects. In the test presented, the lead vehicle is manually driven on the journey given in Figure 4.27. The light gray curve represents the trajectory acquired by GPS receiver Sagitta 02 in LRK centimetric mode. In this experiment, we assume that the vehicles are initially on the reference trajectory and do not necessarily have the same course as the lead vehicle. Each follower vehicle has the trajectory of the lead vehicle and the speed of the preceding vehicle at its disposal via Wifi and in real time. The follower vehicles determine their position by using a LASER telemeter and the position of the preceding vehicle. In Figure 4.27, we can see that the follower vehicles perfectly follow the trajectory of the lead vehicle. Figure 4.28 gives more details on zone which is circled in Figure 4.27.

Figure 4.26. *The LAGIS TRAVEL platform*

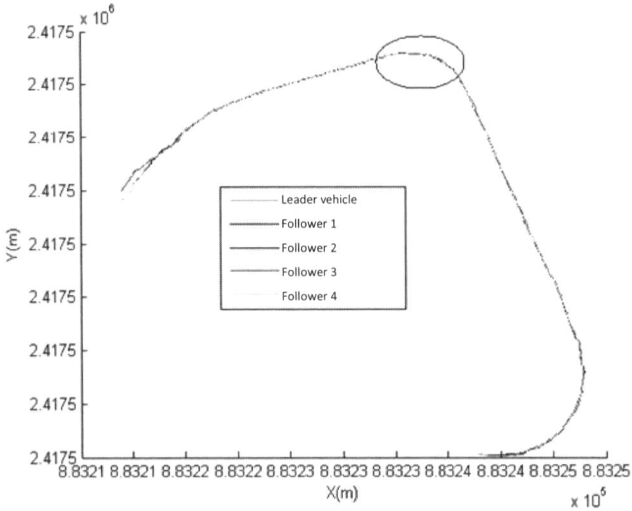

Figure 4.27. *Trajectories of the follower vehicles given by RB*

Another point to take into account in the formation of a vehicle convoy is the speed control of the different vehicles in a convoy to continue to keep the specified gap between vehicles. This command can be set via interdistance or relative time between the vehicles. In our experiments, the interdistance threshold between two vehicles was set to 1.5 meters.

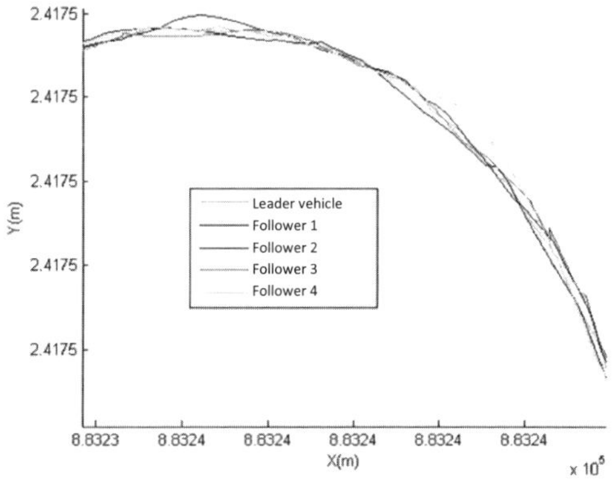

Figure 4.28. *Close-up of the circled area in Figure 4.27*

Figure 4.29 gives the interdistance between the vehicles as a function of their evolution time. We can see that the distance threshold was adhered to. The true speed of the lead vehicle shows several oscillations that influenced the distance curves at the beginning of the test, the longitudinal command (simply proportional) being so elementary. This aspect and the impact of the chosen model were assessed in order to be dealt with in the functional analysis carried out within the FD2S project framework. It is presented in the following section.

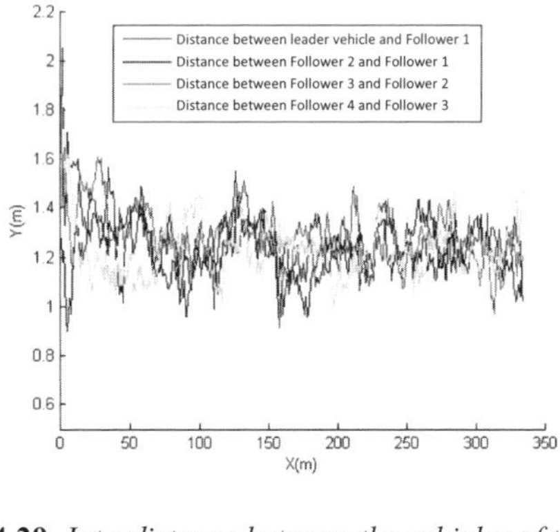

Figure 4.29. *Interdistance between the vehicles of the convoy*

4.5.9. *Functional analysis of models of the convoy*

Within the framework of the FD2S project, a functional study of the chosen convoy models was carried out [GLA 08]. The dynamic functional analysis of the models consists of evaluating a simulation of the functional properties and robustness of the algorithms at stake. By ignoring the material constraints, the study aims to focus on the longitudinal control (1D simulation) and the lateral control (2D simulation) of the convoy. By designing a dedicated simulator, the goal is to explore the following properties:

– the non-collision between vehicles;

– the global stability of the convoy (does it converge towards a stable state or not, and under which hypotheses?); and

– the preservation of the global coherence of the convoy faced with particular perturbating events (the sudden stopping of a vehicle, large interdistance between vehicles, radius of the curvature, etc.).

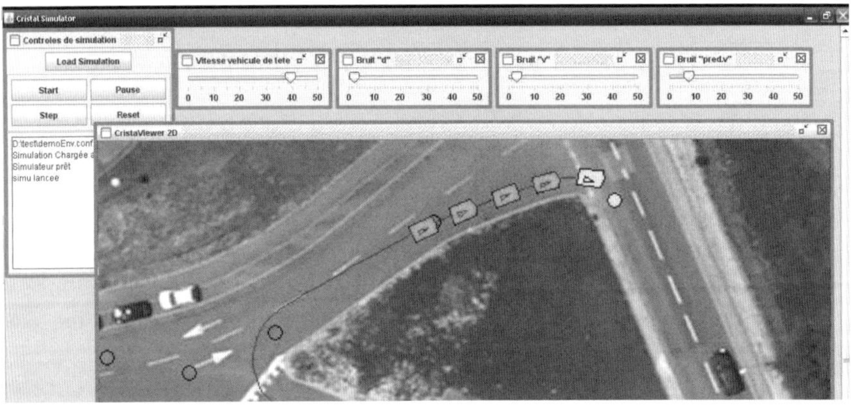

Figure 4.30. *Window of 2D simulation with the integration of aerial views*

As well as testing the robustness of the system using specific scenarios presenting the extreme cases (worst cases), the presence of noise (simulated error) in sensor measurements, and a systematic study of the parameters in order to establish the ranges of application and the optimal values of operation, this procedure lies on several levels of simulation according to an incremental “granularity”:

– a simplistic 1D model of the vehicles (point model), ideal sensors and actuators;

– a 2D tricycle model [BOM 06] of vehicles, the integration of noise on the sensors and actuators; and

– a 2D model connected to a physical motor, with the integration of constraints such as features of the ground (adherence), slope, etc.

The third level of simulation is carried out by running fine models of existing vehicles. Within this framework we aim to use a simulator such as Callas that enables us to link the results with the supervision module.

4.6. Conclusions and perspectives

The pieces of work presented in this chapter have shown the relevance of using Bayesian networks as a tool to merge multisensor data and a decision algorithm through the use of probabilistic graph theory. The Bayesian network proves to be a formalism that is potentially very useful because it can integrate the data fusion, diagnostics, learning and control in the same formalism. The other very interesting criterion that led us to choose this formalism is the reconfiguration aspect of an approach based on Bayesian networks by using specific configurations such as *switching state–space model*, *jump-Markov model* or *SKF* [MUR 02]. In the previous pieces of rearch, it is *SKF* that was put in place.

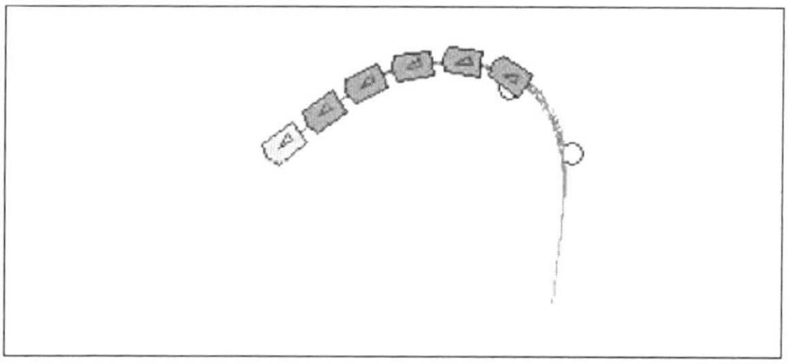

Figure 4.31. *Illustration of the study of lateral control on specific scenarios*

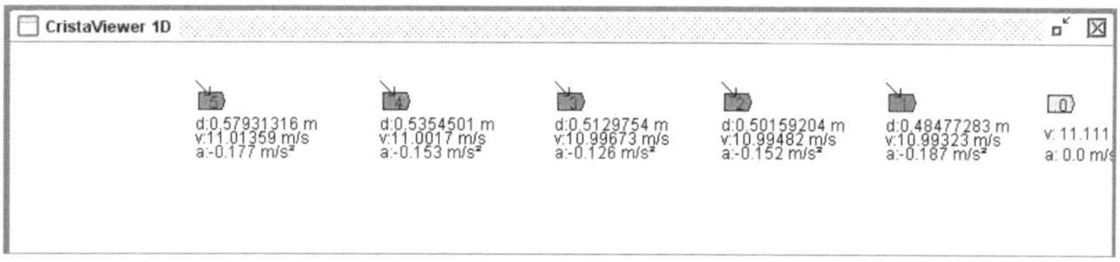

Figure 4.32. *Window of the 1D simulation allowing us to zoom in on the longitudinal control of the convoy*

Within the framework of the GIS 3SGS ongoing FDADI project, we suggest integrating a diagnostic layer and sensor management before merging the data. We suggest adopting a complementary approach with the use of information theory, and in particular Shannon's entropy [DES 98]. Whereas Bayes' criterion relies on the minimization of the error probability, the entropy enables better results in terms of ROC (receiver operating characteristic) curves (where the probability of detection as a function of the probability of false alarms), mainly when the *a priori* probabilities of the hypotheses implemented are unknown or weak. Finally, the method of model construction in the form of trees is generally carried out by using a *top-down* approach which, in the end does not give an optimal model, contrary to the *bottom-up* approach that we are implementing.

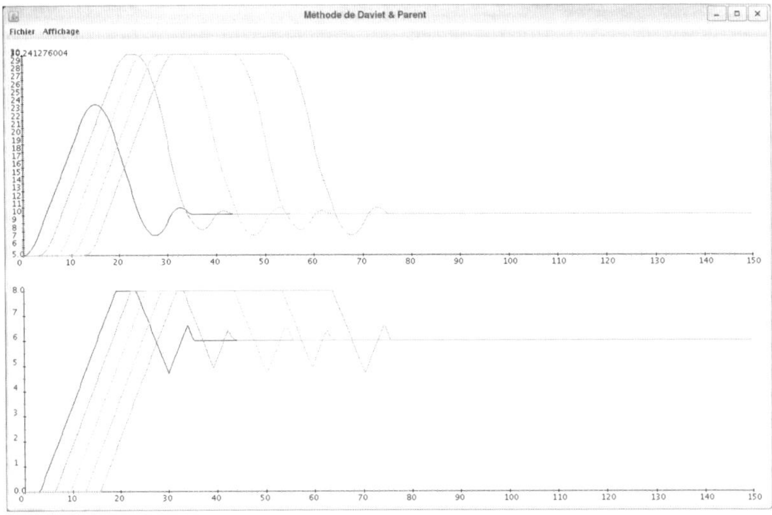

Figure 4.33. *Module to draw curves of interdistances and speeds against time. Illustration of a step-by-step model with six vehicles with an initial interdistance of 5 m. The lead vehicle aims for 6m/s (21.6 km/h), with the command update after 10 ms with convergence in 76 s*

The issue in which we are currently interested within the framework of the FDADI project is the absolute localization of each vehicle in a convoy with an open GPS (which provides raw satellite data). The goal is to tackle the problem upstream with "tight coupling" [BON 05] instead of "loose coupling", as was the case within the framework of the FD2S project.

The localization approach under development comes within the context of studying the contribution of the infrastructure–vehicle interaction. More particularly, it aims to study the contribution of local information drawn from the evolution of the environment on monitoring of the integrity and loss of GPS measurements made in an urban environment.

The goal of our projects is to develop a method of data fusion with an integrated diagnostic of errors that can affect pseudo-distances (satellite–receiver distance) such as the multipath phenomenon of waves. We will develop a tightly coupled approach by looking to fuse the pseudo-distances of a G*lobal navigation satellite system* with measurements built with a LASER telemeter and a 3D cartographic database of the environment. The design of such an approach aims to use the redundancy of information in order to diagnose errors in satellite pseudo-distances for the elimination of failing satellite measurements from the procedure of position calculation.

4.7. Bibliography

[ARU 02] ARULAMPALAM S., MASKELL S., GORDON N., "A tutorial on particle filters for online nonlinear/non-Gaussian Bayesian tracking", *IEEE Transactions on Signal Processing*, vol. 50, pp. 174-188, 2002.

[BAR 82] BAR-SHALOM Y., CHANG K., BLOM H., "Tracking a manoeuvring target using input estimation versus the Interacting Multiple Model Algorithm", *IEEE Transactions on Aerospace and Electronic Systems*, vol. AES-25, no. 2, pp. 296-300, 1982.

[BAR 88] BAR-SHALOM Y., FORTMANN T., *Tracking and Data Association*, Academic Press, New York, 1988.

[BER 99] BERGMAN N., Recursive Bayesian estimation: Navigation and tracking applications, PhD Thesis, Linkoping University, Sweden, 1999.

[BLO 88] BLOM H., BAR-SHALOM Y., "The interacting multiple model algorithm for systems with Markovian switching coefficients", *IEEE Transactions on Auto. Contr.*, vol. AC-33, no. 8, pp. 780-783, 1988.

[BOM 06] BOM J., Etude et mise en œuvre d'un convoi de véhicules urbains avec accrochage immatériel, PhD Thesis, University Blaise Pascal – Clermont II, July 2006.

[BON 05] BONNIFAIT P., Contribution à la localisation dynamique d'automobiles. Application à l'aide à la conduite, HDR Thesis, University of Technology of Compiègne, December 2005.

[CRI 00] CRISAN D., DOUCET A., Convergence of Seqeuntial Monte Carlo Methods, Technical Report *CUED/F-INFENG/TR381*, Signal Processing Group, Departement of Engineering, University of Cambridge, 2000.

[DES 98] DESROUSSEAUX C., Utilisation d'un critère entropique dans les systèmes de détection, Thesis, Lille University I, number 2381, December 1998.

[DIS 01] DISSANAYAKE M.-G., NEWMAN P., CLARK S., DURRANT-WHYTE H., "A solution to the simultaneous localization and map building (SLAM) problem", *IEEE Transactions on Robotics and Automation*, vol. 17, no. 3, pp. 229-241, 2001.

[ELB 03] EL BADAOUI EL NAJJAR M., Localisation dynamique d'un véhicule sur une carte routière numérique pour l'assistance à la conduite, PhD thesis, University of Technology of Compiègne, France, 2003.

[FEA 98] FEARNHEAD P., Sequential Monte Carlo methods in filter theory, PhD thesis, University of Oxford, 1998.

[GLA 08] GLAD A., SIMONIN O., BUFFET O., CHARPILLET F., "Theoretical study of ant-based algorithms for multi-agent patrolling", *ECAI'08 18th European Conference on Artificial Intelligence in Proceedings*, pp. 626-630, 2008.

[GOR 93] GORDON N.-J., SALMOND D.-J., SMITH A.-F.-M., "A novel approach to nonlinear/non-Gaussian Bayesian state estimation", *IEE Proceedings on Radar and Signal Processing*, vol. 140, pp. 107-113, 1993.

[GUS 02] GUSTAFSSON F., GUNNARSSON F., BERGMAN N., FORSSELL U., JANSSON J., KARLSSON R., NORDLUND P., "Particle filters for positioning, navigation, and tracking", *IEEE Transactions on Signal Processing*, vol. 50, no. 2, pp. 425-437, 2002.

[JAB 07] JABBOUR M., Localisation de véhicules en milieu urbain à l'aide d'un lidar et d'une base de données navigables, PhD thesis, University of Technology of Compiègne, France, November 2007.

[JEN 90] JENSEN F., LAURITZEN S., OLSEN K., "Bayesian updating in recursive graphical models by local computations", *Computational Statisticals Quarterly*, vol. 4, pp. 269-282, 1990.

[JUL 97] JULIER S., UHLMANN J., "A new extension of the Kalman filter to nonlinear systems", *International Symposium on Aerospace/Defense Sensing, Simulation and Controls*, pp. 182-193, Orlando, Florida, USA, 1997.

[KAL 60] KALMAN R.-E., "A new approach to linear filtering and prediction problems", *Transaction of the AMSE – Journal of Basic Engineering*, vol. 82, pp. 34-45, 1960.

[LAU 89] LAURITZEN S.-L., WERMUTH N., "Graphical models for associations between variables, some of which are qualitative and some quantitative", *Annals of Statistics*, vol. 17, pp. 31-57, 1989.

[LI 05] LI X., JILKOV V., "Survey of maneuvering target tracking part V: Multiple-model methods", *IEEE Trans. Aero. Elect. Syst.,* vol. 41, no. 4, pp. 112-120, October 2005.

[MAR 05] MARTINET P., THUILOT B., BOM J., "From autonomous navigation to platooning in urban context", *Proceedings of the IARP-Workshop on Adaptive and Intelligent Robots: Present and Future*, vol. 1, no.1, pp. 1-9, 2005.

[MUR 99] MURPHY P., A variational approximation for Bayesian networks with discrete and continuous latent variables, Thesis, UAI, 1999.

[MUR 02] MURPHY P., Dynamic Bayesian networks: representation, inference and learning, PhD Thesis, University of California, Berkeley, 2002.

[NOR 00] NORGAARD M., POULSEN N.-K., RAVN O., Advances in Derivative-free State Estimation for Nonlinear Systems, Technical Report *IMM-REP-1998-15*, Technical University of Denmark, 2000.

[QUD 06] QUDDUS M.A., NOLAND R.B., OCHIENG W.Y., "A high accuracy fuzzy logic based map matching algorithm for road transport", *Journal of Intelligent Transportation Systems: Technology, Planning, and Operations,* vol. 10, no. 3, pp.103-115, 2006.

[SMA 08a] SMAILI C., EL NAJJAR M.-E., CHARPILLET F., "A road matching method for precise vehicle localization using hybrid Bayesian network", *Special Issue of the International Journal of Intelligent Transportation Systems: Technology, Planning, and Operations*, vol. 12, no. 4, pp. 176-188, 2008.

[SMA 08b] SMAILI C., EL NAJJAR M.-E., CHARPILLET F., "Multi-sensor Fusion Method Using Bayesian Network for Precise Multi-vehicle Localization", *The 11th International IEEE Conference on Intelligent Transportation Systems,* Beijing, China, October 12-15, 2008.

[SMA 10] SMAILI C., Fusion des données multi-capteurs à l'aide des réseaux Bayesiens dynamiques pour la navigation d'un mobile en monde extérieur, PhD thesis, University of Nancy 2, 2010.

[THU 02] THUILOT B., CARIOU C., MARTINET P., BERDUCAT M., "Automatic guidance of a farm tractor relying on a single cp-dgps", *Autonomous Robots*, vol. 13, no. 1, pp. 53-71, 2002.

[URI 02] URI L., Hybrid Bayesian networks for reasoning about complex systems, PhD thesis, Stanford University, Department of Computer Science, October 2002.

[ZHA 97] ZHAO Y., *Intelligent Transportation Systems: Vehicle Location and Navigation Systems*, Artech House, Boston, 1997.

Chapter 5

Damage and Forecast Modeling

5.1. Introduction

The underlying issue that justifies a prognosis procedure is the implementation of maintenance policies or decision rules that take into account the history of the system, its environment, the diagnostic on its current state, and possibly the future operational conditions. Hence, we are placed straightaway in a context where online information must be integrated, and where decision rules are condition-based and/or dynamic with respect to this information. Thus, the aim is to exclude as much as possible decision-making from a mean behavior and to consider more delicate situations from the point of view of probabilistic approaches. Besides, we are often placed in a very operational approach, in the sense that we set a dynamic decision rule *a priori* and we look for tools enabling us to apply it by predicting the evolution of the system on a restricted horizon (by calculating the remaining useful lifetime in particular). In this framework, the use of models based on knowledge/learning makes sense and can happen to be better adapted and easier to implement than a probabilistic evolution model. On the contrary, on a more strategic level with an optimization point of view, it can be more useful to have at our disposal a mean performance criterion enabling us to optimize the possible decision rule(s), on an infinite horizon. A probabilistic approach will then certainly enable us to supply the required indicators. Hence, we have chosen to compare the modeling tools referred to as probabilistic – currently developed at Institut Charles Delauney (ICD) (lifetime models, ARMA models, Markov chains, stochastic processes, etc.) – and the non-probabilistic modeling tools currently developed at Centre de Recherche en Automatique de Nancy (CRAN)

Chapter written by Anne BARROS, Eric LEVRAT, Mitra FOULADIRAD, Khanh LE SON, Thomas RUIN, Benoît IUNG, Alexandre VOISIN, Maxime MONNIN, Antoine DESPUJOLS, Emmanuel REMY and Ludovic BÉNÉTRIX.

(neural networks and techniques based on artificial intelligence, etc.) at operational and strategic levels by favoring *a priori* the following two points:

– at the operational level, how are the two approaches complementary or competing?

– at the strategic level, how the results of the prognosis carried out from non-probabilistic approaches can be used to optimize decision rules with a probabilistic decision criterion?

5.1.1. *Operational level*

As shown in the literature review undertaken by Maxime Monnin [MON 09, BAN 09, CHI 99, LIA 06, MAZ 07], it turns out that in practice, the probabilistic models are used to the exclusion of non-probabilistic models (and vice versa), and it appeared interesting to us to challenge this split in DEPRADEM project. If this lack of interaction is explained partly by linking these tools to different scientific communities, it is also justified by the nature of the problems dealt with. The aim of the works developed within Khanh Le Son's master and doctoral theses that started in October 2009 is to identify how the deterioration models developed at CRAN and those developed at ICD are comparable (in particular with respect to their use in maintenance decision-making) and how they can be complementary or competing. We are interested in the type of information that is used to build these models (online and offline in particular), in the kinds of components under consideration (size, structure, quantity of indicators characterizing their state), in the nature of covariables (present and future operational conditions in particular), and in the damage and decision indicators that can be supplied. The involvement of EDF (Electricité de France) in the project must favor the emergence of real study cases or, at least, must enable us to target the work on a few typical problems.

5.1.2. *Strategic level*

At this level, the main question is to know how the two approaches can be complementary. The tools used at CRAN to forecast failures enable us to build a data history regarded as offline feedback data. Based on this latter, we can propose a lifetime distribution or a stochastic process modeling the deterioration of the system under consideration. This modeling can be used to estimate the remaining useful lifetimes and to build prognosis-based maintenance policies.

During the first stage of Khanh Le Son's thesis, we considered the operational level and implemented probabilistic approaches to predict a remaining useful life (RUL). These approaches are based on either the use of lifetime distributions that enable us to model the lifetime of a system or the use of stochastic processes that enable us to model the evolution of the system from the start to the failure. We gave more importance to the second approach that is more sophisticated. The advantage of this second approach depends on the number of states considered for the process (from two to infinity for a continuous process). It should be pointed out that the choice of the approach depends mainly on the type of data we deal with. The goal of this chapter is to give a general idea of these methods. To make a comparison with non-probabilistic approaches, we chose to work on data supplied by the International Conference on Prognostics and Health Management (PHM). The winners of the challenge [HEI 08, PEE 08, WAN 08] use non-probabilistic models based on neural networks or time series. To compare our method with the other non-probabilistic methods, we suggested a probabilistic method to model the deterioration of the system considered by the 2008 PHM data and to estimate the RUL.

The chapter is organized as follows. In section 5.2, the analysis of the sensor measurements of PHM data set is carried out. In section 5.3, a degradation indicator based on principal component analysis is proposed. In section 5.4, the prognostic method is applied with the new degradation indicator by using stochastic processes and the RUL estimation method is presented.

5.2. Preliminary study of data

5.2.1. *Structure of the database*

In the framework of the PHM challenge, two data sets are available: the training data set and the test data set. The PHM data set consists of data collected from 218 components of an unspecified system. There are three operational settings that have a substantial effect on the performance of the units. The data for each cycle of each unit include the unit ID, cycle index, three values for the operational settings, and 21 values for 21 sensor measurements connected to the state-of-the-units. The sensor data are noisy.

5.2.2. *Performance criterion for the prognostic*

The algorithm performance is evaluated by a function called score function, denoted by S, which evaluates the error in predicting the number of remaining time cycles and is defined as follows:

$$S = \sum_{k=1}^{K} S_k \qquad [5.1]$$

where

$$S_k = \begin{cases} e^{-d_k/13} - 1, & d_k \leq 0 \\ e^{d_k/10} - 1, & d_k > 0 \end{cases} \quad k = 1, ..., K \qquad [5.2]$$

and

$$d_k = \text{estimated RUL for component } k - \text{real RUL of component } k \qquad [5.3]$$

Other criteria are used in other articles, in particular the root mean square error (RMSE) is defined as follows:

$$\text{RMSE} = \sqrt{\sum_{i=1}^{N} (d_i)^2} \qquad N = 218 \qquad [5.4]$$

The lower is the score, the better is the proposed RUL estimation algorithm. In the PHM Conference challenge, the following papers obtained the best results. Wang [WAN 08] obtained the total score of 5,636 with a similarity-based approach for estimating the RUL. In [PEE 08], an artificial neural networks mixed with a Kalman filter is used and the obtained mean square error is 984. Felix in [HEI 08] with a recurrent neural networks method achieved the mean square error of 519.8.

5.2.3. *Definition of a deterioration indicator*

In the framework of probabilistic prognosis methods, a stochastic process is used to model the deterioration phenomenon, i.e. an evolution from the operating state to the failure *via* intermediate states. For this purpose, we need to have a scalar indicator that models the deterioration. In the particular case of PHM data, among all the papers that obtained the best scores, we did not find any satisfactory deterioration indicator. This is due to the fact that they use either a neural network approach where a deterioration model is not necessary [HEI 08, PEE 08] or a very simplified model as in [WAN 08].

After an initial study of the training data set, the following features are pointed out:

– A degradation indicator cannot be directly deduced from the 21 sensor paths.

– The three measurements connected to operational settings can be represented by six clusters in a three-dimensional space. This second point is illustrated in Figure 5.1, where the variables OP1, OP2, and OP3 correspond to the measures of the three sensors devoted to operational settings. The six observed dots are actually six highly concentrated clusters. Hereafter, it is supposed that these clusters indicate six discrete operating conditions of the system. Hence, each cycle of a unit can be labeled by a regime ID from 1 to 6, replacing the original three variables of operational settings.

– For each unit, the changes of modes occur frequently and can represent very different sequences.

– If we select only the measurements corresponding to the same regime ID, some of the 21 sensor measurements connected to state-of-the-units will exhibit prominent trend along the life of a unit. In [WAN 08], for each mode, the plot of the collected data from one sensor is used to show the trend of the system's degradation. Some sensors do not show a clear trend due to high noise or due to their low sensitivity to the degradation. Their inclusion in the analysis may lower the accuracy of prediction. According to [WAN 08], using a subset of the available sensors can help to improve the accuracy of RUL estimation; therefore, they propose to use a selection of seven sensors (2, 3, 4, 7, 11, 12 and 15). Hereafter, for the data analysis, we use this selection of seven sensors.

These first analyses brought out the following facts: the deterioration indicator cannot be defined directly from the measurements of 21 sensors, and the operational conditions corresponding to the six modes influence the trajectories of the 21 sensors and must be taken into account in the construction of the deterioration indicator. Furthermore, it is useless to take into account the 21 sensors, and the sensor selection suggested in [WAN 08] can be adopted.

5.3. Construction of the deterioration indicator

To build a deterioration indicator, we should identify a set of trajectories that can be considered as the realization of the same stochastic process and a real trend in these realizations that can be interpreted as the deterioration from the beginning of the lifecycle until the failure. The next sections are devoted to the construction of a deterioration indicator.

5.3.1. *Study of the failure space with PCA*

First, we apply PCA without taking into account the operational modes. We consider the data subset of the failure times for 218 units. Let X be the original data set where its rows indicate the information on individuals (at the failure time) and its columns indicate the information on variables (seven sensors). For the PCA corresponding to the chosen data set, the principal component's coefficients and variances are the eigenvectors and eigenvalues of the covariance matrix of the centering matrix of X. The criterion to choose the principal component is based on the histogram of the variances of principal components. It can be seen in Figure 5.1(a) that the cumulative percentage of the total variation of the first two principal components is more than 98%. Figure 5.1(b) shows that the projections of failure time measurements on the two-dimensional principal component space are divided into six different groups. This could be due to the fact that the operational modes and the presence of six classes complicate the construction of a degradation indicator. To overcome this problem, we separate the operational modes of the failure times to build six data sets and we apply a PCA to data corresponding to each operational mode. Therefore, we obtain six different principal component spaces. The percentages of histogram for the first two principal components in Table 5.1 show that we can use the two-dimensional principal component space, called failure space, to build a degradation indicator for each unit in each mode. Moreover, the high concentration of the positions of different component failure states on the two principal components space depicted in Figure 5.2 leads us to consider only one failure place for each operational mode. We can consider the barycenter of the cluster of failure states on the two principal components space as the unique failure place.

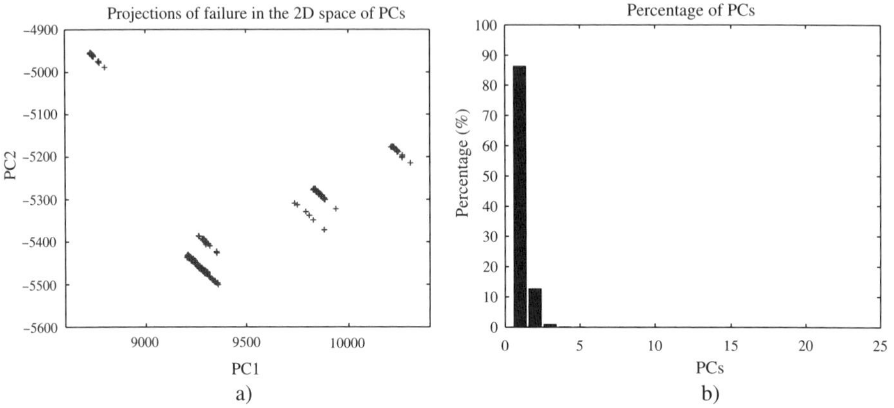

Figure 5.1. *Principal plane of PCA*

	Mode 1	Mode 2	Mode 3	Mode 4	Mode 5	Mode 6
PC1	60.85	72.64	61.45	54.41	58.95	79.65
PC2	38.04	26.75	37.85	44.55	40.07	19.11
PC3	0.66	0.28	0.34	0.56	0.41	0.77

Table 5.1. *Contribution of the first three axes of PCA in each mode*

5.3.2. *Damage indicator defined as a distance*

The degradation indicator can be calculated from this unique failure place at each time t, for each unit, as the distance between the seven selected sensor measurements projected in the failure space of the current unit's mode and the unique failure place of the current mode. To improve the accuracy of the degradation indicator, the distance of each unique failure space is divided by the dispersion. The dispersion of each unique failure space is defined as the standard deviation of the barycenter of the failure places cluster in each mode.

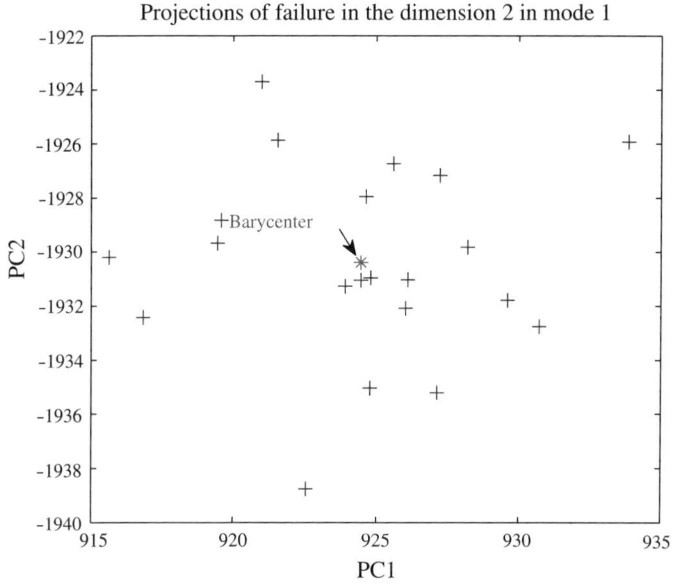

Figure 5.2. *Projection of the recordings of seven sensors in the principal plane of PCA for mode 1*

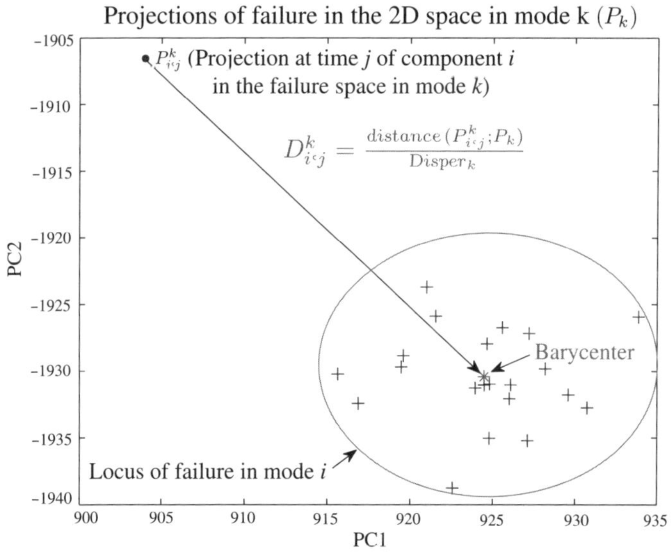

Figure 5.3. *Definition of the damage indicator*

The construction of the indicator can be formalized in the following way:

– let N_k be the number of individuals at the failure time in mode k,

– let $(\bar{x}^k, \bar{y}^k)$ be the barycenter of the projected measurements of the failure places in mode k, $k = 1, ..., 6$,

– let (x_i^k, y_i^k), $i = 1, ..., N_k$, be the ith projected failure place in mode k,

– let (x_{ij}^k, y_{ij}^k) be the projected measurement of the component i at the arbitrary operating time j in mode k.

We then consider the following distance for component i, at time j, in mode k:

$$D_{i,j}^k = \frac{\sqrt{(x_{ij}^k - \bar{x}^k)^2 + (y_{ij}^k - \bar{y}^k)^2}}{\text{Disper}_k} \qquad [5.5]$$

where Disper_k the dispersion of the failure places in mode k is defined in the following way:

$$\text{Disper}_k = \sqrt{\frac{1}{N_k - 1} \sum_{i=1}^{N_k} ((x_i^k - \bar{x}^k)^2 + (y_i^k - \bar{y}^k)^2)} \qquad [5.6]$$

for each unit, the degradation indicator is calculated by equation [5.5]. Thus, the damage indicator on $[0, 1, ..., r_i]$ for r_i time steps is described by the following vector:

$$D_{i,[1:r_i]} = \left[D_{i,1}^{k(1)}, D_{i,2}^{k(2)}, ..., D_{i,r_i}^{k(r_i)}\right] \qquad [5.7]$$

where $k(j)$ refers to the operating mode at time j.

As an example, we give in Figure 5.4(a) the degradation indicator for component 1, then for all the components (Figure 5.4(b)). We can see that in Figure 5.4(b) there is an important variation between the degradation indicators of different units and there is a general nonlinear and decreasing trend (at the beginning of the cycles, the indicator value tends to be around 1, and at the end of the cycles, at the failure times, it tends to zero). We model then the trend of this indicator by considering this characteristic and then we propose the criterion of RUL estimation in the next section.

5.4. Estimation of the residual life span (RUL)

5.4.1. *Simple approach based on the life span*

As a first step, we apply the most basic probabilistic method for the prediction of RUL. This method consists of a simple lifetime approach. This means that we only consider the failure times of the units without taking into account the progressive degradation phenomena that lead us from the initial state to the failure.

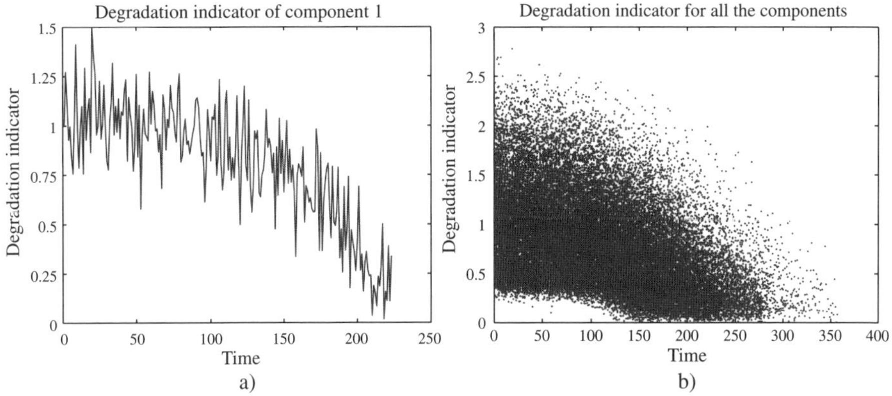

Figure 5.4. *Damage indicator corresponding to component 1 (a) and to all the components (b)*

Let T be the lifetime of a unit. While treating the failure times (or lifetime), the RUL at time t of each unit can be considered as the mean residual lifetime at time t presented as follows:

$$\text{RUL}(t) = \mathbf{E}(T|T > t) = \frac{1}{R(t)} \int_t^{\infty} R(x)dx \qquad [5.8]$$

where $R(x) = \mathbf{P}(T > x)$ is the reliability function.

According to the histogram of the failure times of 218 units presented in Figure 5.5, it seems reasonable to propose a Weibull distribution to model the lifetime distribution. For the lifetime distribution modeling and the mean residual life calculation, a three-parameter Weibull distribution is considered. Since the failure time does not start from zero, it is reasonable to consider a non-zero location parameter. The Weibull density function with three parameters (μ, λ, θ) is defined as follows:

$$f(x, \mu, \lambda, \theta) = \mu\lambda^{-\mu}x^{\mu-1} \exp\left(-\left(\frac{x-\theta}{\lambda}\right)^{\mu}\right) \qquad [5.9]$$

By a nonlinear regression, the estimated parameters are obtained, respectively:

$$(\mu, \lambda, \theta) = (2.606,\ 1/119.2,\ 95.02)$$

The RUL is then estimated by using equation [5.8] on the test data set. The score obtained is S = 9,450. This score is not very convincing in comparison with other non-probabilistic methods where the obtained scores are around 5,636.

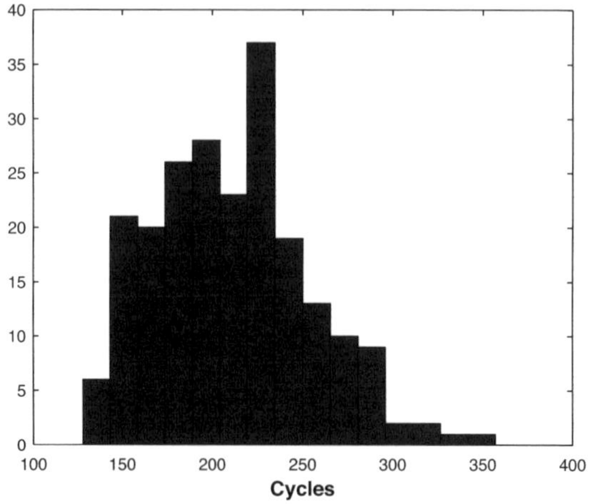

Figure 5.5. *Histogram of breakdown dates for 218 components*

5.4.2. *Stochastic deterioration model*

We suggest here using a Wiener process to model the deterioration and to estimate the RUL.

5.4.2.1. *Estimation of parameters of Wiener process*

Let $D_{i,j}$ be the deterioration indicator of component i at time j. We assume that each $D_{i,[1:r_i]}, r_i \geq 0, i = 1, ..., 218$, is a realization of a Wiener process with three parameters (η, σ, α) and therefore satisfies the following conditions:

1) the increments $D_{-,j} - D_{-,k}, j \neq k$, are independent,

2) for all $0 < s < j$, $D_{-,j+s} - D_{-,j}$ is normally distributed with mean $\eta((j+s)^\alpha - j^\alpha)$ and variance $\sigma^2((j+s) - j) = \sigma^2 s$,

where $D_{-,j}$ is the value of the process at time j. The deterioration increments $\triangle d_{i,j} = D_{i,j+1} - D_{i,j}$, $i = 1, \ldots, N$, for each component follow a Gaussian distribution $\mathcal{N}(\eta((j+1)^\alpha - j^\alpha), \sigma^2)$. We also write observations $\triangle d_i = (\triangle d_{i,1}, \ldots, \triangle d_{i,r_i})$, for $i = 1, \ldots, N$, where r_i is the date of the last recording before failure of component i. We then estimate the parameters of Wiener process (η, σ, α) by maximizing the likelihood function [MOL 04]:

$$l = \log(f_{\eta,\sigma,\alpha}(\triangle d_1, \ldots, \triangle d_N)) = \sum_{i=1}^{N} \log(g_{\eta,\sigma,\alpha}(\triangle d_{i,1}, \ldots, \triangle d_{i,r_i-1}))$$

$$= \sum_{i=1}^{N} \log\left(\prod_{j=1}^{r_i}\left(\frac{1}{\sqrt{2\pi\sigma^2((j+1)^\alpha - j^\alpha)}} e^{\left(\frac{-(\triangle d_{i,j} - \eta((j+1)^\alpha - j^\alpha))^2}{2\sigma^2((j+1)^\alpha - j^\alpha)}\right)}\right)\right)$$

where N is the number of units, $f_{\eta,\sigma,\alpha}$ is the probability density function of all increments, and $g_{\eta,\sigma,\alpha}$ is the probability density function of all increments corresponding to each unit. The estimated parameters are obtained as follows:

$$\hat{\eta} = -0.0016, \ \hat{\sigma} = 1.2551, \ \hat{\alpha} = 0.6707 \quad [5.10]$$

These parameters will be applied to the test data set in the order to estimate the RUL for each unit. The next section discusses the RUL estimation method by Monte Carlo simulations.

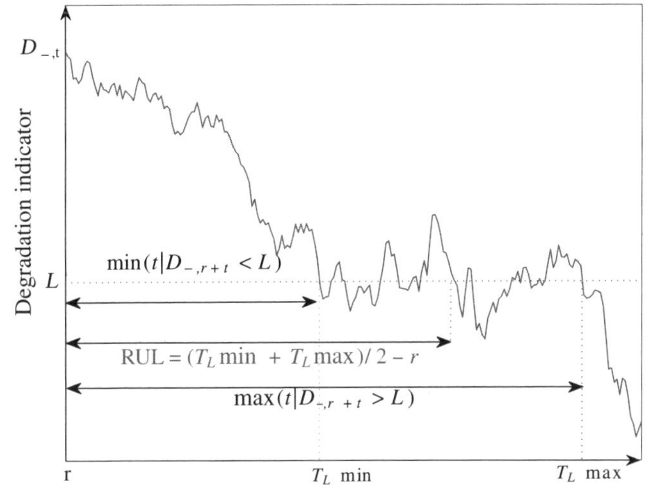

Figure 5.6. *Criterion suggested to calculate RUL from a trajectory of Wiener process*

5.4.2.2. *RUL estimation by simulation of a Wiener process*

The last cycle of the test data set is the time upon which we calculate the RULs. The estimation is carried out as follows: let $r_{i'}$ be the last recording time in the test data set for component i', $D_{i',r_{i'}}^{k(r_{i'})}$ the deterioration indicator at time $r_{i'}$, and L a fixed deterioration level. If we assume that L is the failure threshold, then $L = 0$. The Wiener process is non-monotonous and therefore can cross, several times, the same threshold (see Figure 5.6). Therefore, considering the failure time as the first passage time or as the last passage time to L might overestimate or underestimate the RUL. Here, after experiments on the test data set, we chose to consider that RUL at time $r_{i'}$ is the time that elapses between $r_{i'}$ and the mean of the first and last times of the overrun of level L written, respectively, as $T_L^{i'}\text{min}$ and $T_L^{i'}\text{max}$ (Figure 5.6). We thus define, for a simulated trajectory of Wiener process, the following:

$$\text{RUL}_{\text{simulated}}^{i'} = \frac{(T_L^{i'}\text{min} + T_L^{i'}\text{max})}{2} - r \quad [5.11]$$

where $T_L^{i'}\min = \inf\{j > r_{i'}, D_{i',j}^{k(j)} \leq L\}$ and $T_L^{i'}\max = \inf\{j > r_{i'}, \forall s > 0, D_{i',j+s}^{k(j+s)} \leq L\}$ are, respectively, the first and last times when threshold L is crossed. After n = 10,000 simulations, the estimated RUL, denoted by $\text{RUL}_{\text{estimated}}^{i'}$, is calculated by the empirical mean. For example, the estimated RUL of two units are given in Table 5.2.

Unit i'	$\text{RUL}_{\text{estimated}}^{i'}$	$\bar{\sigma}^{i'}$	$\text{RUL}_{\text{min}}^{i'}$	$\text{RUL}_{\text{max}}^{i'}$
1	89.32	3.02	86.3	92.34
2	78.22	1.80	76.42	80.02

Table 5.2. *Estimated RUL and associated variance*

By using the performance criteria from the PHM Conference, the obtained score S is 5,520, and the RMSE = 438. The performance of our method is thus better than those suggested by the winners of the 2008 PHM Conference data challenge.

5.5. Conclusion

The scores obtained with our method based on the use of Wiener process are, respectively, equal to S = 5,520 and RMSE = 438. These results show that a probabilistic approach of our probabilistic method based on the stochastic process can be more appropriate than other methods. These results tend to show that the procedure that consists of building a damage indicator is meaningful and that the use of a stochastic process can happen to be beneficial to the quality of the prognostic.

We can compare our result with some results obtained as follows.

– Non-probabilistic model with the similarity-based prognostic method of Wang in the 2008 PHM Conference [WAN 08]: the score is 5,636.

– Non-probabilistic model based on the neural networks of Peel [PEE 08] and Heimes [HEI 08] shows, respectively, the mean square error 519.8 and 984.

These results and our obtained score lead us to justify the accuracy of our probabilistic method.

Thus, the probabilistic approaches similar to what we present here seem to be promising. They are generally based on the construction of a deterioration indicator that makes sense with respect to the evolution of a system from a new state to the failure. Furthermore, the use of a stochastic process enables us to finely model the uncertainty observed at the level of this indicator. In contrast to the non-probabilistic approaches like neural networks, the stochastic processes enable us to take into account uncertainties and to give a distribution of RUL, which can be interesting as information to bring to a decision support process.

5.6. Bibliography

[BAN 09] BANJEVIC D., "Remaining useful life in theory and practice", *Metrika*, Springer, vol. 69, no. 2, pp. 337–349, 2009.

[CHI 99] CHINNAM R., "On-line reliability estimation of individual components, using degradation signals", *IEEE Transactions on Reliability*, vol. 48, no. 4, pp. 403–412, 1999.

[HEI 08] HEIMES F., "Recurrent neural networks for remaining useful life estimation", *International Conference on Prognostics and Health Management, PHM 2008*, pp. 1–6, 2008.

[LIA 06] LIAO H., ZHAO W., GUO H., "Predicting remaining useful life of an individual unit using proportional hazards model and logistic regression model", *Proceedings of the Reliability and Maintainability Symposium (RAMS)*, pp. 127–132, 2006.

[MAZ 07] MAZHAR M., KARA S., KAEBERNICK H., "Remaining life estimation of used components in consumer products: life cycle data analysis by Weibull and artificial neural networks", *Journal of Operations Management*, vol. 25, no. 6, pp. 1184–1193, 2007.

[MOL 04] MOLER C., "Numerical computing with MATLAB", *Society for Industrial Mathematics*, 2004.

[MON 09] MONNIN M., Projet DEPRADEM – Modélisation des Dégradation et du Pronostic pour l'Aide à la Décision en Maintenance - Synthèse Bibliographique, Rapport de fin de post doctorat – CRAN - Henri Poincaré University Nancy 1, 2009.

[PEE 08] PEEL L., "Data driven prognostics using a kalman filter ensemble of neural network models", *International Conference on Prognostics and Health Management, PHM 2008*, pp. 1–6, 2008.

[WAN 08] WANG T., YU J., SIEGEL D., LEE J., "A similarity-based prognostics approach for remaining useful life estimation of engineered systems", *International Conference on Prognostics and Health Management, PHM 2008*, pp. 1–6, 2008.

Chapter 6

Diagnosis of Systems with Multiple Operating Modes

The COSMOS (design and observation of systems with multiple safe operating modes) and DICOP (diagnosis of power converters: application to railway transport) projects of GIS 3SGS have involved four academic partners (CRAN[1], CReSTIC[2], LAGIS[3] and LAMIH[4]). These projects come within the general theme of the "dependability of complex systems". The goals were to develop tools and methods for the observation and fault diagnosis (fault detection, isolation and estimation) of systems characterized by several operating modes. These collaborative projects have enabled significant advances in this area of research. This chapter presents a few of the results obtained within this framework.

6.1. Introduction

Several dynamic systems, such as those found in the areas of power electronics, nuclear energy, aerospace or robotics, are considered complex systems due to their

---

Chapter written by Taha BOUKHOBZA, Frédéric HAMELIN, Benoît MARX, Gilles MOUROT, Anca Maria NAGY, José RAGOT, Djamel Eddine Chouaib BELKHIAT, Kevin GUELTON, Dalel JABRI, Noureddine MANAMANNI, Sinuhé MARTINEZ, Nadhir MESSAI, Vincent COCQUEMPOT, Assia HAKEM, Komi Midzodzi PEKPE, Talel ZOUARI, Michael DEFOORT, Mohammed DJEMAI and Jérémy VAN GORP.
1 CRAN (UMR 7039), University of Nancy.
2 CReSTIC (EA3804), University of Reims Champagne-Ardenne.
3 LAGIS (UMR 8219), University of Lille 1.
4 LAMIH (UMR 8201), University of Valenciennes et du Hainaut Cambrésis.

physical configuration, involving interactions between several components with nonlinear behaviors.

The dependability of these systems is a major challenge that researchers and industrialists attempt to address by developing adapted and high-performance tools for monitoring, observation and diagnosis. These tools aim to improve the safety of individuals and equipment, and to increase the availability and profitability of facilities.

System monitoring consists of a procedure aiming to detect any system behaviors that differ from the normal ones. Generally, monitoring methods can be classified into two categories: those not using any mathematical model of the system; and those based on a model describing the behavior of the system that is being monitored. We are interested in methods using behavioral models (known as model-based methods).

Most real-life systems are described by complex models that are very difficult to use for control and diagnosis.

In order to decrease the complexity of such models, two techniques that rely on decomposition into simpler models can be used:

– decomposition of the model according to particular operating modes that correspond in general to a physical configuration of the system by including a switching process between the different modes: we refer to these as switching systems (SSs) [ANT 00, DEC 00, ZAY 01];

– decomposition of the strongly nonlinear dynamic behaviors into a set of linear or weakly nonlinear behaviors: we refer to this as multiple-model representation (MMR) [MUR 97, TAK 85].

SSs are a class of hybrid dynamic systems. A SS consists of an interaction between a set of subsystems (modes) with continuously linear or nonlinear dynamics. The discrete evolution from one mode to another is ruled by a switching law, which can either depend on continuous evolution in the current mode or can result from an external event.

The MMR of systems has generated a great deal of interest in the past few years, particularly because of its ability to accurately represent nonlinear dynamic systems and its aptitude for dealing with the diagnosis of faults and control problems by using tools and techniques mastered for linear systems. The MMR is based on the principle of "divide and rule", which consists of the decomposition of the dynamics of a nonlinear system into a set of subsystems (that are weaker nonlinear or even linear). The behavior of each elementary model is valid in a limited range. The global model, with a wider operating range, is obtained by aggregating the submodels by interpolation with time-varying weights. The generic structure of multiple models includes several representations among which we can cite Takagi-Sugeno's fuzzy models, local model networks, models with sequential gains, linear parameter-varying (LPV) models, statistical mixed models and piecewise affine models.

This chapter deals with the modeling and the use of models for the estimation and diagnosis of dynamic systems whose time evolution of state variables is described either by switches between subsystems (SS), or by the aggregation of a set of systems sharing all or part of their state variables (MMR). The collaborative projects COSMOS and DICOP have enabled significant advances in this area, as much on the development of theoretical knowledge as on algorithmic and methodological developments enabling their practical implementation.

6.2. Detection of faults for a class of switching systems

6.2.1. *Introduction*

In this part, we suggest a synthesis approach of robust hybrid observers for the detection of faults in a class of switched linear system [ANT 00, DEC 00]. The latter is assumed to be subject to undesirable effects due to unknown inputs (measurement noise, external disturbances, etc.) and/or modeling errors. It is also characterized by the presence of state jumps at the switching instants and by the absence of *a priori* knowledge on the active mode of the switched linear system (SLS), a particular class of SS. Thus, in the rest of this part, we consider the SLS to be composed of N linear subsystems and given by:

$$\left\{ \begin{array}{lcl} \dot{x}(t) & = & A_q x(t) + B_q u(t) + B_{d_q} d(t) + B_{f_q} f(t) \\ y(t) & = & C_q x(t) + D_q u(t) + D_{d_q} d(t) + D_{f_q} f(t) \end{array} \right. \qquad [6.1]$$

where:

_ $x \in \mathbb{R}^n$, $u \in \mathbb{R}^m$, $y \in \mathbb{R}^p$, $d \in \mathbb{R}^m$ and $f \in \mathbb{R}^l$ are respectively the state vector (non-measurable or partially measurable), the input vector, the output vector, the vector of unknown inputs (disturbances, unstructured uncertainties, noises) and the vector of faults to be detected;

_ A_q, B_q, B_{d_q}, B_{f_q}, C_q, D_q, D_{d_q}, D_{f_q} are the known matrices of adequate dimensions; and

_ $q \in Q = 1, 2, \ldots, N$ is the index of the active mode at time t.

We assume that there are only a finite number of mode changes, that the SLS and all of the pairs (A_q, C_q) are observable and that vectors d and f are bounded in L_2 norm. Finally, we consider that the switch from one subsystem to another only occurs at specific locations of the state space, denoted $G_{i,j}$, which are described by linear hyper-planes as:

$$G_{i,j} = \{x \in \mathbb{R}^n | g_{i,j}(x) = 0\}, \ (i, j) \in I_g \qquad [6.2]$$

where I_g is a set of *uplets* including all of the possible transitions between the two modes.

6.2.2. *Structure of the residual generator and observer design*

The structure of the proposed residual generator for the class of SLS considered is illustrated in Figure 6.1 [FRA 09, FRI 00, COC 04, OLI 09]. It is made of three blocks: a continuous observer for the estimation of the continuous state; a block for the identification of the active mode; and a weighting gain. The weighting gains and the observer gains are chosen in order to ensure a good compromise between the robustness of the residuals, with respect to the unknown inputs, and their sensitivity to faults [BEL 09, BEL 11d, BEL 10, ZHO 03].

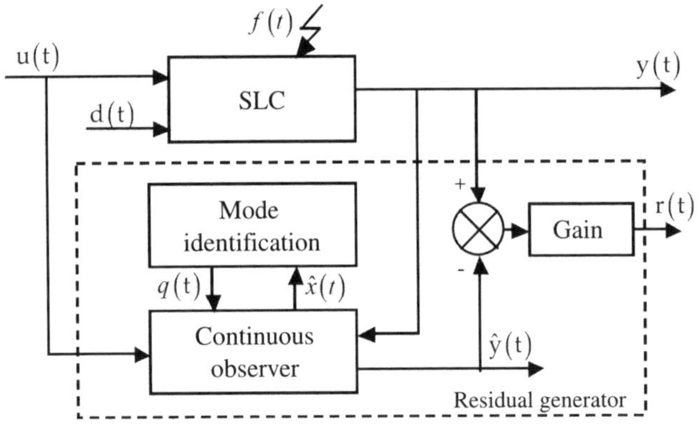

Figure 6.1. *Structure of a residual generator*

For the generation of residuals, let us consider that the observer is given by:

$$\begin{cases} \dot{\hat{x}}(t) &= A_{\hat{q}}\hat{x}(t) + B_{\hat{q}}u(t) + K_{\hat{q}}(y(t) - \hat{y}(t)) \\ \hat{y}(t) &= C_{\hat{q}}\hat{x}(t) + D_{\hat{q}}u(t) \\ r(t) &= V(y(t) - \hat{y}(t)) \end{cases} \qquad [6.3]$$

where $x \in \mathbb{R}^n$ and $y \in \mathbb{R}^p$ are the estimated state and output vectors, respectively, r is the residual vector, and $K_{\hat{q}}$ and V are respectively the gain matrix and the weighting matrix to be determined.

Let q indicate the SLS mode and $\hat{q}$ the observer mode. The dynamics of estimation error $e(x) = x(t) - \hat{x}(t)$ and the residual $r(t)$ are then given by [BEL 11b]:

$$\dot{e} = \bar{A}_{\hat{q}}e(t) + \Delta\bar{A}_{q,\hat{q}}x(t) + \Delta\bar{B}_{q,\hat{q}}u(t) + \bar{B}_{d_q}d(t) + \bar{B}_{f_q}f(t) \quad [6.4]$$

$$r(t) = V\left(C_{\hat{q}}e(t) + \Delta C_{q,\hat{q}}x(t) + \Delta D_{q,\hat{q}}u(t) + D_{d_q}d(t) + D_{f_q}f(t)\right) \quad [6.5]$$

with:

$$\Delta\bar{A}_{q,\hat{q}} = \Delta A_{q,\hat{q}} - K_{\hat{q}}\Delta C_{q,\hat{q}},\ \Delta C_{q,\hat{q}} = C_q - C_{\hat{q}},$$

$$\Delta\bar{B}_{q,\hat{q}} = \Delta B_{q,\hat{q}} - K_{\hat{q}}\Delta D_{q,\hat{q}},\dots$$

Equations [6.4] and [6.5] can be reformulated as follows:

$$e(t) = e_x(t) + e_u(t) + e_d(t) + e_f(t) \quad [6.6]$$

$$r(t) = r_x(t) + r_u(t) + r_d(t) + r_f(t) \quad [6.7]$$

where:

$$e_x(t) = e(t)\,|_{u=0,d=0,f=0}, \qquad r_x(t) = r(t)\,|_{u=0,d=0,f=0},$$
$$e_u(t) = e(t)\,|_{x=0,d=0,f=0}, \qquad r_u(t) = r(t)\,|_{x=0,d=0,f=0},$$
$$e_d(t) = e(t)\,|_{x=0,u=0,f=0}, \qquad r_d(t) = r(t)\,|_{x=0,u=0,f=0},$$
$$e_f(t) = e(t)\,|_{x=0,u=0,d=0}, \qquad r_f(t) = r(t)\,|_{x=0,u=0,d=0}$$

This reformulation shows that the dynamics of the residual is the sum of four components and depends on the fault vector f, an unknown input vector d, state vector x and input vector u.

6.2.2.1. *Observer design for fault detection*

In this section, we express Theorem 6.1 stating the conditions ensuring the boundedness of two components $r_x(t)$ and $r_u(t)$ of the residual.

THEOREM 6.1.– Let us consider SLS [6.1] with observer [(6.3] and assume that for $T_0 > 0$ we have $\sup_{t>T_0} \|x(t)\| \leq x_{max}$ and $\sup_{t>T_0} \|u(t)\| \leq u_{max}$. Let us also assume that there are matrices $P_{\hat{q}}^T = P_{\hat{q}} > 0$, $K_{\hat{q}}$ and scalars $\epsilon > 0$, $\alpha > 0$, $\xi > 0$ and $v_{q,\hat{q}} \geq 0$ and $\mu_{q,\hat{q}} \leq 1$ so that the conditions given hereafter are fulfilled:

– $\forall(q, \hat{q}) \in I_g$, the following matrix inequalities are fulfilled:

$$\alpha I \leq P_{\hat{q}} \leq \xi I \quad \hat{q} \in Q \qquad [6.8]$$

$$\psi_{q,\hat{q}} = \begin{bmatrix} \bar{A}_{\hat{q}}^T P_{\hat{q}} + P_{\hat{q}} A_{\hat{q}} + I + v_{q,\hat{q}} I & \bar{A}_{\hat{q}}^T P_{\hat{q}} & P_{\hat{q}} \Delta \bar{A}_{q,\hat{q}} & P_{\hat{q}} \Delta \bar{B}_{q,\hat{q}} \\ \bar{A}_{\hat{q}}^T P_{\hat{q}} & \bar{A}_{\hat{q}}^T P_{\hat{q}} + P_{\hat{q}} A_{\hat{q}} + I + v_{q,\hat{q}} I & P_{\hat{q}} \Delta \bar{A}_{q,\hat{q}} & P_{\hat{q}} \Delta \bar{B}_{q,\hat{q}} \\ (*) & (*) & -\epsilon^2 v_{q,\hat{q}} I & 0 \\ (*) & (*) & 0 & -\epsilon^2 v_{q,\hat{q}} I \end{bmatrix} \leq 0$$

[6.9]

$$P_q - \mu_{q,\hat{q}} P_{\hat{q}} \quad (\hat{q}, q) \in I_g \qquad [6.10]$$

– $\forall \hat{x} \in G_{q,\hat{q}}$ the hybrid observer state is updated according to:

$$\hat{x}^+ = \left(I - R_{\hat{q}}^{-1}\left(C_{\hat{q}} R_{\hat{q}}^{-1}\right) C_{\hat{q}}\right) \hat{x} + \left(R_{\hat{q}}^{-1}\left(C_{\hat{q}} R_{\hat{q}}^{-1}\right)^{\dagger}\right) y - \left(R_{\hat{q}}^{-1}\left(C_{\hat{q}} R_{\hat{q}}^{-1}\right)^{\dagger} D_{\hat{q}}\right) u \qquad [6.11]$$

Then, the estimation error $e_w = e_x + e_u$ is bounded by:

$$\lim_{t\to\infty} \sup \|e_w(t)\| \leq \sqrt{\frac{\bar{v}}{1+\bar{v}} \frac{\xi}{\alpha} \epsilon^2 \left(x_{max}^2 + u_{max}^2\right)}$$

where $R_{\hat{q}} \in \mathbb{R}^{n \times n}$ is a positive definite symmetrical matrix, $(R)^{\dagger}$ is the pseudo-inverse of R, $\bar{v}$ is the sup of $v_{q,\hat{q}}$, $(q, \hat{q} \in I_g)$ and I_g is a set of *uplets* including the different transitions between two modes that can take place in the SLS.

6.2.2.2. *Robustness with respect to unknown inputs and sensitivity to faults*

In this section, we deal with components $r_f(t)$ and $r_d(t)$ by using the techniques of H_∞. The principle applied consists of minimizing the transfer of unknown inputs $d(t)$ to $r_d(t)$ and maximizing the transfer of fault vector $f(t)$ to $r_f(t)$. Thus, we can write:

$$\|r_d\|_2^2 - \gamma_{\hat{q}}^2 \|d\|_2^2 < 0 \qquad [6.12]$$

$$\|r_f\|_2^2 - \beta_{\hat{q}}^2 \|f\|_2^2 > 0 \qquad [6.13]$$

where $\gamma_{\hat{q}}$ and $\beta_{\hat{q}}$ are two positive scalars.

The following lemma and theorem can be proved [BEL 11b].

LEMMA 6.1.– Let us consider SLS [6.1] with observer [6.3]. The dynamics of estimation error e_d are asymptotically stable and fulfill condition [6.12] if there are matrices $P_{\hat{q}} > 0$, $H_{\hat{q}}$ and V so that the following linear matrix inequality (LMI) is fulfilled:

$$\begin{bmatrix} A_{\hat{q}}^T P_{\hat{q}} + P_{\hat{q}} A_{\hat{q}} - H_{\hat{q}} C_{\hat{q}} - C_{\hat{q}}^T H_{\hat{q}}^T + 2\varphi_1(V, V_c^n) & H_{\hat{q}} D_{f_q} - P_{\hat{q}} B_{f_q} & C_{\hat{q}}^T V^T \\ * & \beta_{\hat{q}}^2 I + 2\varphi_2(V, V_c^n) & D_{f_q}^T V^T \\ * & * & -I \end{bmatrix} \leq 0$$

[6.14]

with:

$V_c^n = V^{n-1} C_{\hat{q}}$, $\quad V_f^n = V^{n-1} F_{f_q}$, $\quad n = 1, 2, \ldots$

$\varphi_1(V, V_c^n) = (V_c^n)^T V_c^n - (V_c^n)^T V C_{\hat{q}} - C_{\hat{q}}^T V^T V_c^n$,

$\varphi_2(V, V_f^n) = (V_f^n)^T V_f^n - (V_c^n)^T V D_{f_q} - D_{f_q}^T V^T V_f^n$ and F_{f_q} is a weighting matrix.

The previous results allow us to express the following final result:

THEOREM 6.2.– Let us consider SLS [6.1] with observer [6.3]. Residual [6.5] is bounded and fulfills conditions [6.12] and [6.13] if:

– matrices $P_{\hat{q}} > 0$, $H_{\hat{q}}$ and V exist as do scalars $\epsilon > 0$, $\alpha > 0$, $\xi > 0$ and $v_{q,\hat{q}} \geq 0$ and $\mu_{q,\hat{q}} \leq 1$ so that LMIs [6.8], [6.10], [6.9] and [6.14] are fulfilled; and

– the estimated state is updated at the switching instants according to condition [6.11].

REMARK 6.1.– Let us note that using an iterative procedure suggested in [BEL 11a, BEL11c], it is possible to optimize the choice of parameters $\gamma_{\hat{q}}$ and $\beta_{\hat{q}}$.

6.2.3. *Simulation and results*

In this example, we consider the switching system with two modes defined as follows:

MODE 1.–

if $S_1 = 0$: $A_1 = \begin{bmatrix} 1.5 & 2 \\ -2 & -0.5 \end{bmatrix}$, $B = \begin{bmatrix} 0 \\ 0 \end{bmatrix}$, $C = \begin{bmatrix} 1 & -2.4 \end{bmatrix}$, $B_{d_1} = \begin{bmatrix} 1 & 0 \\ 1 & 0 \end{bmatrix}$,$B_{f_1} = \begin{bmatrix} 0 \\ 0 \end{bmatrix}$, $D_{d_1} = \begin{bmatrix} 1 & 0 \end{bmatrix}$, $D_{f_1} = 14.72$

MODE 2.–

if $S_2 = 0$: $A_2 = \begin{bmatrix} -2 & 10 \\ -2 & -1 \end{bmatrix}$, $B = \begin{bmatrix} 0 \\ 0 \end{bmatrix}$, $C = [1 \quad -2.4]$, $B_{d_2} = \begin{bmatrix} 1 & 0 \\ 0 & 1 \end{bmatrix}$, $B_{f_2} = \begin{bmatrix} 0 \\ 0 \end{bmatrix}$, $D_{d_2} = [1 \quad 0]$, $D_{f_2} = 0$

where $S_i = a_i x_1 + b_i x_2$, $i \in \{1, 2\}$ are the switching laws with $a_1 = 0,25$, $b_1 = 1$, $a_2 = 1$ and $b_2 = 1$.

We note that the first subsystem is unstable; but the second one is stable. The switching system, which is made of two subsystems, is stable.

Application of theorem 6.2 enables us to find the following result:

$$\gamma_1 = \gamma_2 = 1,9667, \quad \beta_1 = 3,4831, \quad \lambda_1 = \lambda_2 = 5 \quad \text{and} \quad K_1 = \begin{bmatrix} 3,5616 \\ -0,8557 \end{bmatrix},$$
$$K_2 = \begin{bmatrix} -0,0434 \\ 0,0545 \end{bmatrix} \text{and } V = 0,333$$

with $\xi = 4,9$, $\alpha = 1$, $\epsilon = 6,78$, $\gamma_{1_{init}} = \gamma_{2_{init}} = 4$, and $\beta_{1_{init}} = 0$.

In order to prove the efficiency of the suggested approach, the unknown inputs d are considered to be centered Gaussian white noise with an amplitude of 0.2. Fault f is simulated as an impulse with an amplitude equal to 0.05, which happens in mode 2, between time 5.2 s and 5.4 s. The generated residual, which is represented by Figure 6.2, converges towards zero at time 2 s, and moves away from zero after the appearance of the faults.

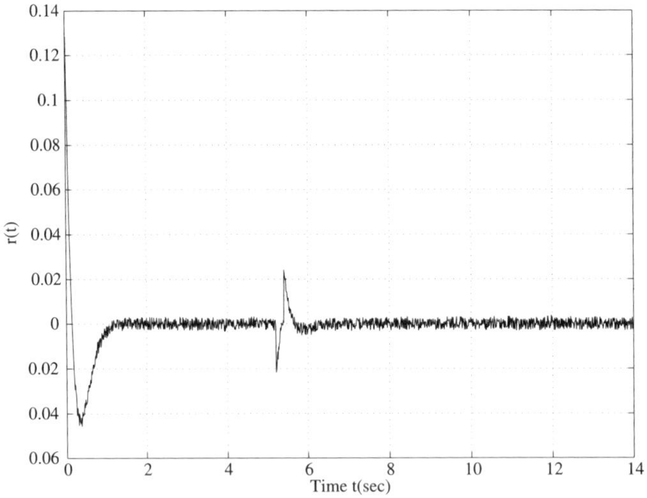

Figure 6.2. *Generated residual*

6.2.4. *Conclusions*

In this part, a robust hybrid observer is designed for a class of SLS. This observer aims to detect faults in a SLS subject to disturbances and/or modeling errors. The suggested approach is based on the use of results developed for robust control. Thus, we formalized the design problem through the minimization of a H_∞ criterion and established the convergence conditions under LMIs. The numerical example clearly illustrates the efficiency of the approach.

6.3. Analytical method to obtain a multiple model

6.3.1. *Introduction*

The issue of complexity in nonlinear dynamic systems appears in several scientific domains including engineering. Numerous model reduction and simplification techniques have been developed in the past few years in order to decrease this complexity to meet with analysis, identification or command goals. In most of the studies that have been carried out, the reduction in complexity has been due to a model order reduction that results in a loss of information due to the variables and dynamics that have been disregarded. Another way to proceed is to rewrite the model in a way that is easier to study, by breaking it down into simpler units without loss of information.

Writing a system in a multiple-model (MM) form is a modeling tool that can be used for nonlinear systems, based on an interpolation of linear submodels [MUR 97]. The global model is a convex combination of these submodels. The major interest in this technique is due to the fact that important properties such as stability, controllability, observability, that are largely studied within the framework of linear time invariant systems can be (at least partially) generalized.

There are different methods that can be used to obtain an MM. The first is based on a linearization of the nonlinear model around one (or several) operating point(s) or around an arbitrary trajectory. The second technique consists of identifying the structure of the MM and its parameters from experimental data. The loss of information is the first drawback of these techniques. A second drawback is the choice of either the points or trajectory of linearization and premise variables expressing the different nonlinearities of the system. These choices are very difficult to make.

In what follows, we present a systematic procedure for transforming a model of a nonlinear system in MM form by avoiding these major drawbacks: the transformation is carried out without loss of information and without linearization;

and the choice of variables is made more systematically. From a general nonlinear model, several quasi-linear parameter varying (quasi-LPV) models can be extracted. Each model is defined by matrix polytopes: the matrices are time-varying convex combinations of constant matrices that define the vertices of each polytope. The latter are obtained by the convex polytopic transformation of all of the variables associated with a representation. Each vertex defines a linear submodel, the nonlinearities of the system being included in the weighting functions. Thus a MM is obtained.

Hence, the problem involves choosing a quasi-LPV representation from among all of those possible, knowing that they are formally equivalent. We thus suggest defining the criteria for choosing the model in accordance with the observer design based on the MM structure. Most of the results concerning MM observer design involve solving LMI from Lyapunov's method. From the analysis of sufficient multi-observer conditions and the complexity of the LMI to be solved, we suggested a certain number of criteria to guide the choice of quasi-LPV form to be kept. The method suggested generalizes the nonlinearity approach [TAN 01] by taking into account the goals of the study (such as the stability or performance analysis, the controller or observer design).

Initially, the formalism of the MMR is recalled. Then, we present the method enabling the transformation of a nonlinear model into a MMR and the criteria kept for the choice of quasi-LPV.

6.3.2. *Setting the problem*

Let us consider a nonlinear dynamic system described by the following differential equations:

$$\left\{ \begin{array}{l} \dot{x}(t) = f(x(t), u(t)) \\ y(t) = g(x(t), u(t)) \end{array} \right. \quad [6.15]$$

where $x \in \mathbb{R}^n$ is the state vector, $u \in \mathbb{R}^m$ is the input vector and $y \in \mathbb{R}^l$ is the output vector.

The MM is defined as a convex combination of r submodels:

$$\left\{ \begin{array}{lcl} \dot{x}(t) & = & \sum\limits_{i=1}^{r} \mu_i(z) \; [A_i \, x(t) + B_i \, u(t)] \\ y(t) & = & \sum\limits_{i=1}^{r} \mu_i(z) \; [C_i \, x(t) + D_i \, u(t)] \end{array} \right. \quad [6.16]$$

Here, A_i, B_i, C_i and D_i, are matrices with constant coefficients of adequate dimensions and $z \in \mathbb{R}^p$ is the vector of premise variables. Function $\mu_i(z)$ represents the weighting of submodel i, represented, in the global model by $\{(A_i, B_i, C_i, D_i)\}$. Weighting functions $\mu_i(z), i = 1, \ldots, r$ have the following property:

$$\sum_{i=1}^{r} \mu_i(z) = 1 \qquad \mu_i(x,u) \geq 0, \forall (z) \in^p \qquad [6.17]$$

The following section describes how a nonlinear model [6.15] can be transformed into an MMR [6.16].

6.3.3. *Transformation in multiple-model form*

The general form of a nonlinear model is first rewritten in the quasi-LPV form. Each quasi-LPV representation has a corresponding set of premise variables, which will be partitioned in two by using the convex polytopic transformation. The MM will thus be a convex combination of linear submodels.

6.3.3.1. *General method*

Assuming that $f(x(t), u(t))$ and $g(x(t), u(t))$ are continuous and bounded in $U \subseteq \mathbb{R}^n$ with $f(0, \cdot) = 0$ and $g(0, \cdot) = 0$ by a factorization procedure, system [6.15] can be represented in a quasi-LPV form:

$$\begin{bmatrix} \dot{x}(t) \\ y(t) \end{bmatrix} = \begin{bmatrix} A(\rho(x,u)) & B(\rho(x,u)) \\ C(\rho(x,u)) & D(\rho(x,u)) \end{bmatrix} \begin{bmatrix} x(t) \\ u(t) \end{bmatrix} \qquad [6.18]$$

where

$A(\rho(x,u)) \in \mathbb{R}^{n\times n}$, $B(\rho(x,u)) \in \mathbb{R}^{n\times m}$, $C(\rho(x,u)) \in \mathbb{R}^{l\times n}$, $D(\rho(x,u)) \in \mathbb{R}^{l\times m}$, $\rho(x,u) \in C^1(\mathbb{R}^{n+m}, \mathbb{R}^{n_1+m_1})$

with $n_1 + m_1 \leq n + m$.

In the quasi-LPV form of a nonlinear system, state matrices *A, B, C* and *D* depend on the state and input variables of the system. This quasi-LPV form of a nonlinear system is affine in the state and input variables.

In this case, the matrix $A(\rho(x,u))$ is given by:

$$A(\rho(x,u)) = \begin{bmatrix} a_{1,1}(\rho(x,u)) & \cdots & a_{1,n}(\rho(x,u)) \\ \vdots & & \\ a_{n,1}(\rho(x,u)) & \cdots & a_{n,n}(\rho(x,u)) \end{bmatrix} \qquad [6.19]$$

Matrices $B(\rho(x,u))$, $C(\rho(x,u))$ and $D(\rho(x,u))$ are built in the same way as $A(\rho(x,u))$.

From the quasi-LPV representation coming from [6.15] and given in explicit form in [6.18], we can define the set of premise variables V_z in the following way:

$$\begin{aligned} V_z &= \{a_{i,q}(\rho(x,u))\,|a_{i,q} \neq K,\; i=\overline{1,n},\quad q=\overline{1,n}\;\} \\ \cup &\; \{b_{i,s}(\rho(x,u))\,|b_{i,s} \neq K,\; i=\overline{1,n},\quad s=\overline{1,m}\;\} \\ \cup &\; \{c_{j,q}(\rho(x,u))\,|c_{j,q} \neq K,\; j=\overline{1,l},\quad q=\overline{1,n}\;\} \\ \cup &\; \{d_{j,s}(\rho(x,u))\,|d_{j,s} \neq K,\; j=\overline{1,l},\quad s=\overline{1,m}\;\} \end{aligned}$$

where $\overline{1,n} = \{1,...,n\}$ and K refer to any constant.

In what follows, this set is simply referred to as:

$$V_z = \{z_1(\rho(x,u)),\ldots,z_p(\rho(x,u)) \mid p \leq (n+l)(n+m)\}$$

where p represents the dimension of the set of premise variables V_z and at the same time the number of nonlinearities identified by using the quasi-LPV form of system [6.15]. Quantities $z_1(\rho(x,u))$, . . ., $z_p(\rho(x,u))$ are the premise variables.

REMARK 6.2.– $(n+l)\times(n+m)$ represents the maximum number of premise variables.

REMARK 6.3.– The number of submodels that make up the MM is equal to $z_j(x,u) = F_{j,1}(z_j)\,z_{j,1} + F_{j,2}(z_j)\,z_{j,2}$.

The 2^p submodels of a MM are generated by the application of the convex polytopic transformation to p premise variables [TAN 01]:

$$z_j(x,u) = F_{j,1}(z_j)\,z_{j,1} + F_{j,2}(z_j)\,z_{j,2} \qquad [6.20]$$

with:

$$\begin{aligned} F_{j,1}(z_j) &= \frac{z_j(x,u) - z_{j,2}}{z_{j,1} - z_{j,2}} \\ F_{j,2}(z_j) &= \frac{z_{j,1} - z_j(x,u)}{z_{j,1} - z_{j,2}} \end{aligned} \qquad [6.21]$$

where scalars $z_{j,1}$ and $z_{j,2}$ are defined in [6.22]:

$$\begin{aligned} z_{j,1} &= \max_{x,u} \{z_j(x,u)\} \\ z_{j,2} &= \min_{x,u} \{z_j(x,u)\} \end{aligned} \qquad \forall j = 1, ..., p \qquad [6.22]$$

Submodel i	σ_i	A_i	$\mu_i(z)$
1	(1,1,...,1)	$A(z_{1,1}, z_{2,1}, ..., z_{p,1})$	$\prod_{j=1}^{p} F_{j,\sigma_1^j}$
2	(1,1,...,2)	$A(z_{1,1}, z_{2,1}, ..., z_{p,2})$	$\prod_{j=1}^{p} F_{j,\sigma_2^j}$
$\vdots$	$\vdots$	$\vdots$	$\vdots$
2^{p-1}	(1,2,...,2)		
$2^{p-1}+1$	(2,1,...,1)	$A(z_{1,2}, z_{2,1}, ..., z_{p,1})$	$\prod_{j=1}^{p} F_{j,\sigma_{2^{p-1}+1}^j}$
	(2,1,...,2)		
$\vdots$	$\vdots$	$\vdots$	
	(2,2,...,1)		
2^p	(2,2,...,2)	$A(z_{1,2}, z_{2,2}, ..., z_{p,2})$	$\prod_{j=1}^{p} F_{j,\sigma_{2^p}^j}$

Table 6.1. *Obtaining an MM with* p *premise variables*

Each submodel i, represented in Table 6.1 by line i, has a corresponding p-uplet, σ_i, which codes the partitions of the premise variables involved in the weighting function corresponding to this submodel. A table gathering together all the partitions of premise variables can be built. By multiplying between each of the functions that describe these partitions, we obtain weighting function $\mu_i(z)$:

$$\mu_i(z) = \prod_{j=1}^{p} F_{j,\sigma_i^j}(z_j(x,u)) \qquad [6.23]$$

where σ_i^k represents the index appearing at k^{th} position in p-uplet σ_i. Then $\sum_{i=1}^{r} \mu_i(z) = 1$ and $\mu_i(z) \geq 0$ $(i = 1, ..., r)$, because:

$$F_{j,1}(z_j(x,u)) + F_{j,2}(z_j(x,u)) = 1, \qquad \forall j = 1, ..., p$$

It must be noted that matrix functions $A(\rho(x,u))$, $B(\rho(x,u))$, $C(\rho(x,u))$ and $D(\rho(x,u))$ involve the premise variables z_j for $j = 1, ..., p$. Thus, matrices A_i, B_i, C_i and D_i $(i = 1, ..., r)$ of equation [6.16] are determined by evaluating matrices A, B, C and D at the vertices of the polytope defined by the partitions of the premise variables.

In [NAG 10], we show that the MM [6.16], which is a convex combination of linear subsystems $\{(A_i, B_i, C_i, D_i)\}$ $(i = 1, ..., r)$, is equivalent to system [6.18], with $r = 2^p$ where p is the number of premise variables.

The following section provides tools that can be used to choose the most suitable MM structure for observer/controller design.

6.3.3.2. *Criteria for the choice of quasi-LPV form*

The results from stability analysis and the controller/observer design of a MM [AKH 04, BER 01, CHA 01, GUE 07, ICH 09, TAN 01, WAN 96] are based on solving LMIs resulting from Lyapunov's stability method. The convex property produced by weighting functions means that LMIs are only evaluated at the vertices of the polytope (A_i, B_i, C_i and D_i) and the weighting functions are not included when solving these LMIs. Only matrices A_i, B_i, C_i and D_i contribute to the solution. It is important to note that the LMI conditions used to build a controller or observer based on a MM are sufficient conditions but are not necessary.

In general, the quasi-LPV representation [6.18] for a nonlinear model with the form given in [6.15] is not unique; each quasi-LPV representation has a corresponding set of premise variables V_z; thus, the choice of quasi-LPV form is equivalent to choosing set V_z. The choice of set V_z is important because it influences the number of submodels and the structure of the global model. This degree of freedom is used to facilitate commandability, observability and stability analysis studies.

First, to ensure the observability/commandability of the global system represented in a MM form (i.e. to ensure a solution for the LMIs is associated with the MM in order for the controller/observer to be found) the observability/commandability of each submodel is necessary [GUE 07]. Thus, the quasi-LPV forms that generate submodels that do not possess these properties must be discarded. It must be noticed that this is not a sufficient condition to ensure the observability/commandability of the MM [MUR 02].

Furthermore, as mentioned at the beginning of this section, in the stability studies of MMs, the stability conditions of observers are given in terms of LMIs. Different techniques are suggested to decrease the number of LMIs, which leads to fewer calculations and favors the existence of a solution [BER 01]. In conclusion, the number of LMIs is linear or polynomial in the number of submodels [TAN 01]. This is the reason it is desirable that a MM contains the minimum number of submodels. This number is linked to the number of premise variables; hence a quasi-LPV form that has a minimum set of premise variables will be preferred.

Finally, the construction of an observer/controller for an MM with premise variables that depend on states is harder than in the case where these premise variables are known [ICH 09]. As a consequence, an MM with premise variables that depend on a minimum number of state variables is preferable.

Taking these remarks into account, the following rules can be followed to provide the simplest possible structure:

– discard the quasi-LPV forms for which matrix B has zero columns and C has zero rows, since these matrices do not fulfill the rank conditions of observability/commandability;

– choose the quasi-LPV form that contains the minimum number of premise variables from among the quasi-LPV forms for which the premise variables depend on the same number of state variables;

– identify the decompositions containing premise variables that are common to different state and/or output equations in [6.18], and therefore decrease the total number of premise variables; and

– choose the quasi-LPV form for which the premise variables depend on the minimum number of state variables from among the quasi-LPV forms that have the same number of premise variables.

6.3.4. *Conclusion*

In this section, we have proposed a general analytical method that can be used to simplify the complexity of a nonlinear model by replacing it with a set of submodels with simpler, linear structures and a set of adequate aggregation functions to aggregate these submodels into a MM. This way of rewriting models will be used in the following section to generate fault indicators for the operating diagnosis of SSs whose modes are nonlinear.

6.4. Detection of switching and operating mode recognition without the explicit use of model parameters

6.4.1. *Introduction*

SSs have several modes, which can be linear or nonlinear. Switching from one mode to another is ruled by a commutation function that can depend on inputs, outputs or other variables that are internal or external to the system (see [PEK 07]). Methods for fault detection and isolation in SSs are suggested in the literature but most focus on systems with linear modes only [COC 04, DOM 07].

In order to detect and diagnose failures, to command or predict the behavior of a SS, it is critical to determine the time at which one mode switches to another (the switching time) and to recognize the active mode at each point in time. This information will be even more critical if some modes of SS correspond to failure.

This section gives two methods that can be used to generate fault indicators or residuals for such systems. The residuals generated enable us to detect when one operating mode switches to another and to recognize the active mode. The first method [HAK 11] is based on the use of data from the system and is dedicated to SSs with linear modes. The second one [ZOU 11a, ZOU 11b] addresses SSs with nonlinear modes, which will be represented by MMs.

6.4.2. *Diagnosis of SSs with linear modes*

This part deals with a SS with only linear modes. The goal is to estimate the switching time and to recognize the current mode. This method is derived from a data-based fault detection method as suggested in [RAG 04]. The general principle of the method is to use indicator signals, referred to as residuals, that are generated by the projection of measurements in a subspace created by the inputs on a given time window. The advantage of this residual generation method is that it does not use the values of model parameters, and only uses the structure of the modes, which are assumed to be linear. The residuals enable us to detect the commutations, if the conditions of distinguishability between operating modes are fulfilled. An extension of the method is proposed that can be used to recognize the active mode by using data collected offline and labeled along with data gathered online.

6.4.2.1. *Formulation of the problem*

Let us consider the online measurements of inputs $u_k \in \mathbb{R}^m$, which are assumed to be deterministic, and outputs $y_k \in \mathbb{R}^\ell$, which are affected by centered and colored noise $w_k \in \mathbb{R}^\ell$ coming from the following SS:

$$\left\{ \begin{array}{l} x_{k+1} = A_{\sigma(k)} x_k + B_{\sigma(k)} u_k \\ y_k = C_{\sigma(k)} x_k + D_{\sigma(k)} u_k + w_k \end{array} \right. \qquad [6.24]$$

$$\sigma(k) : k \rightarrow \{1, 2, ..., d\}$$

where $x_k \in \mathbb{R}^n$ are the states, (d is the number of modes) and $\sigma(k)$ refers to the *commutation signal*. Here we refer to the *dwell-time* of mode γ ($1 \leq \gamma \leq d$), as Δk_γ.

Where $\forall \sigma(k) \in \{1, 2, ..., d\}$, we assume that:

– each mode is stable;

– the dwell-time of the system is bounded: $\Delta k_\gamma \geq L + i + 1$ where L and i are two integers;

– matrices $A_{\sigma(k)} \in \mathbb{R}^{n \times n}, B_{\sigma(k)} \in \mathbb{R}^{n \times m}$, $C_{\sigma(k)} \in \mathbb{R}^{\ell \times n}$, and $D_{\sigma(k)} \in \mathbb{R}^{\ell \times m}$ are unknowns; and

– the only online information available contains the values taken from the inputs and outputs.

Our goal is *to estimate the switching times* and *to recognize the current mode* using only the collected and recorded/labeled data describing the behavior of the system in each mode.

6.4.2.2. *Residual calculation for the estimation of switching time*

The goal here is to calculate residuals for the detection of switching. The only information on the system we have at our disposal relates to the linearity and stability of the modes. For a fixed value of $\sigma(k)$, and by repeated substitutions of equation [6.24], we obtain:

$$y_k = C_{\sigma(k)} A^i_{\sigma(k)} x_{k-i} + H_{\sigma(k),i} \bar{u}_{k,i} + w_k \qquad [6.25]$$

with:

$$\bar{u}_{k,i} = \begin{pmatrix} u^T_{k-i} & u^T_{k-i+1} & \cdots & u^T_k \end{pmatrix}^T \in R^{m.(i+1) \times 1}$$

If a value of i is chosen that is sufficiently large, the stability condition implies that $\left\| A^i_{\sigma(k)} \right\|_2 \to 0, \forall \sigma(k) \in \{1, 2, ..., d\}$. Thus $A^i_{\sigma(k)} x_{k-i}$ tends towards zero and becomes insignificant with respect to the other terms of equation [6.25], hence we can write:

$$y_k \simeq H_{\sigma(k),i} \bar{u}_{k,i} + w_k \qquad [6.26]$$

where:

$$H_{\sigma(k),i} = \begin{bmatrix} C_{\sigma(k)} A^{i-1}_{\sigma(k)} B_{\sigma(k)} & \cdots & C_{\sigma(k)} B_{\sigma(k)} & D_{\sigma(k)} \end{bmatrix}$$

is the matrix of Markov's parameters for the ith order system.

By piling equation [6.26] into a window of size L, we obtain:

$$Y_k \simeq H_{\sigma(k),i} U_k + W_k \qquad [6.27]$$

where $Y_k = [y_{k-L+1} \cdots y_k] \in \mathbb{R}^{\ell \times L}$ is a matrix of the outputs and W_k is built in a similar manner to Y_k. U_k is the extended Hankel's matrix defined by:

$$U_k = \begin{bmatrix} \bar{u}_{k-L+1,i} & \bar{u}_{k-L+2,i} & \cdots & \bar{u}_{k,i} \end{bmatrix} \in R^{m(i+1) \times L} \qquad [6.28]$$

We consider projection matrix Π_k, which defines the kernel to the right of matrix U_k, to be:

$$\Pi_k = I - U_k^T(U_k U_k^T)^{-1}U_k \quad [6.29]$$

and we have:

$$U_k \Pi_k = 0 \quad [6.30]$$

Π_k needs to be non zero when:

$$L > mi$$

This implies that matrix U_k has more independent columns than independent rows. We have L to be an *even* integer to warrant the existence of integer $\frac{L}{2}$, which is used in the following developments.

The suggested data-based residual is obtained by the projection of Y_k onto Π_k:

$$\epsilon_k = Y_k \Pi_k Z \in R^{\ell} \quad [6.31]$$

$$\epsilon_k \simeq \underbrace{H_i U_k \Pi_k Z}_{=0} + W_k \Pi_k Z \quad [6.32]$$

where $Z = \begin{pmatrix} 0 & \dots & 0 & 1 \end{pmatrix}^T \in R^{\ell \times 1}$ is the selection vector. The form of calculation for this residual, given by equation [6.31], depends solely on the inputs and outputs of the system. Since the measurement noise is assumed to be centered and the inputs are deterministic signals, the expected value of the residual is given by:

$$E[\epsilon_k] \simeq E[W_k \Pi_k Z] = 0 \quad [6.33]$$

6.4.2.3. *Sensitivity of the residual to changes in mode*

Let us assume that a single mode change happens during interval $k - L + 1$ and k. Let τ be the commutation time: $k - L + 1 \leq \tau \leq k$. By repeated substitutions in equation [6.24], we obtain the following input–output matrix relationship:

$$Y_k = \begin{bmatrix} y_{k-L+1} & \cdots & y_\tau & \cdots & y_k \end{bmatrix}$$

We obtain:

$$\begin{aligned} Y_k &= H_{1,i}\begin{bmatrix} \bar{u}_{k-L+1,i} | & \cdots | & \bar{u}_{\tau-1,i} | & \bar{u}_{\tau,i} | & 0_{m.(i+1),k-\tau} \end{bmatrix} \\ &+ \chi + \begin{bmatrix} H_{(1,2),1} | & \cdots | & H_{(1,2),i-1} | & H_{(1,2),i} \end{bmatrix} U^* \\ &+ H_{2,i}\begin{bmatrix} 0_{m.(i+1),L-k+\tau+i} | & \bar{u}_{\tau+i+1,i} | & \cdots | & \bar{u}_{k-1,i} | & \bar{u}_{k,i} \end{bmatrix} + W_k \end{aligned} \quad [6.34]$$

with:

$$\chi = \begin{bmatrix} C_1 A_1^i x_{k-L-i+1} | & C_2 A_1^i x_{\tau-i+1} | & C_2 A_2 A_1^{i-1} x_{\tau-i+2} | & \cdots \\ C_2 A_2^{i-2} A_1^2 x_{\tau-1} | & C_2 A_2^{i-1} A_1 x_\tau | & \cdots | & C_2 A_2^i x_{k-i} \end{bmatrix}$$

$$U^* = \begin{bmatrix} 0_{mi.(i+1),\tau-k+\frac{L}{2}} | & U^\bullet | & 0_{mi.(i+1),k-\tau-i+\frac{L}{2}} \end{bmatrix} \text{ and}$$

$$U^\bullet = \begin{bmatrix} \bar{u}_{\tau+1} & 0_{m.(i+1),1} & \cdots & \cdots & 0_{m.(i+1),1} \\ 0_{m.(i+1),1} & \bar{u}_{\tau+2} & \ddots & \ddots & \vdots \\ \vdots & \ddots & \ddots & \ddots & \vdots \\ \vdots & \ddots & \ddots & \ddots & 0_{m.(i+1),1} \\ 0_{m.(i+1),1} & \cdots & \cdots & 0_{m.(i+1),1} & \bar{u}_{\tau+i} \end{bmatrix}$$

$$H_{(1,2),1} = \begin{bmatrix} C_2 A_1^{i-1} B_1 | & C_2 A_1^{i-2} B_1 | & \cdots | & C_2 B_1 | & D_2 \end{bmatrix}$$

$$\vdots$$

$$H_{(1,2),i-1} = \begin{bmatrix} C_2 A_2^{i-2} A_1 B_1 | & C_2 A_2^{i-2} B_1 | & \cdots | & C_2 B_2 & | D_2 \end{bmatrix}$$

$$H_{(1,2),i} = \begin{bmatrix} C_2 A_2^{i-1} B_1 | & C_2 A_2^{i-2} B_2 | & \cdots | & C_2 B_2 | & D_2 \end{bmatrix}$$

For two matrices, A_1 and A_2, with $0 \le j \le i$, we have the following inequality:

$$\left\| A_1^j A_2^{i-j} \right\| < \left\| A_1^j \right\| \left\| A_2^{i-j} \right\| < (\max(\|A_1\|, \|A_2\|))^i$$

If the choice of i is sufficiently large, the term $\max(\|A_1\|, \|A_2\|)^i$ can be disregarded. As a consequence, $\left\| A_1^j A_2^{i-j} \right\|$ can also be ignored, as can the contribution of the initial state:

$$\chi = \begin{bmatrix} C_1 A_1^i x_{k-L-i+1} | & C_2 A_1^i x_{\tau-i+1} | & C_2 A_2 A_1^{i-1} x_{\tau-i+2} | & \cdots \\ C_2 A_2^{i-2} A_1^2 x_{\tau-1} | & C_2 A_2^{i-1} A_1 x_\tau | & \cdots | & C_2 A_2^i x_{k-i} \end{bmatrix}$$

By adding and subtracting the term:

$$H_{1,i} \begin{bmatrix} 0_{m.(i+1),\tau-k+L-1} | & \bar{u}_{\tau+1,i} | & \cdots | & \bar{u}_{k-1,i} | & \bar{u}_{k,i} \end{bmatrix}$$

in equation [6.34], we obtain:

$$\begin{aligned} Y_k = H_{1,i} U_k + [\ H_{(1,2),1} - H_{1,i} | \ \cdots | \ H_{(1,2),i-1} - H_{1,i} | \ H_{(1,2),i} - H_{1,i}\ ] U^* \\ + (H_{2,i} - H_{1,i}) [\ 0_{m.(i+1),L-k+\tau+i} | \ \bar{u}_{\tau+i+1,i} | \ \cdots | \ \bar{u}_{k-1,i} | \ \bar{u}_{k,i}\ ] + W_k \end{aligned} \quad [6.35]$$

Finally, by projecting relationship [6.35] onto the kernel of input matrix U_k with the help of Π_k, we obtain:

$$\begin{aligned} \epsilon_k = Y_k \Pi_k Z = [\ H_{(1,2),1} - H_{1,i} | \ \cdots | \ H_{(1,2),i-1} - H_{1,i} | \ H_{(1,2),i} - H_{1,i}\ ] U^* \Pi_k \\ + (H_{2,i} - H_{1,i}) [\ 0_{m.(i+1),L-k+\tau+i} | \ \bar{u}_{\tau+i+1,i} | \ \cdots | \ \bar{u}_{k-1,i} | \ \bar{u}_{k,i}\ ] \Pi_k + W_k \Pi_k \end{aligned}$$

If $u_k \neq 0$, residual ϵ_k is sensitive to the switching from mode $H_{1,i}$ to distinguishable mode $H_{2,i}$ ($H_{2,i} \neq H_{1,i}$). When a switching happens, residual ϵ_k is not zero.

REMARK 6.4.– The distinguishability condition used above expresses the fact that the Markov's parameters of the two modes are different, which means that their minimal realizations are different.

6.4.2.4. *Residual based on data for recognizing the current mode*

In this section, a residual calculated from data that are labeled offline in each mode is used to recognize the mode. For a given mode γ, the data are piled up in matrices as follows:

$$\underline{U}_{\gamma,k} = \begin{bmatrix} \overline{u}_{1,i} & \cdots & \overline{u}_{\frac{L}{2},i} \end{bmatrix} \in \mathbb{R}^{m.(i+1)\times \frac{L}{2}},$$
$$\underline{Y}_{\gamma,k} = \begin{bmatrix} \underline{y}_{1} & \cdots & \underline{y}_{\frac{L}{2}} \end{bmatrix} \in \mathbb{R}^{\ell\times \frac{L}{2}}, 1 \leq \gamma \leq d$$

These data can be collected offline, by constraining the system to operate in each mode or by using feedback from failing behaviors.

The offline labeled data are used with online collected data $(U_{s,k}, Y_{s,k})$. The input and output matrices used to calculate the residuals are:

$$\tilde{U}_k(s,\gamma) = \begin{bmatrix} \underline{U}_{\gamma,k} & U_{s,k} \end{bmatrix} \text{and } \tilde{Y}_k(s,\gamma) = \begin{bmatrix} \underline{Y}_{\gamma,k} & Y_{s,k} \end{bmatrix} \in \mathbb{R}^{\ell\times L}$$

with:

$$U_{s,k} = \begin{bmatrix} \bar{u}_{\tau+i,i} & \cdots & \bar{u}_{\frac{L}{2}+\tau+i-1,i} \end{bmatrix} \in \mathbb{R}^{m.(i+1)\times \frac{L}{2}}$$

and:

$$Y_{s,k} = \begin{bmatrix} y_{\tau+i} & \cdots & y_{\frac{L}{2}+\tau+i-1} \end{bmatrix} \in \mathbb{R}^{\ell\times \frac{L}{2}}$$

Here, the same amount of offline (labeled) and online data is used for mode recognition. This enables us to balance the importance of information collected offline and online.

By using the same procedure as in the previous section, we obtain:

$$\tilde{Y}_k(s,\gamma) = H_{s,i} \begin{bmatrix} 0_{m.(i+1),\frac{L}{2}} & U_{s,k} \end{bmatrix} + H_{\gamma,i} \begin{bmatrix} \underline{U}_{\gamma,k} & 0_{m.(i+1),\frac{L}{2}} \end{bmatrix} + W_k \quad [6.36]$$

where $0_{s1,s2}$ is a matrix with $s1$ rows and $s2$ columns containing only zeros.

Equation [6.36] is equivalent to:

$$\tilde{Y}_k(s,\gamma) = H_{\gamma,i}\left[\begin{array}{cc} \underline{U}_{\gamma,k} & U_{s,k} \end{array}\right] + (H_{s,i} - H_{\gamma,i})\left[\begin{array}{cc} 0_{m.(i+1),\frac{L}{2}} & U_{s,k} \end{array}\right] + W_k \quad [6.37]$$

By projecting equation [6.37] in the same way as we have previously, we obtain:

$$\tilde{\epsilon}_{(s,\gamma),k} = \tilde{Y}_k(s,\gamma)\tilde{\Pi}_k(s,\gamma)Z = (H_{s,i} - H_{\gamma,i})\left[\begin{array}{cc} 0_{m.(i+1),\frac{L}{2}} & U_{s,k} \end{array}\right]\tilde{\Pi}_k(s,\gamma)Z + W_k\tilde{\Pi}_k(s,\gamma)Z$$

with $\tilde{\Pi}_k(s,\gamma)$ built in the same way as Π_k.

PROPOSITION 6.1.– The current mode is identified as γ (i.e. $s = \gamma$) if and only if $E\left[\tilde{\epsilon}_{(s,\gamma),k}\right] \cong 0$.

PROOF.– If $s = \gamma$, we have $H_{s,i} = H_{\gamma,i}$, then:

$$\tilde{Y}_k(s,\gamma) = H_{\gamma,i}\left[\begin{array}{cc} \underline{U}_{\gamma,k} & U_{s,k} \end{array}\right] + W_k$$

which implies that $E[\tilde{\epsilon}_{(s,\gamma),k}] = E[W_k\tilde{\Pi}_k(s,\gamma)Z] \cong 0$.

6.4.2.5. *Tuning the method*

Two time windows are used to calculate the residuals:

1) the first one is of size i and enables us to express output y_k at time k as a function of inputs collected on a horizon of length i and initial state x_{k-i}. The choice of integer i must be sufficiently large that the influence of the initial state weighted by A^i can be disregarded (A^i tends toward zero for a sufficiently large value of i); and

2) the second window of size L chosen is large enough that the right kernel of Hankel's matrix of inputs U_k has a dimension greater than $\ell + mi$. The choice of integer L depends on i and ℓ because inequality $L \geq (mi + \ell)$ must be fulfilled.

To find a good compromise between algorithm complexity, the size of matrices to be handled and sensitivity to the initial condition, we determine the minimum value i_0 of i that enables us to limit the effect of the initial condition as much as possible. For this we consider the following criterion:

$$\begin{aligned} J(i) &= min_i \left\|(Y_k)_{wf}\Pi_k\right\|_2 \\ &= min_i \left\|C_{\sigma(k)}A^i_{\sigma(k)}X_k\Pi_k + w_k\Pi_k\right\|_2 \end{aligned} \quad [6.38]$$

where $X_k = [x_{k-L+1}...x_{k-1}x_k]$ and index wf indicates that the outputs are collected when the system is faultless.

This criterion can be calculated in a tuning phase prior to the first use of the system by using persistent inputs that enable us to trigger all of the system's dynamics.

6.4.2.6. *Summary of the method*

The following algorithm details the different steps that enable us to detect the switching and recognize the successor mode.

Procedure

Step 1: initialization:

– *tuning the method*: choice of i and L;

– *construction of the reference database*: $(\underline{U}_\gamma, \underline{Y}_\gamma)$ is collected, with a known corresponding label $\gamma, 1 \leq \gamma \leq d$ for each set of data and for all γ.

Step 2: recognition of the current mode:

– *similarity detector*: calculate $\tilde{\epsilon}_{(s,\gamma),k} = \tilde{Y}_k(s,\gamma)\tilde{\Pi}_k(s,\gamma)$ with $1 \leq \gamma \leq d$;

– *label of the current mode*: if $E[\tilde{\epsilon}_{(s,\gamma),k}] = 0$ then $s = \gamma$;

– *unknown mode*: if for all γ, $E[\tilde{\epsilon}_{(s,\gamma),k}] \neq 0$, trigger an alarm.

Step 3: estimation of the switching time:

– *switching indicator*: at each time, calculate $\epsilon_k = Y_k\Pi_k$;

– *switching detection*: if $E[\epsilon_k] \neq 0$ after execution of *step 2* at least after collection of $L/2$ samples of current data.

6.4.3. *Diagnosis of a switching system with uncertain nonlinear modes*

This section deals with SS with nonlinear modes containing uncertain parameters. An MM of Takagi-Sugeno type [TAK 85] is used to represent the nonlinear dynamics of each mode and create a projection using the activation function to generate the residuals. As previously in the linear case, this method is based on partial knowledge of the models in each operating mode, here partial knowledge of MM, i.e. knowledge of activation functions only.

It is not necessary to have knowledge of local model parameters. The general principle of the method is to project the output matrix onto the kernel of weighted inputs Hankel's matrix [ZOU 11a, ZOU11b].

For several applications, even if a behavioral model that is relatively accurate can be obtained from the laws of physics or by using parametric identification techniques, these models do not exactly reflect the behavior of the system. This is because there is imperfect knowledge of the value of model parameters or the normal variation of these parameters during the running. These parametric uncertainties influence the residuals *a priori*, which makes it difficult to detect mode changes and to identify the active mode. Robust methods of diagnosis are generally suggested in the literature that can be used to monitor such systems [ISE 05, GER 97, PAT 91].

The suggested method uses partial knowledge of the system and is intrinsically robust to parameter values.

6.4.3.1. *Model of a switching system with nonlinear modes*

Let us consider the SS written in the form:

$$\Sigma_\sigma : \left\{ \begin{array}{l} x_{k+1} = f_\sigma(x_k, u_k) \\ y_k = g_\sigma(x_k, u_k) \end{array} \right. \qquad [6.39]$$

$$\sigma : [0, \infty[\rightarrow \{1, 2, \ldots, s\}$$

where $u_k \in \mathbb{R}^m$ and $y_k \in \mathbb{R}^l$ are the input and output vectors, respectively, and $x_k \in X \subset \mathbb{R}^n$ is the state vector. Function $\sigma(k) \in \{1, 2, \ldots, s\}$ indicates that the index of the active mode and f_σ and g_σ are nonlinear functions. A mode can represent either normal behavior or a failing behavior of the system. Each model Σ_σ ($\sigma \in \{1, 2, \ldots, s\}$) is represented by a Takagi-Sugeno MM in what follows.

REMARK 6.5.– In order to simplify the notations, the current value of index $\sigma(k)$ will be represented by j (for instance $j = \sigma(k)$).

Takagi-Sugeno's MM for a given mode, j, represented by h_j local models, consists of a set of rules $R_{\gamma,j}$ given by $R_{\gamma,j}$:

$$\left\{ \begin{array}{l} If\ z_k^1\ in\ \alpha_{\gamma,1}\ and\ \ldots\ and\ z_k^{g_j}\ in\ \alpha_{\gamma,g_j} \\ Then \left\{ \begin{array}{l} x_{k+1}^{\gamma,j} = \left(A^{\gamma,j} + \Delta A^{\gamma,j}\right) x_k^{\gamma,j} + \left(B^{\gamma,j} + \Delta B^{\gamma,j}\right) u_k \\ y_k^{\gamma,j} = \left(C^{\gamma,j} + \Delta C^{\gamma,j}\right) x_k^{\gamma,j} + \left(D^{\gamma,j} + \Delta D^{\gamma,j}\right) u_k \end{array} \right. \\ 1 \leq \gamma \leq h_j \end{array} \right. \qquad [6.40]$$

where $x_k^{\gamma,j} \in \mathbb{R}^{n_{\gamma,j}}$ and $y_k^{\gamma,j} \in \mathbb{R}^\ell$ are respectively the state vector and the output vector of local model γ of mode j. Here, $\left(A^{\gamma,j}, B^{\gamma,j}, C^{\gamma,j}, D^{\gamma,j}\right)$ are nominal matrices of state representation that we have at our disposal for h_j local models of mode j. Matrices $\left(\Delta A^{\gamma,j}, \Delta B^{\gamma,j}, \Delta C^{\gamma,j}, \Delta D^{\gamma,j}\right)$ represent the behavioral gap in the real system with respect to the nominal model of h_j local models of mode j. All of the local models are assumed to be stable.

Integers $\alpha_{\gamma,q}, q \in \{1, \ldots, g_j\}$ represent the modality obtained by premise variables z_k^i $(1 \leq i \leq g_j)$ in rule $R_{\gamma,j}$.

The output of the Takagi-Sugeno model for mode *j* is given by:

$$y_k^j = \sum_{\gamma=1}^{h_j} \omega^{\gamma,j} (\underline{z}_k) y_k^{\gamma,j} \qquad [6.41]$$

Weighting functions $\omega^{\gamma,j}$ are defined as being the normalized validities of rules $R_{\gamma,j}$ and fulfill the following properties:

$$\begin{cases} \sum_{\gamma=1}^{h_j} \omega_k^{\gamma,j} (\underline{z}_k) = 1 \quad \forall k \\ 0 \leq \omega_k^{\gamma,j} (\underline{z}_k) \leq 1 \quad \forall k, 1 < \gamma \leq h_j \end{cases} \qquad [6.42]$$

where premise variables $\underline{z}_k = \begin{pmatrix} z_k^1 & z_k^2 & \ldots & z_k^{g_j} \end{pmatrix}^T \in \mathbb{R}^{g_j}$ depend on unknown and/or known variables such as the input or output signals.

In what follows, we will assume that these premise variables are known at all times. For mode j represented by h_j local models, the Takagi-Sugeno representation [ORJ 08] is thus given by:

$$\begin{cases} \begin{bmatrix} x_{k+1}^{1,j} \\ \vdots \\ x_{k+1}^{h_j,j} \end{bmatrix} = \begin{bmatrix} A^{1,j} & & 0 \\ & \ddots & \\ 0 & & A^{h_j,j} \end{bmatrix} \begin{bmatrix} x_k^{1,j} \\ \vdots \\ x_k^{h_j,j} \end{bmatrix} + \begin{bmatrix} B^{1,j} \\ \vdots \\ B^{h_j,j} \end{bmatrix} u_k \\ y_k^j = \sum_{\gamma=1}^{h_j} \omega^{\gamma,j} (z_k) \left(C^{\gamma,j} x_k^{\gamma,j} + D^{\gamma,j} u_k \right) \end{cases} \qquad [6.43]$$

6.4.3.2. *Residual generation*

The residuals that will be used as detectors of mode changes and indicators of the current mode are generated by using the input–output data. The input–output relationship is established in a given time window and the projection matrix is then calculated in order to eliminate the unknown state and the model parameters [CHE 09].

The following input–output matrix is obtained if mode j is active [ZOU 11a]:

$$Y_k \simeq \bar{H}^j \Phi_k^j \qquad [6.44]$$

where $Y_k = \begin{pmatrix} y_k & y_{k+1} & \cdots & y_{k+L-1} \end{pmatrix} \in \mathbb{R}^{\ell \times L}$ is the output matrix.

$\bar{H}^j \in \mathbb{R}^{\ell \times m(i+1)h_j}$ is the matrix of Markov's parameters:

$$\bar{H}^j = \begin{bmatrix} \bar{H}^{1,j} & \dots & \bar{H}^{h_j,j} \end{bmatrix}$$

and:

$$\bar{H}^{\gamma,j} = \left[\overline{C}^{\gamma,j} \left(\overline{A}^{\gamma,j}\right)^{i-1} \overline{B}^{\gamma,j} \quad \dots \quad \overline{C}^{\gamma,j} \overline{B}^{\gamma,j} \quad \overline{D}^{\gamma,j} \right]$$

where $\overline{A}^{\gamma,j} = A^{\gamma,j} + \Delta A^{\gamma,j}$ and $\overline{B}^{\gamma,j}$, $\overline{C}^{\gamma,j}$ and $\overline{D}^{\gamma,j}$are similarly defined.

$\Phi_k^j \in \mathbb{R}^{h_j m(i+1) \times L}$ is the weighted input–output matrix:

$$\Phi_k^j = \begin{bmatrix} \Phi_k & & 0 \\ & \ddots & \\ 0 & & \Phi_k \end{bmatrix} \underbrace{\begin{bmatrix} \Omega^{1,j} \\ \Omega^{2,j} \\ \vdots \\ \Omega^{h_j,j} \end{bmatrix}}_{\Omega^j} \qquad [6.45]$$

where $\Omega^j \in \mathbb{R}^{h_j L \times L}$ is the weighting matrix of all the local models of mode j and $\Phi_k \in \mathbb{R}^{m(i+1) \times L}$ is Hankel's matrix defined by:

$$\Phi_k = \begin{bmatrix} \bar{u}_{k-i} & \bar{u}_{k-i+1} & \dots & \bar{u}_{k-i+L-1} \end{bmatrix}$$

$$\bar{u}_{k-i} = \begin{pmatrix} u(k-i)^T & \dots & u(k)^T \end{pmatrix}^T$$

and for $1 \leq r \leq h_j, \Omega^{r,j} \in \mathbb{R}^{L \times L}$:

$$\Omega^{r,j} = \begin{bmatrix} \omega_k^{r,j} & 0 & \dots & 0 \\ 0 & \omega_{k+1}^{r,j} & \dots & 0 \\ \vdots & \vdots & \dots & \vdots \\ 0 & 0 & \dots & \omega_{k+L-1}^{r,j} \end{bmatrix}$$

The residuals are obtained by projecting input–output relationship [6.44] onto the kernel to the right of weighted input matrix Φ_k^j. For this, we introduce the projection matrix $\left(\Phi_k^j\right)^{\perp} \in \Re^{L \times L}$:

$$\left(\Phi_k^j\right)^{\perp} = I - \left(\Phi_k^j\right)^T \left(\Phi_k^j \left(\Phi_k^j\right)^T\right)^{-1} \Phi_k^j \qquad [6.46]$$

By right-multiplying equation [6.44] by $\left(\Phi_k^j\right)^{\perp}$, the residual expression is the following:

$$r_k^j = Y_k\left(\Phi_k^j\right)^{\perp} = 0 \qquad [6.47]$$

Projection matrix $\left(\Phi_k^j\right)^{\perp}$ is not zero for all u_k and $\omega^{\gamma,j}\left(\underline{z}_k\right)$ if the number of independent rows of matrix Φ_k^j is less than that of its independent columns. This is translated into the following condition:

$$L > h_j(mi+1)$$

If the above condition is fulfilled and mode j is active, residuals r_k^j have a value of zero. If this is not the case, these residuals indicate a change of mode.

The projection method does not take into account system parameters $\bar{H}^j$ and requires only the input–output measurements Y_k and weighting functions Φ_k^j, which do not depend on uncertainties $\left(\Delta A^{\gamma,j}, \Delta B^{\gamma,j}, \Delta C^{\gamma,j}, \Delta D^{\gamma,j}\right)$. This proves the robustness of the residuals to these uncertainties on the state matrices of local models.

6.4.4. *Conclusions*

Two data-based methods were suggested for the detection of commutations and recognition of the active mode. The first one is aimed at SSs made of linear modes and uses only input–output data for the switching detection and offline labeled data for the recognition of the active mode. The second method is aimed at SSs exhibiting nonlinear modes and uses Takagi-Sugeno type MM modeling for each mode. The switching detection and recognition of the active mode is based on partial knowledge of the model, particularly the activation functions. These two methods have shown their efficiency in the simulation of academic examples [ZOU 11a, ZOU 11b, HAK 11]. The application to a real system is ongoing.

6.5. Modeling, observation and monitoring of switching systems: application to a multicellular converter

6.5.1. *Introduction*

This section presents results on the modeling, observation and monitoring of a power converter (multicellular converter). Multicellular converters, by their nature, exhibit a hybrid behavior and therefore are a specific type of SS. The hybrid aspect is caused by the presence of discrete variables (i.e. switches) and continuous variables (voltages and currents in the passive components, such as resistors, inductors and capacitors).

The study presented in this section was carried out within the framework of the DICOP project using an experimental platform available at LAMIH (Laboratoire d'Automatique, de Mecanique et d'Informatique Industrielles et Humaines de Valenciennes).

The platform enables us to validate the theoretical developments obtained around the modeling and observation of SS.

This project focused on producing solutions to the following scientific hurdles:

– fault detection methods for a class of SSs;

– binary command for a SS; and

– design of a reconfigurable fault tolerant control law.

6.5.2. *Multicellular converter with two arms or four quadrants*

The goal was to define a topology working in four quadrants on the basis of a multicellular converter in series in order to control a direct current machine. The idea was to place two multicellular converters in series (see Figure 6.3). The system consisted of a reversible chopper that was based on a two-arm structure of multicellular converters with three cells. This type of structure considerably reduced the voltage at the terminals of the command switches to that of a conventional chopper.

From its structure, six command signals were able to act on the state of the converter. Let us note that the command of each arm was able to be carried out independently. Thus, each arm needed to supply the necessary voltage to have the requested voltage at the output of the converter. For this structure, we had the following cases:

– if the output voltage of the first arm was greater than that of the second arm, the current was therefore positive (the engine rotated in one direction);

– if the output voltage of the first arm was less than that of the second arm, the current was therefore negative (the engine rotated in the opposite direction); and

– if the output voltage of the first arm equaled that of the second arm, the engine slowed down.

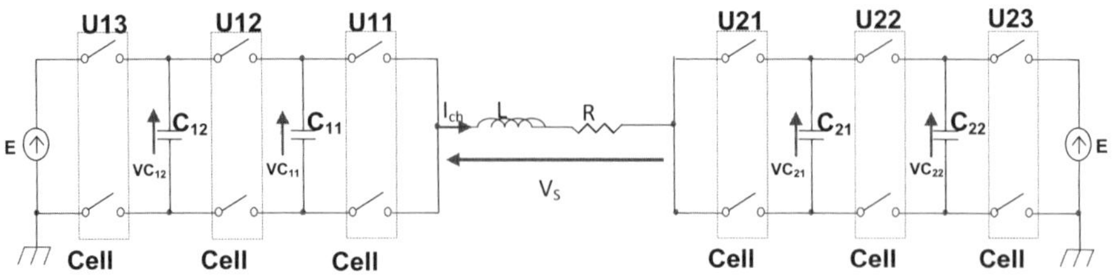

Figure 6.3. *Structure of a multicellular converter with three cells and two arms*

The modeling of this system gave the following system

$$\begin{aligned} \dot{v}_{c11} &= \tfrac{u_{12}-u_{11}}{C_{11}} I_{ch} & \dot{v}_{c12} &= \tfrac{u_{13}-u_{12}}{C_{12}} I_{ch} \\ \dot{v}_{c21} &= \tfrac{u_{22}-u_{21}}{C_{21}} I_{ch} & \dot{v}_{c22} &= \tfrac{u_{23}-u_{22}}{C_{22}} I_{ch} \\ \dot{I}_{ch} &= -\tfrac{R}{L} I_{ch} + \tfrac{1}{L} V \end{aligned}$$

The output voltage was given by:

$$V = (u_{13}-u_{23})E+(u_{11}-u_{12})v_{c11}+(u_{12}-u_{13})v_{c12}-(u_{21}-u_{22})v_{c21}-(u_{22}-u_{23})v_{c21}$$

Thus, the output voltage could have seven levels: $-E$, $-2E/3$, $-E/3$, 0, $E/3$, $2E/3$, and E. This enabled us to command a direct current machine in engine mode or in generator mode.

6.5.3. *Diagnosing faults in the four quadrant converter*

6.5.3.1. *Overview of the converter faults*

A fault is defined as a deviation of at least one property or characteristic parameter of the system from the normal operating conditions that are allowed [ISE 97]. In the literature, faults are classified according to their localization, thus there are three types of faults: actuator faults, sensor faults and system or component faults.

Before starting the diagnosis, we must initially list the different faults that can occur in our multicellular converter, which has several commutation cells. These are made of two semiconductor power components, a passive component (capacitor) and a junction between the command card and the semiconductor components.

By taking into account this setup, we can list the possible faults:

– *cell faults*: the commutation cell remains stuck in one state. The actuator will then take a fixed command value of either $u = 0$ or $u = 1$. This fault does not appear because of the failure of semiconductor components but because of the junction between command card and semiconductors. Thus, the command signal cannot reach the trigger of the components and the cell remains stuck in one state;

– *capacitor faults*: this includes either a sudden functional failure or ageing; and

– *switch faults*: there are two modes of failure, and these are [VAL 03]: a switch in open circuit and a switch in closed circuit (the component remains in on-state).

6.5.3.2. *Diagnosis based on an observer*

We are interested in the diagnosis of cell faults for the four-quadrant converter. Most methods of diagnosis are broken down into two steps [JAM 07]: residual generation and decision making. We will use the observer method to generate the residuals, which are the fault-indicator signals. Figure 6.4 illustrates the general principle of the diagnosis based on an observer.

The only measurement is the load current. We use the following sliding mode observer [SLO 91] to estimate the unmeasured variables:

$$\begin{array}{llllll} \dot{\hat{v}}_{c11} & = & \frac{u_{12}-u_{11}}{C_{11}}\hat{I}_{ch} - \lambda_1 sign(S) & \dot{\hat{v}}_{c12} & = & \frac{u_{13}-u_{12}}{C_{12}}\hat{I}_{ch} - \lambda_2 sign(S) \\ \dot{\hat{v}}_{c21} & = & \frac{u_{22}-u_{21}}{C_{21}}\hat{I}_{ch} - \lambda_3 sign(S) & \dot{\hat{v}}_{c22} & = & \frac{u_{23}-u_{22}}{C_{22}}\hat{I}_{ch} - \lambda_4 sign(S) \end{array}$$

$$\begin{array}{lll} \dot{\hat{I}}_{ch} & = & -\frac{R}{L}\hat{I}_{ch} + \frac{E}{L}(u_{13} - u_{23}) + \frac{1}{L}(u_{11} - u_{12})\hat{v}_{c11} \\ & & +\frac{1}{L}(u_{12} - u_{13})\hat{v}_{c12} - \frac{1}{L}(u_{21} - u_{22})\hat{v}_{c21} - \frac{1}{L}(u_{22} - u_{23})\hat{v}_{c21} - \lambda_5 sign(S) \end{array}$$

where $S = \hat{I}_{ch} - I_{ch}$ represents the sliding surface; v_{c11}, v_{c12}, v_{c21} and v_{c22} are the estimated values of voltages at the terminals of the capacitors; and λ_1, λ_2, λ_3, λ_4 and λ_5 represent the corrective gains of observation.

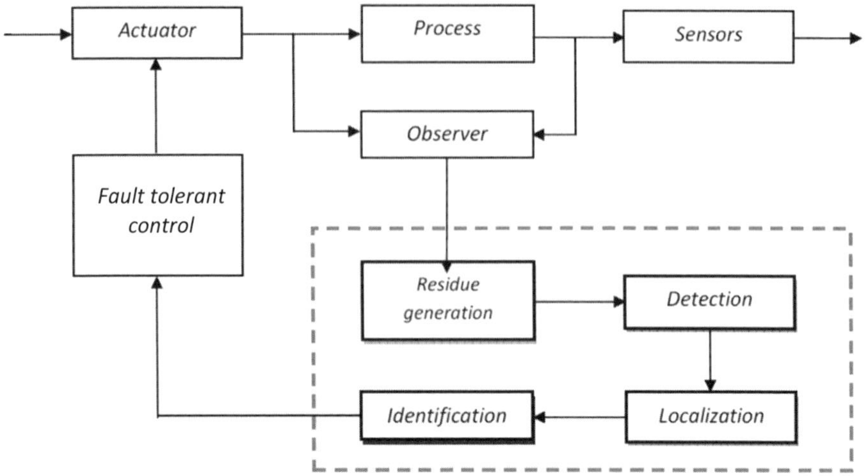

Figure 6.4. *Diagnosis based on an observer*

The convergence conditions were developed for a multicellular converter with p cells in [BEN 01]. By adapting them to the four-quadrant converter, the corrective gains become:

$$\begin{cases} \lambda_1 = -\frac{\lambda_5}{\tau} L(u_{12} - u_{11}) & , \quad \lambda_2 = -\frac{\lambda_5}{\tau} L(u_{13} - u_{12}) \\ \lambda_3 = \frac{\lambda_5}{\tau} L(u_{22} - u_{21}) & , \quad \lambda_4 = \frac{\lambda_5}{\tau} L(u_{23} - u_{22}) \\ \lambda_5 > |\Delta f_5|_{max} & , \quad \Delta f = f(\hat{x}, u) - f(x, u) \end{cases}$$

τ is the parameter that must be accurately adjusted because it sets the dynamics of the observer during the sliding mode.

To define the residual, we must initially know the consequences of the different cells stuck on the state variables of the system. For this, we establish the faulty operation signatures. We illustrate this in Table 6.2.

The chosen residual has the following form:

$$R = \left\| \frac{\Delta v_{c11}}{v_{c11ref}} + \frac{\Delta v_{c12}}{v_{c12ref}} + \frac{\Delta v_{c21}}{v_{c21ref}} + \frac{\Delta v_{c22}}{v_{c22ref}} + \frac{\Delta I_{ch}}{I_{ch,ref}} \right\|$$

Blocked state	V_{c11}	V_{c12}	V_{c21}	V_{c22}	I_{ch}
$u_{11} = 1$	↘	↘	–	–	↗
$u_{11} = 0$	↗	↗	–	–	↘
$u_{12} = 1$	↗	↘	–	–	↗
$u_{12} = 0$	↘	↗	–	–	↘
$u_{13} = 1$	↗	↗	–	–	↗
$u_{13} = 0$	↘	↘	–	–	↘
$u_{21} = 1$	–	–	↗	↗	↘
$u_{21} = 0$	–	–	↘	↘	↗
$u_{22} = 1$	–	–	↘	↗	↘
$u_{22} = 0$	–	–	↗	↘	↗
$u_{23} = 1$	–	–	↘	↘	↘
$u_{23} = 0$	–	–	↗	↗	↗

Table 6.2. *Signature of faulty operation of the four-quadrant converter (– no influence, ↗ increase, ↘ decrease)*

We are now interested in the detection and isolation of faults. Analysis of the effects of different faults shows that the comparison of signals that are measured and estimated with the help of the nominal model does not enable us to isolate the faults. The detection stage is carried out because the residual is not zero in the presence of a fault. The next stage uses the fault signature where a decision logic that uses the measurement of the current and intermediate voltages enables the failure to be isolated. A filter decreasing the oscillations is used. The advantage of this method is that it can be generalized if we know the consequences of the different faults on the system variables.

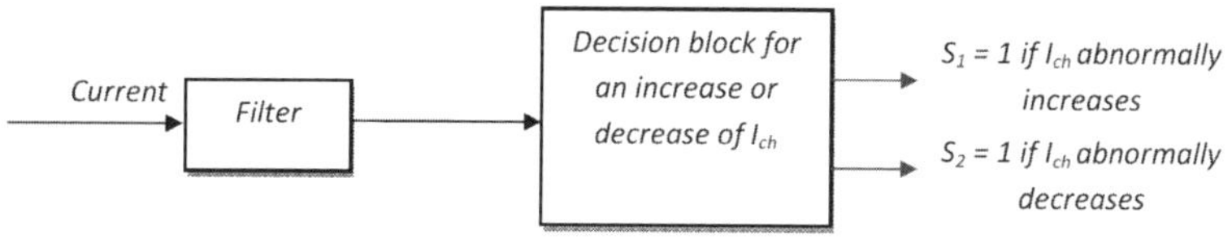

Figure 6.5. *Filtering and decision on the current*

6.5.3.3. *Simulation results*

In order to validate the procedure, simulations are carried out. The parameters of the converter are: E = 1500 V, R = 10 Ω, L = 0.5 mH, $C_{11} = C_{12} = C_{21} = C_{22} = 40\ \mu F$ and $I_{ch\_ref} = 60$ A.

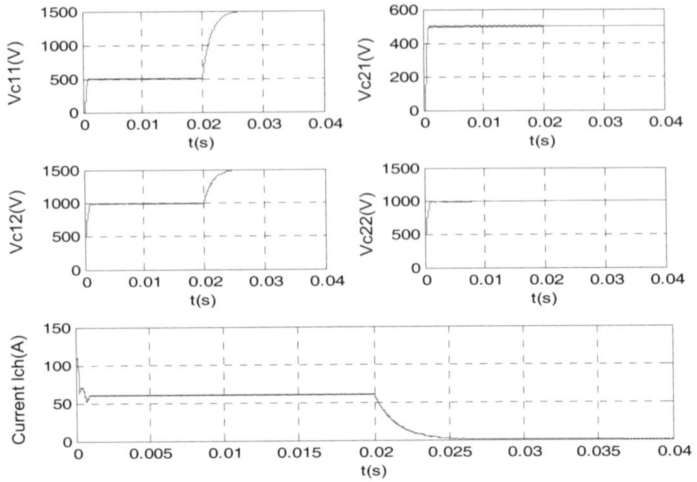

a) Evolution of state variables of the four-quadrant converter in case of stuck switching

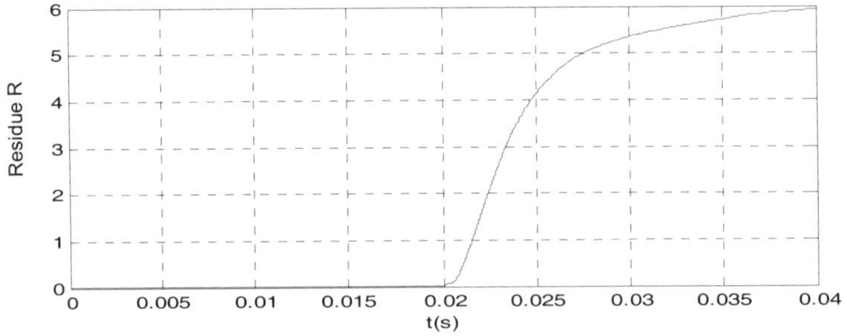

b) Evolution of the residual in case of stuck switching ($u_{11} = 0$)

Figure 6.6. *Results of test 1: we introduce a stuck switching of $u_{11} = 0$ at t= 0.02 sec*

Figures 6.6 and 6.7 show the simulation results for some instances where a cell is stuck. The phenomenon also happens when switches are stuck, in accordance with the voltages at the terminals of the capacitors.

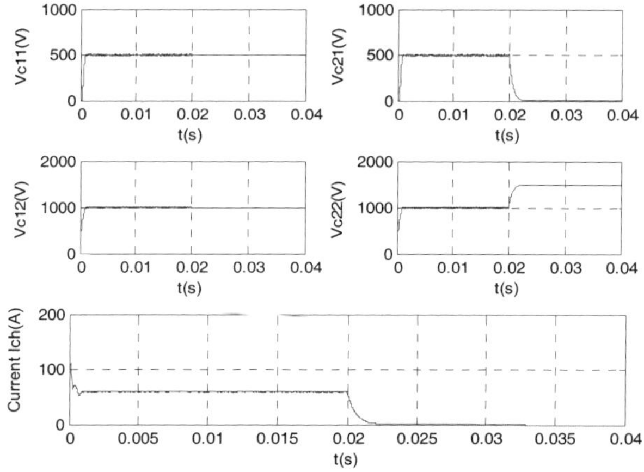

a) Evolution of state variables of the four-quadrants converter in case of stuck switching

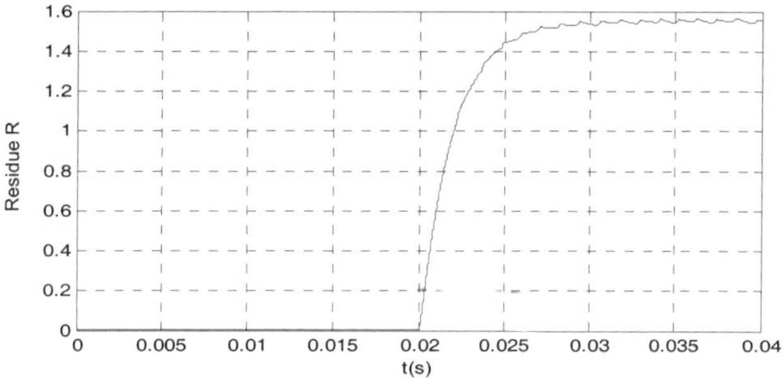

b) Evolution of the residual in case of stuck switching of $u_{22}=1$

Figure 6.7. *Results of test 2: we introduce a stuck switching of $u_{22} = 1$ at $t= 0.02$ sec*

We notice that the residual is close to zero in the absence of a fault. When there is a stuck cell fault, the residual departs from zero, thus ensuring the fault is detected. The detection takes 10^{-4} second and the isolation varies according to the filter used in the decision block but does not go beyond 10^{-3} second. This type of diagnosis based on a sliding mode observer is therefore a high-performance method.

6.5.4. *Experimental benchmark for validation*

To experimentally validate the command, observation and diagnosis techniques, we are going to use a benchmark available at LAMIH. The latter is made of multicellular arms in series with three cells. The switches of each arm are made of MOSFET transistors assembled in antiparallel with diodes. The MOSFET transistors are associated with command modules managing the dead times and protecting against short circuits. The command signals of MOSFET transistors come from a DS1103 card from dSPACE[5], based on a processor dedicated to signal processing, TMS320C31 from Texas Instruments, and several analog-digital and digital-analog converters. A picture of this experimental benchmark is given in Figure 6.8.

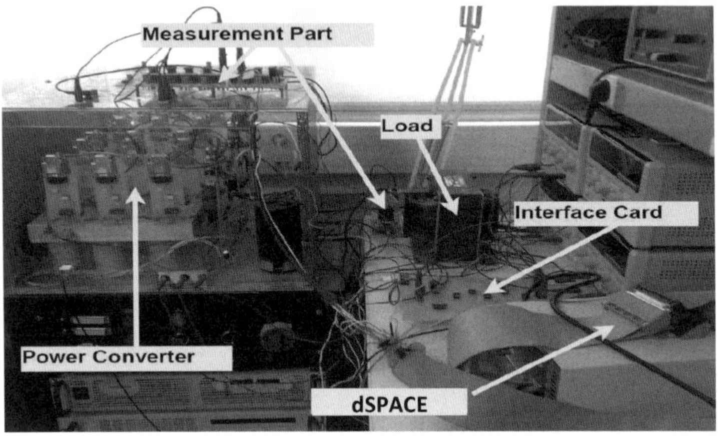

Figure 6.8. *Picture of the multicellular converter in series*

Figure 6.9 shows a synopsis of the benchmark carried out. This diagram is split into four parts:

– the power part that includes:

- the multicellular converter with three cells, switches and capacitors,

- the command and switching cards, and

- the load created by a direct current machine;

– the measurement part, made of voltage sensors (to measure the voltages of the floating capacitors and the voltage of the tachymeter) and a current sensor to measure the load current;

– the command part, made of a DS1103 control card (based on a DSP TMS320C31 device) and computer driven. The computer enables us to generate the object code defined from the command algorithm; and

5For more information go to http://www.dspace.com/en/pub/start.cfm

– the interface part, which is the layer of optical insulation and protects the outputs of the dSPACE card.

The experimental validations are under development.

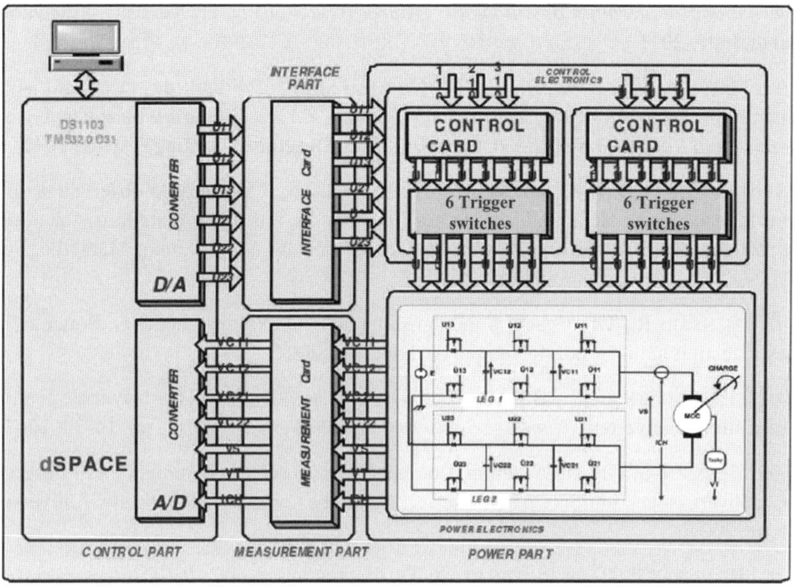

Figure 6.9. *Synopsis of the benchmark*

6.6. Bibliography

[AKH 04] AKHENAK A., CHADLI M., RAGOT J., MAQUIN D., "Estimation of state and unknown inputs of a nonlinear system represented by a multiple model", *11th IFAC Symposium on Automation in Mining, Mineral and Metal processing*, MMM, 8 – 10 September, Nancy, France, 2004.

[ANT 00] ANTSAKLIS P.-J., "Special issue on hybrid systems", *Theory and Applications*, vol. 43, 2000.

[BEL 09] BELKHIAT D.-E.-C., MESSAI N., MANAMANNI N., "Robust fault detection based-observer for linear switched systems", *3rd IFAC Conference on Analysis and Design of Hybrid Systems (ADHS 09)*, 16 – 19 September, Zaragoza, Spain, 2009.

[BEL 10] BELKHIAT D.-E.-C., MESSAI N., MANAMANNI N., "Synthèse d'observateur robuste pour la détection de défauts d'une classe de systèmes à commutations", *6e Conférence Internationale Francophone d'Automatique (CIFA 2010)*, 2 – 4 June, Nancy, France, 2010.

[BEL 11a] BELKHIAT D.-E.-C., MESSAI N., MANAMANNI N., "Design of a robust fault detection based observer for linear switched systems with external disturbances", *Nonlinear Analysis: Hybrid Systems*, vol. 5, no. 2, pp. 206-219, 2011.

[BEL 11b] BELKHIAT D.-E.-C., MESSAI N., MANAMANNI N., "Design of robust detection observer for discrete-time linear switched system", *18th World Congress of the International Federation of Automatic Control (IFAC 2011)*, 28 August – 2 Septemeber, Milan, Italy, 2011.

[BEL 11c] BELKHIAT D.-E.-C., MESSAI N., MANAMANNI N., "A fault detection and isolation scheme for discrete time switched linear systems", *9th IEEE International Conference on Control and Automation (IEEE ICCA'11)*, 19 –21 December, Santiago, Chile, 2011.

[BEL 11d] BELKHIAT D.-E.-C., MESSAI N., MANAMANNI N., "Synthèse d'observateur robuste pour la détection de défauts des systèmes linèaires discrets à commutations", *4e Journées Doctorales/ Journées Nationales MACS (JD-JN-MACS)*, 9 – 10 June, Marseille, France, 2011.

[BEN 01] BENSAID R., Observateurs des tensions aux bornes des capacités flottantes, PhD thesis, Institut national polytechnique de Toulouse, 2001.

[BEN 07] BENMANSOUR K., BENALIA A., DJEMAÏ M., DELEON J., "Hybrid control of multicellular converter", *Nonlinear Analysis: Hybrid Systems*, vol. 1, pp. 16-29, 2007.

[BEN 09] BENMANSOUR K., Réalisation d'un banc d'essai pour la commande et l'observation des convertisseurs multicellulaires série: Approche hybride, PhD thesis, University of Cergy, 2009.

[BER 01] BERGSTEN P., PALM R., DRIANKOV D., "Fuzzy observers", *IEEE International Fuzzy Systems Conference*, pp. 700-703, 2 – 5 December, Melbourne, Australia,2001.

[CHA 01] CHADLI M., MAQUIN D., RAGOT J., "On the stability analysis of multiple model systems", *Proceedings of the 6th European Conference Control (ECC)*, 4 – 7 September, Porto, Portugal, 2001.

[CHE 09] CHENIKHER S., CASSAR J.-P., PEKPE K.-M., "Fault detection and isolation from an identifed mimo Takagi-Sugeno model of a bioreactor", *7th IFAC Symposium on Fault Detection, Supervision and Safety of Technical Processes*, 30 June – 3 July, Barcelona, Spain, 2009.

[COC 04] COCQUEMPOT V., MEZYANI T.-EL., STAROSWIECKI M., "Fault detection and isolation for hybrid systems using structured parity residuals", *5th Asian Control Conference, 2004*, 20 – 23 July, vol. 2, pp. 1204-1212, 2004.

[DEC 00] DECARLO R., BRANICKY M., PETTERSSON S., LENNARTSON B., "Perspectives and results on the stability and stabilizability of hybrid systems", *Proceeding of the IEEE*, vol. 88, pp. 1069-1082, 2000.

[DEF 11] DEFOORT M., DJEMAÏ M., FLOQUET T., PERRUQUETTI W., "Robust finite time observer design for multicellular converter", *International Journal of Systems Sciences, 2010, Special Issue on VSS methods for Hybrid Systems*, Vol. 11, pp. 1464-5319, January 2011.

[DJE 11] DJEMAÏ M., BUSAWON K., BENMANSOUR K., MAAROUF A., "High order sliding mode control of a DC motor drive via a switched controlled multi-cellular converter", *International Journal of Systems Sciences, 2010, Special Issue on VSS methods for Hybrid Systems*, vol. 11, pp. 1464-5319, January 2011.

[DOM 07] DOMLAN E.-A., RAGOT J., MAQUIN D., "Switching systems: Active mode recognition, identification of the switching law", *Journal of Control Science and Engineering*, Vol. 2007, ID50796, 2007.

[FRA 97] FRANK P.-M., DIN X., "Survey of robust residual generation and evolution methods in observer-based detection systems", *Journal of Process Control*, vol. 7, no. 6, pp. 403-424, 1997.

[FRI 00] FRISK E., "Residual generator design for nonlinear, polynomial systems: a Grobner basis approach", *Proceedings of the IFAC Fault Detection, Supervision and Safety for Technical Processes*, Budapest, Hungary, pp. 979-984, 2000.

[GER 97] GERTLER J., "Fault detection and isolation using parity relations", *Control Engineering Practice*, vol. 5, no. 5, pp. 653-661, 1997.

[GUE 07] GUERRA T., KRUSZEWSKI A., BERNAL M., "Control law proposition for the stabilization of discrete Takagi-Sugeno models", *IEEE Transactions on Fuzzy Systems*, Vol. 14, no. 3, pp. 724-731, 2008.

[HAK 11] HAKEM A., PEKPE K.-M., COCQUEMPOT V., "Sensor fault diagnosis for bilinear systems using data-based residuals", *50th IEEE Conference on Decision and Control and European Control Conference: IEEE CDC/ECC 2011*, Orlando, Florida, December 12-15, 2011.

[ICH 09] ICHALAL D., MARX B., RAGOT J., MAQUIN D., "State estimation of nonlinear systems using multiple model approach", *Proceedings of the 2009 American Control Conference*, St. Louis, Missouri, USA, pp. 4636-4641, 2009.

[ISE 97] ISERMANN R., BALLÉ P., "Trends in the application of model-based fault detection and diagnosis of technical processes", *Control Engineering Practice*, vol. 5, no. 5, pp. 709- 719, 1997.

[ISE 05] ISERMANN R., "Model-based fault-detection and diagnosis – status and applications", *Annual Reviews in Control*, vol. 29, no. 1, pp. 71-85, 2005.

[JEM 07] JAMOULI H., Génération de résidus directionnels pour le diagnostic des systèmes linéaires stochastiques et la commande tolérante aux défauts, PhD thesis, Henri Poincaré University - Nancy 1, 2007.

[LIE 06] LIENHARDT A.-M., Etude de la commande et de l'observation d'une nouvelle structure de conversion d'énergie de type SMC (convertisseur multicellulaire superposé), Doctoral thesis, Institut national polytechnique de Toulouse, 2006.

[MUR 97] MURRAY-SMITH R., JOHANSEN T., *Multiple Model Approaches to Modelling and Control*, Taylor & Francis, London, 1997.

[MUR 02] MURPHEY T, BURDICK J., "Nonsmooth controllability theory and an example", *Proceedings of the 41st IEEE Conference on Decision and Control*, pp. 370-376, 2002.

[NAG 10] NAGY A.M., MOUROT G., MARX, B., SCHUTZ G., RAGOT J., "Systematic multi-modeling methodology applied to an activated sludge reactor model", *Industrial & Engineering Chemistry Research*, vol. 49, no. 6, pp. 2790-2799, 2010.

[OLI 09] OLIVIER-MAGET N., ETREUX G.-H., LE LANN J.-M., LE LANN M.V., "Model-based fault diagnosis for hybrid systems: Application on chemical process", *Computers and Chemical Engineering*, vol. 33, pp. 1617-1630, 2009.

[ORJ 08] ORJUELA R., MARX B., RAGOT J., MAQUIN D., "Proportional-integral observer design for nonlinear uncertain systems modelled by a multiple model approach", *47th IEEE Conference on Decision and Control*, CDC, pp. 3577-3582, 2008.

[PAT 91] PATTON R.-J., CHEN J., "A robust parity space approach to fault diagnosis based on optimal eigenstructure assignment", *Control 1991. International Conference on Control'91*, vol. 2, pp. 1056-1061, 1991.

[PEK 07] PEKPE K.-M., CASSAR J.-P., CHENIKHER S., "Identification of MIMO Takagi-Sugeno model of a bioreactor", *FUZZ-IEEE*, IEEE Fuzzy Systems Conference, 23 – 26 July, pp. 1-6, 2007.

[RAG 04] RAGOT J., PEKPE K.-M., MOUROT G., "Subspace method for sensor fault detection and isolation-application to grinding circuit monitoring", *11th IFAC Symposium on automation in Mining, Mineral and Metal processing*, 8 – 10 September, Nancy, France, 2004.

[SAA 06] SAADAOUI H., MANAMANNI N., DJEMAI M., BARBOT J.-P., FLOQUET F., "Exact differentiation and sliding mode observer for switched mechanical systems", *Nonlinear Analysis*, vol. 65, no. 5, pp. 1050-1069, 2006.

[SLO 91] SLOTINE J.-J.E., LI W., *Applied Nonlinear Control*, Prentice Hall, New Jersey, 1991.

[TAK 85] TAKAGI T., SUGENO M., "Fuzzy identification of systems and its applications to modeling and control", *IEEE Trans. Syst. Man. & Cybern*, SMC-15, vol. 1, pp. 116-132, 1985.

[TAN 01] TANAKA K., WANG H., *Fuzzy Control System Design and Analysis, a Linear Matrix Inequality Approach*, John Wiley & Sons, New York, 2001.

[VAL 03] VALLON J., Introduction à l'étude de la fiabilité des cellules de commutation à IGBT sous fortes contraintes, PhD thesis, Institut national polytechnique de Toulouse, 2003.

[WAN 96] WANG H.O., TANAKA K., GRIFFIN M., "An approach to fuzzy control of nonlinear systems: stability and design issues", *IEEE Transactions on Fuzzy Systems*, vol. 4, pp. 14-23, 1996.

[ZAY 01] ZAYTOON J., *Modélisation, Analyse et Commande des Systèmes Dynamiques Hybrides*, Hermès, Paris, 2001.

[ZHO 03] ZHONG M.-X., DING S., LAM J., WANG H., "A LMI approach robust fault detection filter for uncertain LTI systems", *Automatica*, vol. 39, pp. 543-550, 2003.

[ZOU 11a] ZOUARI T., PEKPE K.M., COCQUEMPOT V., KSOURI M., "Diagnosis des systèmes à commutations: Reconnaissance du mode de fonctionnement modélisé sous forme Takagi-Sugeno à paramètres locaux incertains", *JD-JN MACS*, Marseille, France, June 2011.

[ZOU 11b] ZOUARI T., PEKPE K.M., COCQUEMPOT V., KSOURI M., "Data projection method for sensor fault detection and isolation on nonlinear systems based on Takagi Sugeno model", *50th IEEE Conference on Decision and Control and European Control Conference: IEEE CDC/ECC 2011*, Orlando, Florida, December 12-15, 2011.

Chapter 7

Multitask Learning for the Diagnosis of Machine Fleet

7.1. Introduction

Nowadays, in manufacturing or production systems, condition-based maintenance is preferred over preventive maintenance because it is technically achievable and financially advantageous [CAM 09]. It is based on the surveillance of the state of the system considered from information collected on it, i.e. on the ability to detect and diagnose its dysfunctions in order to be able to plan maintenance actions for the "equipment" involved. Since the beginning of 2000, with the rise of information and communication technologies (Web, mobile, etc.), new forms of maintenance such as remote condition-based maintenance (e-CBM) have emerged [MUL 08, HIG 04].

This evolution is translated at the industrial level into the development of new solutions and new services, in particular the possibility to monitor a fleet of equipment (*fleet-wide monitoring*). The main idea is, through an overall vision of the operation of the set of machines, to improve the efficiency of the monitoring of the machines of this fleet. We find examples of these applications in the domain of transport: a fleet of cars [YOU 05], or planes or boats; in the energy sector (see, for instance, the websites of the companies Smartsignal[1] or EPRI[2]): a fleet of thermal, hydraulic, or nuclear power plants [KUN 03], of wind turbines, etc. Many of these solutions mainly offer a hardware and software architecture enabling communication between the different

---

Chapter written by Xiyan HE, Gilles MOUROT, Didier MAQUIN, José RAGOT, Pierre BEAUSEROY, André SMOLARZ and Edith GRALL-MAËS.

1 http://www.smartsignal.com.

2 http://my.epri.com.

levels or agents defining this remote diagnostic center. Others detail a bit more the "surveillance" function of their solutions. However, we can note from publications or patents associated with these solutions that the diagnostic methods used are quite elementary (threshold crossing of a signal, for instance) and do not integrate the recent developments coming from scientific research into the area. Furthermore, none present methods enabling us to develop these monitoring systems by integrating the concept of the fleet [FAM 03, HAS 04, ABU 07, YU 10]. However, this aspect appears to be the main key to this approach.

Overall, for the diagnosis, we can expect several beneficial points of this fleet diagnosis. On the one hand, it is clear that the redundancy inherent to a fleet of similar machines or systems enables us to have much more information and many more measurements relative to the operating modes and dysfunction of the systems considered. On the other hand, the similarity of the systems should or could also facilitate the design and implementation of a diagnosis system on any system of a similar nature.

Initially, we are going to define more precisely the issue at stake. We consider a set of equipment or machines supposed to be identical, for instance, a fleet of pumps of nuclear power plants, a fleet of rolling mill electric engines, and a fleet of plane turbines. Without large restrictions, we will hypothesize that this equipment can go through the same states of operation and dysfunction, and that the set of variables observed enabling us to characterize the states is identical on each piece of equipment throughout the fleet. Even though assumed to be identical, this equipment can have different behaviors, in particular, because of different operating conditions and disturbances. For instance, the same turbine is not subjected to identical operating conditions according to whether it equips a medium-haul plane covering flights in Africa or a long-haul plane covering flights between Nordic countries. It results therefrom that the values of the monitoring variables coming from an equipment and their distribution function in the observation space depend not only on its operating state, but also on the equipment itself and eventually on the history of its use.

Generally, the statistical models of observations are not known and the system is only described by samples. Hence, we assume that we have at our disposal only observation samples. Each observation is represented by a vector of attributes, extracted from each equipment of the fleet. The predefined attributes are able to characterize the states of the equipment. However, the size of the collected samples can significantly vary from one piece of equipment to another and the states corresponding to the observations available are not necessarily the same on each piece of equipment. Furthermore, these observations correspond most often to measurements made in normal operating conditions while not many, if not any, describe abnormal situations.

Among the approaches to the diagnosis of operating systems, we were particularly interested in approaches referred to as no *a priori* model, based on data (pattern recognition). In fact, the implementation of classification methods for learning requires us to have at our disposal a labeled data set. All the data are used according to different modalities to determine a decision rule and to estimate the performances of the rule. The latter is then applied to new observations coming from the same system in order to label them. The performances of the rule depend on both the available learning set and the learning method.

In the literature, this kind of problem is sometimes tackled as a two-class classification problem [KAS 09]:

– a target class also referred to as a positive class, whose data are available;

– another class whose data are difficult or impossible to obtain. This class corresponds to the failure class or negative class. Several methods were suggested within this framework. The general principle consists of artificially generating the data of the negative class, and then employing conventional learning algorithms with two classes. These methods are based on the hypothesis that the proportion of positive examples in the non-labeled set is very weak, which is not always correct. The extraction of negative examples is thus poorly reliable and requires careful attention. Besides, the non-labeled set must be sufficiently large in order to obtain the best classifier.

The classification restricted to one class is another approach for this kind of problem; in this case, learning only takes into account positive examples, the main idea being to learn the normal class by adjusting the model to positive data. The new observations are classified by comparing their score, given by the model, at a decision threshold. There exist two categories of methods of this kind [CAM 08]: the statistical methods and the methods relying on the optimization of a decision function among a class of functions, for instance neural networks or approaches based on support vector machines (SVMs). The 1-SVM method, suggested by B. Schölkopf, is one of them [SCH 01, MAN 01].

The learning methods have proved very efficient to infer monitoring rules adapted to an equipment. However, in several practical applications, we must design a common monitoring system for several similar systems, or rebuild a decision rule for each system. This is the case, for instance, for diagnosis problems when the system has several identical pieces of equipment *a priori*, but which work in different conditions. Here, the term "equipment" can refer to a simple machine (an engine, a pump, etc.), a system (a car, a plane, etc.), and by extension to an industrial installation (a nuclear power plant, etc.). The vehicles of a rental company are an example of this. In the industrial sphere, it is common to find several identical operational sites *a priori* in the domain of construction or energy production. In all these cases, learning of a monitoring rule can be assimilated to a task and it seems relevant to transfer or mutualize the information coming from similar tasks [PAN 10]. These reasons explain undoubtedly the growing interest, for several years, in multitask learning (MTL) [BI 08, ZHE 08, GU 09, BIR 10].

While most learning methods focus on the processing of a task, multitask methods aim to improve the generalization performances of the monitoring system by taking advantage of the links between the tasks. The main idea is to share what is learned from different tasks (for instance a common representation space or some parameters of the model that are close to each other), whereas the tasks are learned in parallel [CAR 97]. Previous works showed that MTL can lead to more efficient learning models with better performances [CAR 97, HES 00, BEN 03, EVG 05, BEN 08] than those obtained from single-task learning techniques.

In recent years, SVMs (support vector machines) [BOS 92, VAP 95] have been successfully used for MTL [EVG 04, JEB 04, EVG 05, WID 10, YAN 10]. The SVM method was initially developed for the classification of data of two different classes. When the data are linearly separable, SVMs determine the separating hyperplane that maximizes the distance to the closest observations (hyperplane with maximum margin). When the data are not linearly separable, "the kernel trick" is used. The basic idea is to use the original data in a space of greater dimension (transformed space) and then to solve a linear problem in that space. The good properties of the kernel functions can also be used within the framework of MTL.

In this chapter, we present a new approach to MTL that relies on one-class support vector machines (1-SVMs). 1-SVMs, suggested by [SCH 01], are adapted to the solving of problems of novelty or outlier detection. These problems appear each time as knowledge of the system to monitor is restricted to one class, or that knowledge of the other classes is too insufficient. In fact, in several applications, it is very difficult to collect a sufficiently large number of observations corresponding to all the abnormal behaviors of the system. The main advantage of a 1-SVM, with respect to other one-class classification methods [TAR 95, RIT 97, ESK 00, SIN 04], is that it focuses only on the estimation of the envelope of a region containing the samples of the target class, the envelope being sufficient for classification. The estimation of the envelope is done by separating a certain proportion of the target samples from their origin by a maximum margin hyperplane (in general, in a transformed space).

Recently, Yan *et al.* [YAN 10] suggested using a multitask approach for learning one-class classifiers. The basic idea is to constrain the classifiers of similar tasks to be close to each other. However, they solved this problem by resorting to conic programming [KEM 08] that happens to be complex. In this chapter, inspired by works of [EVG 04], we introduce an easier MTL framework that relies on 1-SVM method. We hypothesize, as in [EVG 04], that the parameter values of the model of the different tasks considered are close to a certain mean value. This hypothesis is reasonable because the tasks being similar to each other, usually their models are quite close. Then, we determine a classifier for each task and all the 1-SVMs are learned simultaneously. The suggested multitask approach is simple to implement because it only requires a minor modification to the formulation of a the optimization problem solved by 1-SVM. Experimental results prove its efficiency.

This chapter is organized as follows. In section 7.2, the formulation of a one-class SVM algorithm is briefly described. The suggested MTL method is then described in section 7.3. Section 7.4 presents the experimental results. We conclude the chapter with a few comments and suggestions for future work in section 7.5.

7.2. Single-task learning of one-class SVM classifier

Different versions of the methods of an SVM kind were specially adapted to classification in the case of a single class [SCH 01, MAN 01]. 1-SVM approach introduced by Schölkopf *et al.* (2001) relies on the transformation of original positive data from the input space to a space of greater dimension due to a kernel function. In the transformed space, the positive data are separated from the origin with a maximum margin. Besides the original parameters of SVM, 1-SVM requires setting *a priori* the value of a parameter indicating the percentage of positive data allowed to lie outside the description of the positive class. This makes 1-SVM more tolerant to noise in positive data (i.e. outliers). However, the choice of an appropriate value for the aforementioned parameter is not always intuitive and has a major influence on the performance of 1-SVM.

Briefly, the solution of 1-SVM corresponds to the estimation of the border of a region that includes most of the learning samples. If a new observation is a member of this region, it is classified as normal; otherwise, it is recognized as an outlier.

Let $\mathcal{A}_m = \{\boldsymbol{x}_i\}, i = 1, \ldots, m$, be the learning set with $m \in \mathbb{N}$ the total number of observations. $\boldsymbol{x}_i$ is an observation in space $\mathcal{X} \subseteq \mathbb{R}^d$ of dimension d. Let ϕ be a nonlinear transformation of attributes so that the result of the scalar product of ϕ can be obtained indirectly by a simple kernel K (semi-definite positive kernel on $\mathbb{R}^d$ fulfilling Mercer's conditions [BOS 92, VAP 95]):

$$K(\boldsymbol{x}_i, \boldsymbol{x}_j) = \langle \phi(\boldsymbol{x}_i), \phi(\boldsymbol{x}_j) \rangle \qquad [7.1]$$

for instance the Gaussian kernel:

$$K(\boldsymbol{x}_i, \boldsymbol{x}_j) = e^{-\frac{\|\boldsymbol{x}_i - \boldsymbol{x}_j\|^2}{2\sigma^2}} \qquad [7.2]$$

where σ is the kernel bandwidth.

The data must be projected onto a space induced by the kernel used and a proportion $1 - \nu$ of it must be separated from the origin by the hyperplane furthest away from the origin – we refer to it as the hyperplane of maximum margin. For a new observation $\boldsymbol{x}$, the value of the decision function is determined by evaluating on which side of the hyperplane the observation lies in the transformed space. To determine this hyperplane, we must find its normal vector $\mathbf{w}$ and a threshold ρ, for solutions to the following problem:

$$\begin{cases} \min_{\mathbf{w},\boldsymbol{\xi},\rho} & \frac{1}{2}\|\mathbf{w}\|^2 + \frac{1}{\nu m}\sum_{i=1}^{m}\xi_i - \rho \\ \text{under the constraints} & \langle \mathbf{w}, \phi(\boldsymbol{x}_i)\rangle \geq \rho - \xi_i, \quad \xi_i \geq 0, \quad i = 1, \ldots, m \end{cases} \quad [7.3]$$

where $\|\cdot\|$ is the norm associated with scalar product $\langle\cdot,\cdot\rangle$, $\mathbf{w}$ is the vector normal to the separating hyperplane, ρ is the offset parameter, and ξ_i is the slack variables. As a result, the distance between the hyperplane and the origin is $\rho/\|\mathbf{w}\|$.

It is emphasized in [SCH 01] that parameter $\nu \in [0, 1]$ is an upper bound of the rate of anomalies (the learning observations that lie outside the estimated positive region). Slack variables ξ_i, associated with each observation, are introduced. They enable us, in the case where the data are not linearly separable in the transformed space, to use some learning examples that can be misclassified when the maximum margin of the origin is obtained, thus giving some robustness to the method. The optimization algorithm consists of finding a compromise between the maximization of margin $\rho/\|\mathbf{w}\|$ and the minimization of the average of the slack variables. The distance between an anomaly and the hyperplane is $\xi_i/\|\mathbf{w}\|$. The learning observations that lie on the wrong side of the separating hyperplane, those with $\xi_i > 0$, are classified as anomalies. When we use the Gaussian kernel [7.2], the observations are projected onto the surface of the hypersphere in the space transformed by ϕ. Figure 7.1 shows a view in two dimensions where the hypersphere is a circle around the origin. In this case, 1-SVM is equivalent to the *support vector data description* method [TAX 99], which consists of looking for the minimum sphere of radius R, including most or all the positive data.

Let $\alpha_i, \beta_i \geq 0$ be the Lagrange multipliers associated with problem [7.3]. The Lagrangian is written as:

$$L(\mathbf{w}, \boldsymbol{\xi}, \rho, \boldsymbol{\alpha}, \boldsymbol{\beta}) = \frac{1}{2}\|\mathbf{w}\|^2 + \frac{1}{\nu m}\sum_{i=1}^{m}\xi_i - \rho$$

$$- \sum_{i=1}^{m}\alpha_i(\langle \mathbf{w}, \phi(\boldsymbol{x}_i)\rangle - \rho + \xi_i) - \sum_{i=1}^{m}\beta_i\xi_i \quad [7.4]$$

By annulling the derivatives of Lagrangian with respect to primal variables $\mathbf{w}, \boldsymbol{\xi}_i, \rho$, we obtain:

$$\mathbf{w} = \sum_{i=1}^{m}\alpha_i\phi(\boldsymbol{x}_i) \quad [7.5]$$

$$\alpha_i = \frac{1}{\nu m} - \beta_i, \quad \sum_{i=1}^{m}\alpha_i = 1 \quad [7.6]$$

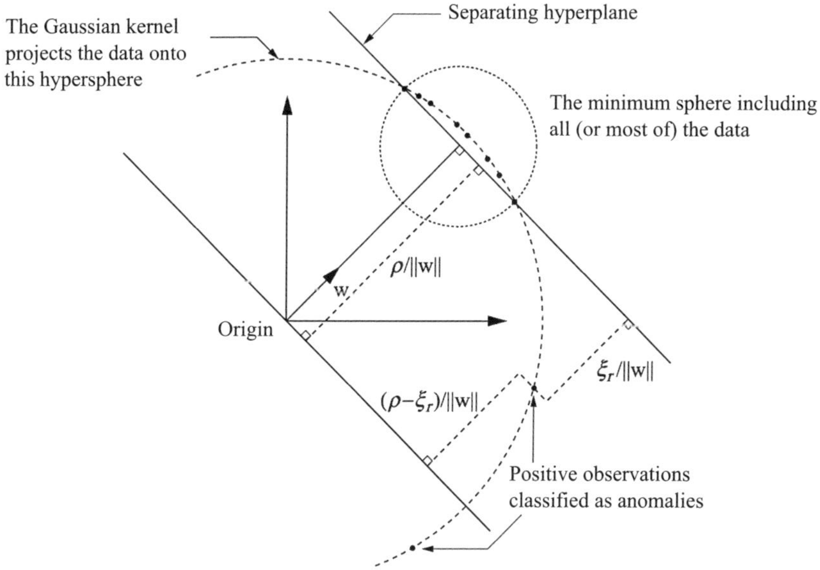

Figure 7.1. *View in two dimensions of* 1*-SVM classifier and its parameters*

Given these expressions, the values of α are solutions to the dual problem given by:

$$\begin{cases} \min_{\boldsymbol{\alpha}} & \frac{1}{2}\sum_{i,j=1}^{m} \alpha_i\alpha_j \left\langle \phi(\boldsymbol{x}_i), \phi(\boldsymbol{x}_j)\right\rangle \\ \text{under the constraints:} & 0 \leq \alpha_i \leq \frac{1}{\nu m}, \quad \sum_{i=1}^{m} \alpha_i = 1 \end{cases} \qquad [7.7]$$

It is not necessary to explicitly calculate the nonlinear transformation for each observation due to the kernel trick. The latter consists of defining directly a scalar product using a function fulfilling Mercer's conditions [VAP 95]:

$$\left\langle \phi(\boldsymbol{x}_i), \phi(\boldsymbol{x}_j)\right\rangle = k(\boldsymbol{x}_i, \boldsymbol{x}_j) \qquad [7.8]$$

By introducing this kernel during the solving of the dual problem, we show that the final decision is expressed only as a function of the scalar product and that it is useless to be explicit about the transformation $\phi(\boldsymbol{x})$:

$$f(\boldsymbol{x}) = \text{sign}\left(\sum_{i=1}^{m} \alpha_i k(\boldsymbol{x}_i, \boldsymbol{x}) - \rho\right) \qquad [7.9]$$

We can cite as an example the Gaussian kernel $k_\sigma(\boldsymbol{x}_i, \boldsymbol{x}_j) = e^{-\frac{\|\boldsymbol{x}_i - \boldsymbol{x}_j\|^2}{2\sigma^2}}$, which is widely used.

7.3. Multitask learning of 1-SVM classifiers

In this section, we introduce the 1-SVM method in the case of MTL. In this context, we consider T tasks defined on the same space $\mathcal{X}$, with $\mathcal{X} \subseteq \mathbb{R}^d$. For each task, we have at our disposal m samples $\{\boldsymbol{x}_{1t}, \boldsymbol{x}_{2t}, \ldots, \boldsymbol{x}_{mt}\}$. The goal is to learn a decision function (a hyperplane) $f_t(\boldsymbol{x}) = \text{sign}(\langle \mathbf{w}_t, \phi(\boldsymbol{x}) \rangle - \rho_t)$ for each task t. Inspired by the method suggested by [EVG 04], we hypothesize that when the tasks are linked to one another, vector $\mathbf{w}_t$ normal to the hyperplane can be represented by the sum of a mean vector $\mathbf{w}_0$ and of a vector $\mathbf{v}_t$ specific to each task:

$$\mathbf{w}_t = \mathbf{w}_0 + \mathbf{v}_t \qquad [7.10]$$

7.3.1. *Formulation of the problem*

By taking into account the hypothesis introduced above, we can generalize the 1-SVM method to the MTL problem. The primal optimization problem can be described as follows:

$$\min_{\mathbf{w}_0, \mathbf{v}_t, \xi_{it}, \rho_t} \frac{1}{2} \sum_{t=1}^{T} \|\mathbf{v}_t\|^2 + \frac{\mu}{2} \|\mathbf{w}_0\|^2 + \sum_{t=1}^{T} \left(\frac{1}{\nu_t m} \sum_{i=1}^{m} \xi_{it} \right) - \sum_{t=1}^{T} \rho_t \qquad [7.11]$$

for all $i \in \{1, 2, \ldots, m\}$ and $t \in \{1, 2, \ldots, T\}$, under the constraints:

$$\begin{aligned} \langle (\mathbf{w}_0 + \mathbf{v}_t), \phi(\boldsymbol{x}_{it}) \rangle &\geq \rho_t - \xi_{it} \\ \xi_{it} &\geq 0 \end{aligned} \qquad [7.12]$$

where ξ_{it} are variables enabling us to minimize the constraints associated with each sample and $\nu_t \in [0, 1]$ is a specific parameter of 1-SVM of each task. To control the level of similarity between the tasks, we introduce a positive regularization parameter μ in the primal optimization problem. With this formulation, a large value of μ tends to force the system to learn tasks independently from one another while a small value of μ will compel the solutions for each task to converge toward a common solution. Like in the case of the conventional 1-SVM, the Lagrangian is defined as follows:

$$\begin{aligned} L(\mathbf{w}_0, \mathbf{v}_t, \xi_{it}, \rho_t, \alpha_{it}, \beta_{it}) = & \frac{1}{2} \sum_{t=1}^{T} \|\mathbf{v}_t\|^2 + \frac{\mu}{2} \|\mathbf{w}_0\|^2 + \sum_{t=1}^{T} \left(\frac{1}{\nu_t m} \sum_{i=1}^{m} \xi_{it} \right) \\ & - \sum_{t=1}^{T} \rho_t - \sum_{t=1}^{T} \sum_{i=1}^{m} \alpha_{it} [\langle (\mathbf{w}_0 + \mathbf{v}_t), \phi(\boldsymbol{x}_{it}) \rangle - \rho_t + \xi_{it}] \\ & - \sum_{t=1}^{T} \sum_{i=1}^{m} \beta_{it} \xi_{it} \end{aligned} \qquad [7.13]$$

where $\alpha_{it}, \beta_{it} \geq 0$ are the Lagrange multipliers. At the optimum, the partial derivatives of the Lagrangian with respect to the variables are equal to zero and we obtain the following equations:

$$\mathbf{w}_0 = \frac{1}{\mu}\sum_{t=1}^{T}\sum_{i=1}^{m}\alpha_{it}\phi(\boldsymbol{x}_{it}) \quad [7.14a]$$

$$\mathbf{v}_t = \sum_{i=1}^{m}\alpha_{it}\phi(\boldsymbol{x}_{it}) \quad [7.14b]$$

$$\alpha_{it} = \frac{1}{\nu_t m} - \beta_{it} \quad [7.14c]$$

$$\sum_{i=1}^{m}\alpha_{it} = 1 \quad [7.14d]$$

By combining equations [7.10], [7.14a], and [7.14b], we obtain the following:

$$\mathbf{w}_0 = \frac{1}{\mu}\sum_{t=1}^{T}\mathbf{v}_t \quad [7.15]$$

$$\mathbf{w}_0 = \frac{1}{\mu + T}\sum_{t=1}^{T}\mathbf{w}_t \quad [7.16]$$

With these equations, the primal optimization problem can be rewritten as a function of only $\mathbf{w}_t$ according to:

$$\min_{\mathbf{w}_t,\xi_{it},\rho_t} \frac{\lambda_1}{2}\sum_{t=1}^{T}\|\mathbf{w}_t\|^2 + \frac{\lambda_2}{2}\sum_{t=1}^{T}\left\|\mathbf{w}_t - \frac{1}{T}\sum_{r=1}^{T}\mathbf{w}_r\right\|^2 + \sum_{t=1}^{T}\left(\frac{1}{\nu_t m}\sum_{i=1}^{m}\xi_{it}\right) - \sum_{t=1}^{T}\rho_t \quad [7.17]$$

with:

$$\lambda_1 = \frac{\mu}{\mu + T} \quad \text{and} \quad \lambda_2 = \frac{T}{\mu + T} \quad [7.18]$$

This new expression of the original primal problem [7.11] enables us to emphasize that, in the context of MTL, the solution is obtained by making compromises between obtaining a 1-SVM model specific to each task and obtaining a unique model for all the tasks.

7.3.2. *Dual problem*

The primal problem [7.11] can be solved more easily by using its dual Lagrangian that is expressed by:

$$\max_{\alpha_{it}} -\frac{1}{2}\sum_{t=1}^{T}\sum_{r=1}^{T}\sum_{i=1}^{m}\sum_{j=1}^{m}\alpha_{it}\alpha_{jr}\left(\frac{1}{\mu}+\delta_{rt}\right)\langle\phi(\boldsymbol{x}_{it}),\phi(\boldsymbol{x}_{jr})\rangle \qquad [7.19]$$

with the constraints:

$$0 \leq \alpha_{it} \leq \frac{1}{\nu_t m}, \quad \sum_{i=1}^{m}\alpha_{it} = 1 \qquad [7.20]$$

where δ_{rt} is Kronecker's symbol:

$$\delta_{rt} = \begin{cases} 1, & \text{if} \quad r = t \\ 0, & \text{if} \quad r \neq t \end{cases} \qquad [7.21]$$

The main difference between this dual problem [7.19] and that of the conventional 1-SVM [7.7] is the introduction of expression $\left(\frac{1}{\mu}+\delta_{rt}\right)$ within the framework of MTL. Let us assume that we define a kernel function such as the one defined by equation [7.8]:

$$k(\boldsymbol{x}_{it},\boldsymbol{x}_{jr}) = \langle\phi(\boldsymbol{x}_{it}),\phi(\boldsymbol{x}_{jr})\rangle \qquad [7.22]$$

where r and t are indices of tasks associated with each sample. Given the properties of the kernels used, the product of two kernels $\delta_{rt}k(\boldsymbol{x}_{it},\boldsymbol{x}_{jr})$ is a valid kernel. Therefore, the following function:

$$\begin{aligned} G_{rt}(\boldsymbol{x}_{it},\boldsymbol{x}_{jr}) &= \left(\frac{1}{\mu}+\delta_{rt}\right)k(\boldsymbol{x}_{it},\boldsymbol{x}_{jr}) \\ &= \frac{1}{\mu}k(\boldsymbol{x}_{it},\boldsymbol{x}_{jr}) + \delta_{rt}k(\boldsymbol{x}_{it},\boldsymbol{x}_{jr}) \end{aligned}$$

is a linear combination of two valid kernels using positive coefficients $\frac{1}{\mu}$ and 1. We can thus solve the optimization problem associated with MTL [7.11] like the one of a conventional 1-SVM with a new kernel function $G_{rt}(\boldsymbol{x}_{it},\boldsymbol{x}_{jr})$. The decision function for each task is given by:

$$f_t(\boldsymbol{x}) = \text{sign}\left(\sum_{r=1}^{T}\sum_{i=1}^{m}\alpha_{ir}G_{rt}(\boldsymbol{x}_{ir},\boldsymbol{x}) - \rho_t\right) \qquad [7.23]$$

7.4. Experimental results

In this section, we present results obtained on two examples by means of this approach: an academic example of small dimension and the analysis of texture images. To evaluate the performances of the method suggested (MTL-OSVM), we compare it with two other methods. The first method consists of carrying out the learning of a classifier 1-SVM, independently for each task (we write it T-OSVM), and the second method consists of learning a single classifier 1-SVM for all the tasks by considering them as a single task (we write it as 1-OSVM). To obtain reliable results in terms of estimation, all the tests were repeated 20 times by using different sets drawn at random.

In the different tests, the Gaussian kernel [7.2] was used for the methods T-OSVM and 1-OSVM. For the suggested method MTL-OSVM, the kernel is built from the kernel specified in equation [7.23].

The MTL-OSVM method requires us to tune different parameters: parameters ν_t and σ_t $(t = 1, \ldots, T)$ of the 1-SVM classifier for each task and regularization parameter μ. There are, in principle, $2 \times T + 1$ parameters to tune, but in order to avoid a combinatorial explosion (ν_t, σ_t, μ) being examined, we hypothesized that all the tasks share the same pair (ν, σ). We then suggest a procedure for the determination of parameters in two stages:

– Determination of the value of the two parameters ν and σ of 1-SVM classifier by cross-validation by considering the set of tasks as a single task. The cross-validation procedure consists of carrying out the learning of the classifier for different combinations of these two parameters defined *a priori* and choosing the one that gives the smallest classification error rate on a validation set [WU 09]. This classification error rate is the sum of the rates of non-detection of "abnormal" observations and false alarm ones ("healthy" observations considered as "abnormal").

– Determination of the regularization parameter μ also by cross-validation, the values of the two parameters ν and σ being fixed.

7.4.1. *Academic nonlinear example*

We have initially tested the suggested method in an example with four tasks. The data sets are obtained as follows. For the first task, each observation $\mathbf{x}_i = [x_{i,1}\ x_{i,2}\ x_{i,3}\ x_{i,4}]^{\mathrm{T}}, i = 1, \ldots, m$, is generated from the following model:

$$x_{i,j} \sim \mathcal{U}(0,1), \quad j = 1, \ldots, 3$$
$$x_{i,4} = x_{i,1} + 2x_{i,2} + x_{i,3}^2 \qquad [7.24]$$

The data samples corresponding to the three other tasks are generated by adding Gaussian white noise of different magnitude to the data set from the first task. The noises added are, respectively, weak for the second task with an amplitude of 1% of the range of the first data set, moderate for the third task with an amplitude of 8%, and high for the fourth task with an amplitude of 15%. To evaluate the rate of non-detection, we have also generated a set of negative samples that are composed of four uniformly distributed variables [7.24]. The training set of each task contains only "healthy" data ($m = 200$) whereas the test set contains both "healthy" and "abnormal" data (200 observations of each kind).

The optimal parameters obtained for 1-SVM are $(\nu, \sigma) = (0.01, 0.5)$ for this example. Figures 7.2–7.5 represent, for each of the four tasks, the variations of the average false positive, false negative, and total error rates for the three methods (MTL-OSVM, T-OSVM, and 1-OSVM) as a function of the regularization parameter μ. We observe that for small values of μ, the performances of the MTL-OSVM method coincide with those of an 1-OSVM method, which is coherent since we consider then all the tasks as a single task. When the value of μ is very high, the performances of the MTL-OSVM method are comparable to those of the method learning independent tasks T-OSVM. When the value of μ increases, the behavior of the MTL-OSVM method for the first three classes is similar. The rate of false alarm for this method decreases with the increase of μ while the rate of non-detection increases. The MTL-OSVM method clearly gives better results than the other two methods for a large range of values of the regularization parameter within interval [0.05, 1]. However, for the fourth task (Figure 7.3), the results obtained are very different from the previous three. In fact, the rates of non-detection and of false alarm of the MTL-OSVM method are no longer included between those of the two other methods. This result is certainly due to the high level of noise added to the initial data set. Furthermore, contrary to the three other tasks, while remaining better, the performances of the MTL-OSVM method are no longer as significantly different from those of the two other methods.

7.4.2. *Analysis of textured images*

We have also tested the suggested approach on images that contain Markovian textures [SMO 97]. Analyzing an image in terms of texture consists of classifying the pixels of the image by labeling them. Representation space $\mathcal{X}$ is of dimension $d = 25$. The variables of the attribute vector characterizing a pixel to classify are attached to the sites belonging to the 5×5 square neighborhood of the pixel to classify as well as the one attached to the central pixel (pixel to classify). Their values are defined by the gray level of each pixel. Like for the previous example of section 7.4.1, four tasks have been generated from an initial textured image. The data set associated with task 1 is made up of samples of dimension 25 that are randomly selected from the textured source image. The samples for the three other tasks are also selected from the same source image as task 1, but they are contaminated by noises of, respectively, a weak (task 2), middle (task 3), and strong magnitude (task 4). "Abnormal" observations generated from a different textured image are also added to the test set. Figure 7.6 represents the different textured images used to generate the data sets. For each test, the learning set of each task is made up of 200 "healthy" observations and the test set contains, besides 200 "healthy" observations, 200 "abnormal" observations. The common values for the parameters of classifier 1-SVM are $(\nu, \sigma) = (0.01, 300)$.

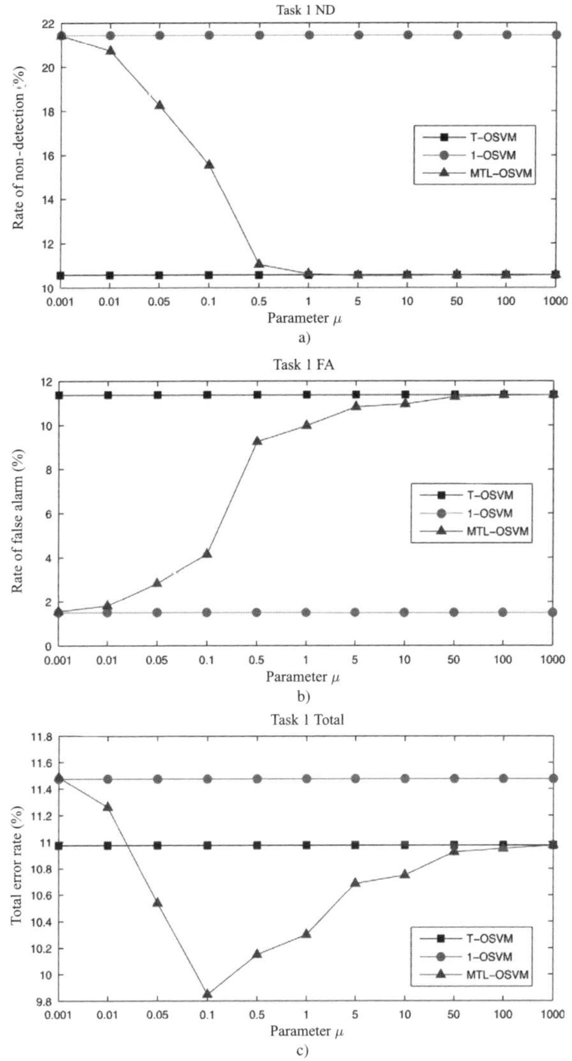

Figure 7.2. *Variation of the error rates (a) non-detection, (b) false alarm, and (c) total classification error for task 1 according to the value of regularization parameter μ (nonlinear example)*

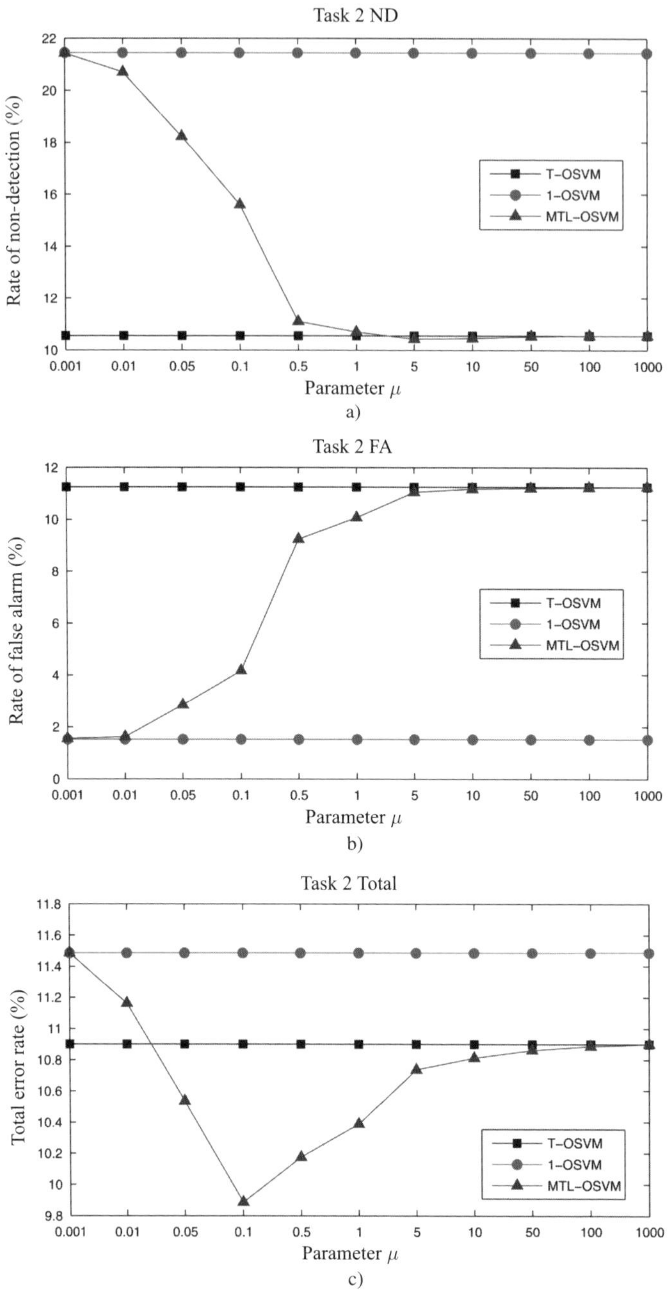

Figure 7.3. *Variation of the error rates (a) non-detection, (b) false alarm, and (c) total classification error for task 2 according to the value of regularization parameter μ (nonlinear example)*

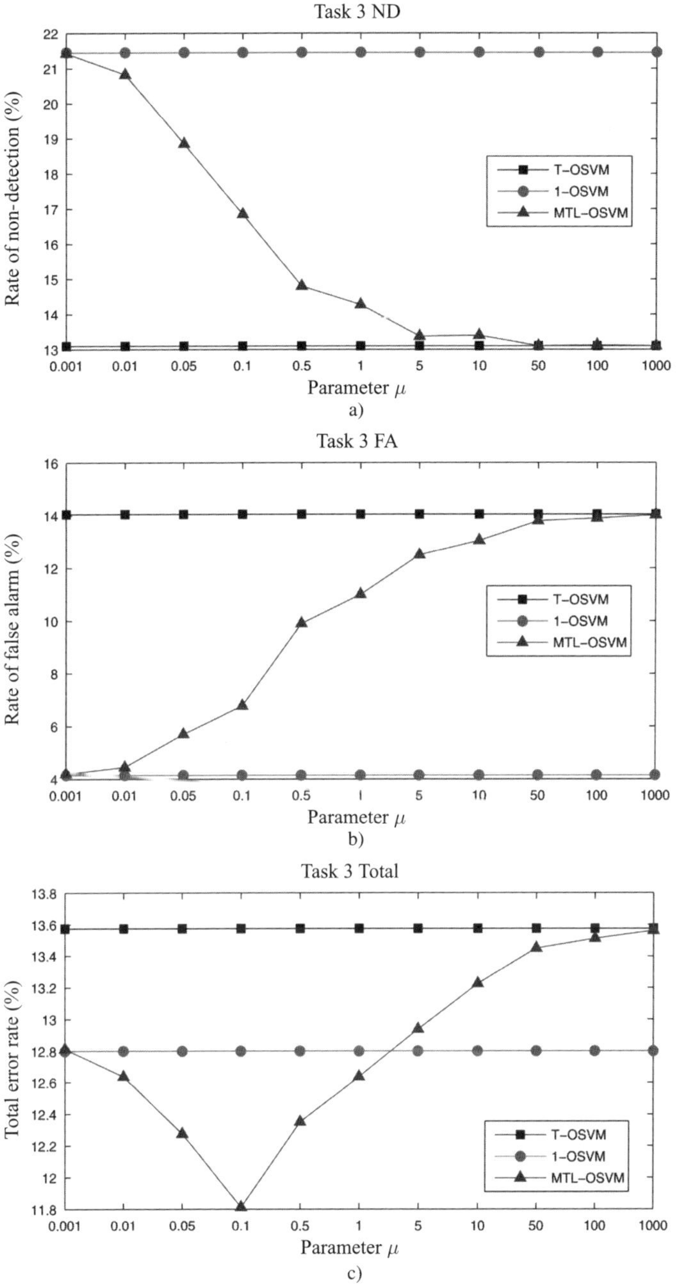

Figure 7.4. *Variation of the error rates (a) non-detection, (b) false alarm, and (c) total classification error for task 3 according to the value of regularization parameter μ (nonlinear example)*

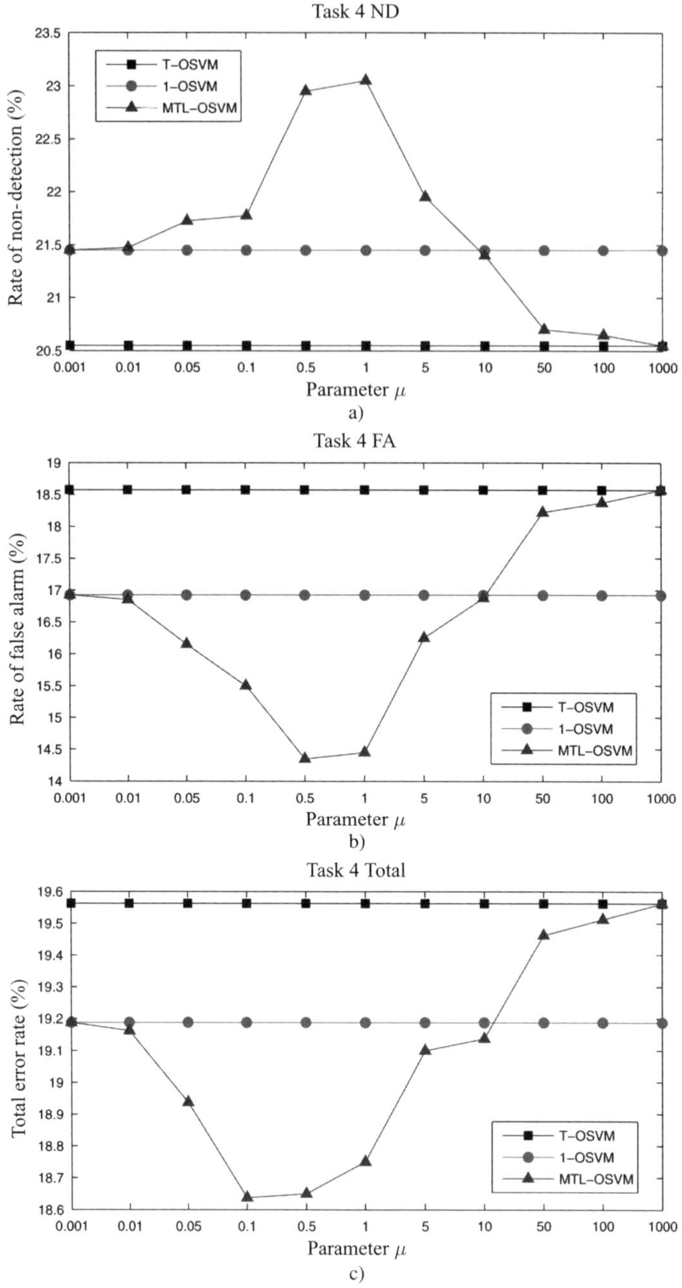

Figure 7.5. *Variation of the error rates (a) non-detection, (b) false alarm, and (c) total classification error for task 4 according to the value of regularization parameter μ (nonlinear example)*

Figure 7.6. *Textured images used to generate the different data sets. (a) Textured source image for task 1. (b) Textured image for task 2 (weak noise). (c) Textured image for task 3 (middle noise). (d) Textured image for task 4 (high noise). (e) Textured source image to generate the "abnormal" observations*

Table 7.1 summarizes the results obtained in this example for the three methods. We observe that the individual learning method T-OSVM generates the smallest rates of non-detection of "abnormal" observations with the highest false alarm rates as a counterpart. Conversely, the overall learning method of all the tasks 1-OSVM gives the smallest rates of false alarm but the highest rates of non-detection. The suggested MTL method MTL-OSVM gives better overall performances by compromising between non-detection and false alarm rates.

	Task 1		
	ND	FA	Total
T-OSVM	**3.62** $\pm$ 1.18	27.0 $\pm$ 3.0	15.3 $\pm$ 1.4
1-OSVM	27.1 $\pm$ 2.6	**2.70** $\pm$ 1.59	14.9 $\pm$ 1.5
MTL-OSVM	14.4 $\pm$ 2.2	9.52 $\pm$ 3.19	**12.0** $\pm$ 1.8
	Task 2		
	ND	FA	Total
T-OSVM	**4.27** $\pm$ 1.11	28.2 $\pm$ 3.6	16.3 $\pm$ 1.6
1-OSVM	27.1 $\pm$ 2.6	**3.27** $\pm$ 2.21	15.2 $\pm$ 1.8
MTL-OSVM	14.6 $\pm$ 2.2	10.3 $\pm$ 3.7	**12.4** $\pm$ 2.0
	Task 3		
	ND	FA	Total
T-OSVM	**7.05** $\pm$ 1.41	28.6 $\pm$ 3.9	17.8 $\pm$ 2.0
1-OSVM	27.1 $\pm$ 2.6	**6.62** $\pm$ 3.73	16.8 $\pm$ 2.2
MTL-OSVM	15.7 $\pm$ 2.3	14.4 $\pm$ 4.1	**15.0** $\pm$ 2.2
	Task 4		
	ND	FA	Total
T-OSVM	**27.4** $\pm$ 3.0	34.4 $\pm$ 8.7	30.9 $\pm$ 4.5
1-OSVM	**27.1** $\pm$ 2.6	34.0 $\pm$ 8.7	30.5 $\pm$ 4.5
MTL-OSVM	29.4 $\pm$ 2.6	**29.0** $\pm$ 9.2	**29.2** $\pm$ 4.5

Table 7.1. *Classification error rate (%) of three methods for each task made up of textured images. ND: non-detection rate, FA: false alarm rate, Total: total classification error rate*

Figures 7.7 – 7.10 represent, for each of the four tasks, the variations in the average false positive, false negative, and total error rates for the three methods (MTL-OSVM, T-OSVM, and 1-OSVM) according to the regularization parameter μ. Overall, we find the same behavior for the three methods as for the previous example. The suggested method MTL-OSVM shows better performances than the two other methods for all the tasks. Hence, we will note the advantage of the MTL-OSVM method that balances between individual learning of tasks and learning of all the tasks gathered as a single task, this compromise being determined from data for each example.

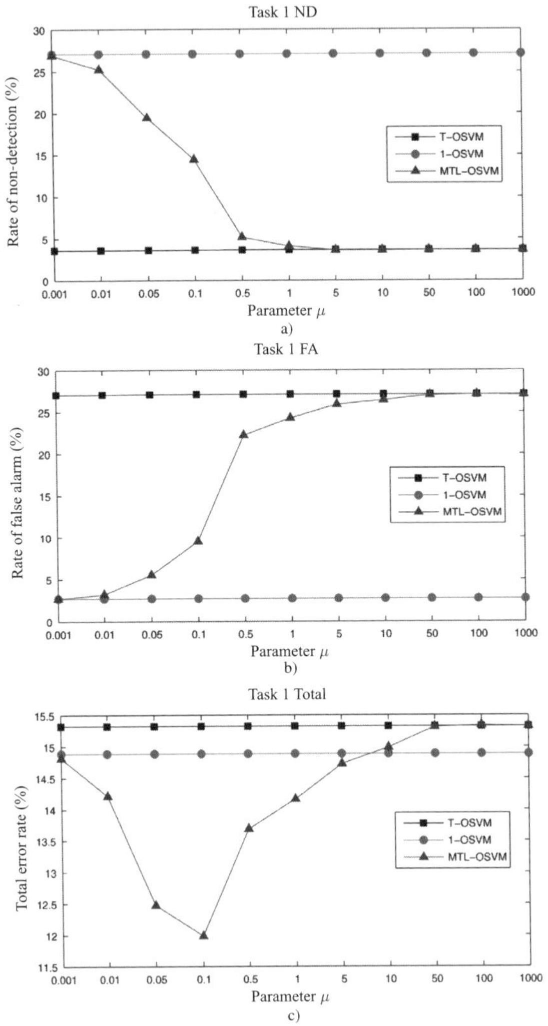

Figure 7.7. *Variation of classification error rates (a) non-detection, (b) false alarm, and (c) total classification error rate for task 1 according to the value of regularization parameter μ (textured images)*

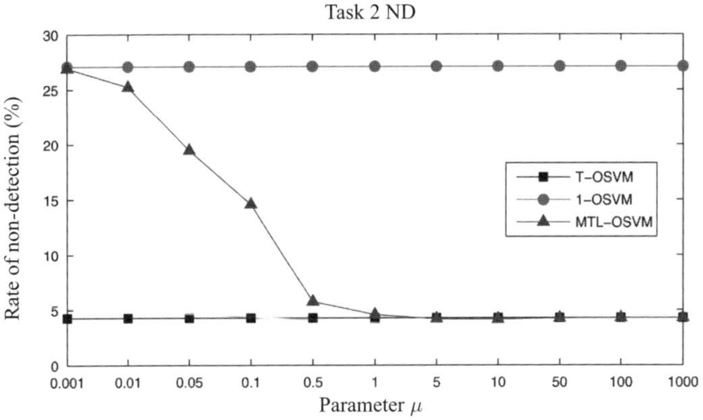

a)

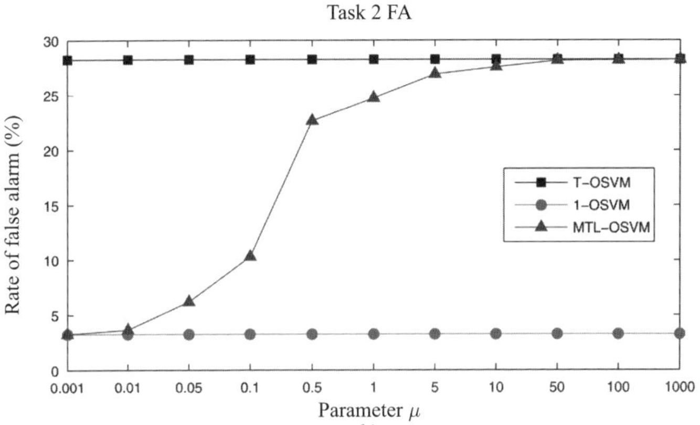

b)

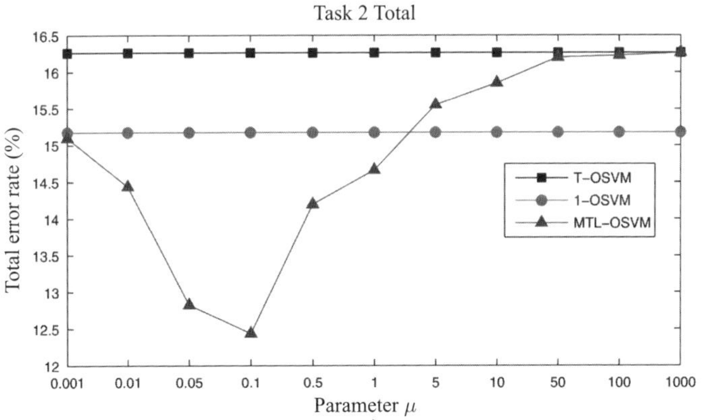

c)

Figure 7.8. *Variation of classification error rates (a) non-detection, (b) false alarm, and (c) total classification error rate for task 2 according to the value of regularization parameter μ (textured images)*

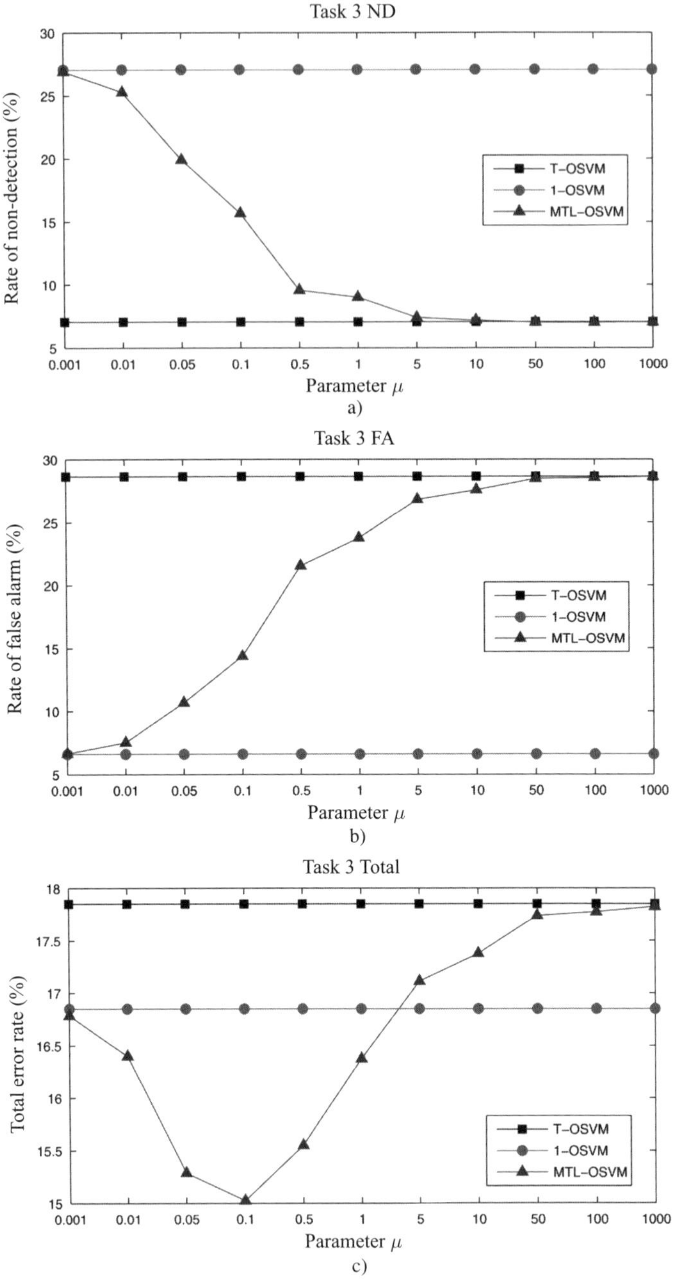

Figure 7.9. *Variation of classification error rates (a) non-detection, (b) false alarm, and (c) total classification error rate for task 3 according to the value of regularization parameter μ (textured images)*

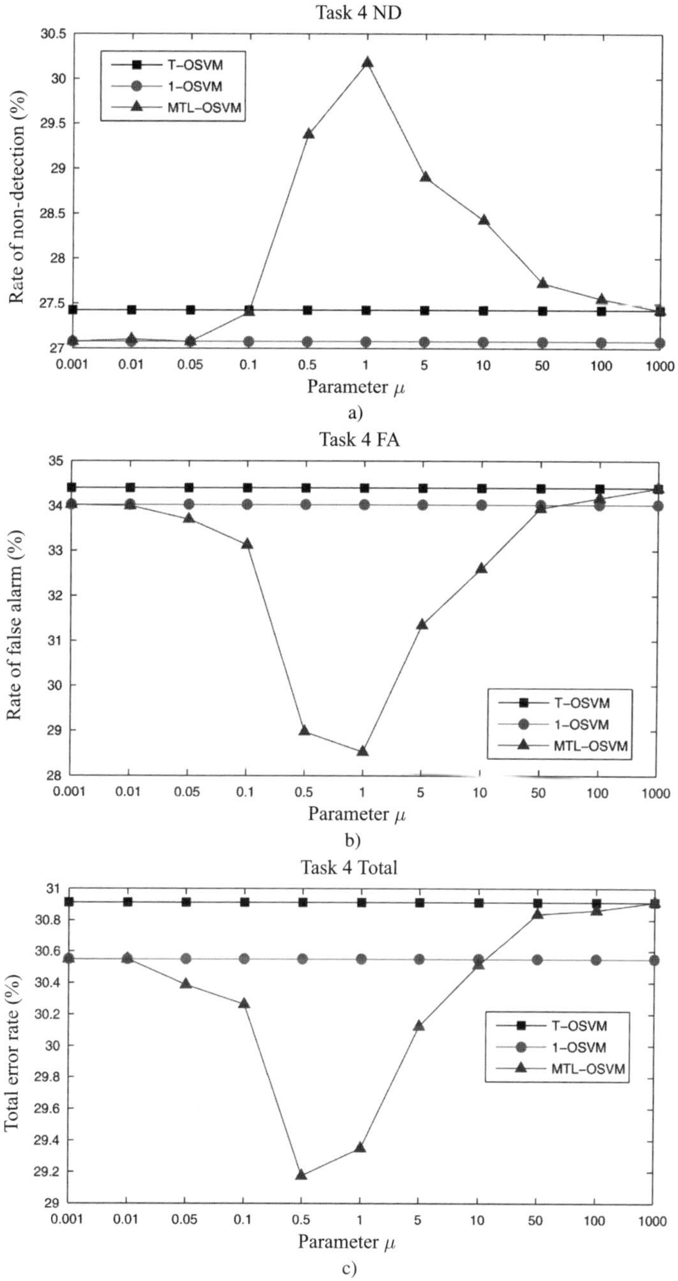

Figure 7.10. *Variation of classification error rates (a) non-detection, (b) false alarm, and (c) total classification error rate for task 4 according to the value of regularization parameter* μ *(textured images)*

The determination of regularization parameter μ is thus essential. Like for the previous example, by analyzing curves of the classification error rates according to this parameter, we notice two important points. First, the tasks quite similar (tasks 1–3) have optimal values very close to each other for μ, which is not the case for task 4, for which the value of this parameter is higher. This behavior, is quite consistent (the higher the value of μ, the more the learning tends toward individual learning), reveals the tasks that differ from the others. Then, we also observe that there exists not a single optimal value, but an interval of values for which the performances are comparable. The method shows robustness according to the setting of this parameter.

7.5. Conclusion

In this chapter, we introduced the one-class SVM within the framework of MTL by hypothesizing that the solutions obtained for related tasks are close to a mean value. A regularization parameter was used in the optimization process in order to control the balance between maximization of the margin for each 1-SVM model and proximity of each 1-SVM to the mean model. The design of new kernels within the multitask framework, based on the kernel properties, significantly facilitates the implementation of the suggested method. The experimental validation was carried out using a single-class data set generated artificially. The results show that the simultaneous learning of several tasks enables us to improve the performances with respect to an independent learning of each class.

We used the same hyperparameters for all the tasks. A conceivable improvement consists of using different hyperparameter values for the different tasks. From this point of view, the composition properties of the kernels should lead to a large array of additional possibilities for the definition of new kernels within the framework of MTL. A more fundamental aspect concerns the use of classification models previously established to characterize a new machine. Generally, this MTL problem can find several applications in the healthcare sector and in the environmental domain.

7.6. Acknowledgments

This project was carried out within the framework of PARDI project with the following partners: EDF, ANDRA, CRAN, and ICD.

7.7. Bibliography

[ABU 07] ABU-EL-ZEET Z.H., PATEL V., Method of condition monitoring, US Patent 7275018, 2007.

[BEN 03] BEN-DAVID S., SCHULLER R., "Exploiting task relatedness for multiple task learning", *Proceedings of the 16th Annual Conference on Computational Learning Theory and the 7th Kernel Workshop*, Washington, DC, pp. 567–580, 2003.

[BEN 08] BEN-DAVID S., BORBELY R.S., "A notion of task relatedness yielding provable multiple-task learning guarantees", *Machine Learning*, vol. 73, pp. 273–287, 2008.

[BI 08] BI J., XIONG T., YU S., DUNDAR M., RAO R.B., "An improved multi-task learning approach with applications in medical diagnosis", *Proceedings of the 2008 European Conference on Machine Learning and Knowledge Discovery in Databases – Part I*, Antwerp, Belgium, pp. 117–132, 2008.

[BIR 10] BIRLUTIU A., GROOT P., HESKES T., "Multi-task preference learning with an application to hearing aid personalization", *Neurocomputing*, vol. 73, pp. 1177–1185, 2010.

[BOS 92] BOSER B.E., GUYON I.M., VAPNIK V.N., "A training algorithm for optimal margin classifiers", *Proceedings of the 5th Annual Workshop on Computational Learning Theory*, Pittsburgh, PA, pp. 144–152, 1992.

[CAM 08] CAMCI F., CHINNAM R., "General support vector representation machine (GSVRM) for stationary and non-stationary classes", *Pattern Recognition*, vol. 41, pp. 3021–3034, 2008.

[CAM 09] CAMPOS J., "Development in the application of ICT in condition monitoring and maintenance", *Computers in Industry*, vol. 60, no. 1, pp. 1–20, 2009.

[CAR 97] CARUANA R., "Multitask learning", *Machine Learning*, vol. 28, pp. 41–75, 1997.

[ESK 00] ESKIN E., "Anomaly detection over noisy data using learned probability distributions", *Proceedings of the 17th International Conference on Machine Learning*, San Francisco, CA, pp. 255–262, 2000.

[EVG 04] EVGENIOU T., PONTIL M., "Regularized multi-task learning", *Proceedings of the 10th ACM SIGKDD International Conference on Knowledge Discovery and Data Mining*, Seattle, WA, pp. 109–117, 2004.

[EVG 05] EVGENIOU T., MICCHELLI C.A., PONTIL M., "Learning multiple tasks with kernel methods", *Journal of Machine Learning Research*, vol. 6, pp. 615–637, 2005.

[FAM 03] FAMILI A., LETOURNEAU S., O'BRIEN C., Method of identifying anormal behaviour in a fleet of vehicles, US Patent 2003/0149550, 2003.

[GU 09] GU Q., ZHOU J., "Learning the shared subspace for multi-task clustering and transductive transfer classification", *Proceedings of the 9th IEEE International Conference on Data Mining*, Miami, FL, pp. 159–168, 2009.

[HAS 04] HASIEWICZ J., HERZOG J., MARCELL R., Equipment health monitoring architecture for fleets of assets, US Patent 2004/0243636, 2004.

[HES 00] HESKES T., "Empirical bayes for learning to learn", *Proceedings of the 17th International Conference on Machine Learning*, San Francisco, CA, pp. 367–374, 2000.

[HIG 04] HIGGS P.A., PARKIN R., JACKSON M., "A survey of condition monitoring systems in industry", *Proceedings of the 7th Biennal ASME Conference Engineering Systems Design and Analysis*, Manchester, UK, pp. 163–178, 2004.

[JEB 04] JEBARA T., "Multi-task feature and kernel selection for SVMs", *Proceedings of the 21st International Conference on Machine Learning*, Banff, Alberta, Canada, 2004.

[KAS 09] KASSAB R., ALEXANDRE F., "Incremental data-driven learning of a novelty detection model for one-class classification with application to high-dimensional noisy data", *Machine Learning*, vol. 74, no. 2, pp. 191–234, 2009.

[KEM 08] KEMP C., GOODMAN N., TENENBAUM J., "Learning and using relational theories", *Advances in Neural Information Processing Systems 20*, pp. 753–760, Cambridge, MA, 2008.

[KUN 03] KUNZE U., "Condition telemonitoring and diagnosis of power plants using web technology", *Progress in Nuclear Energy*, vol. 43, nos. 1–4, pp. 129–136, 2003.

[MAN 01] MANEVITZ L., YOUSEF M., "One-class SVMs for document classification", *Journal of Machine Learning Research*, vol. 2, pp. 139–154, 2001.

[MUL 08] MULLER A., MARQUEZ A.C., IUNG B., "On the concept of e-maintenance: review and current research", *Reliability Engineering & System Safety*, vol. 93, no. 8, pp. 1165–1187, 2008.

[PAN 10] PAN S.J., YANG Q., "A survey on transfer learning", *IEEE Transactions on Knowledge and Data Engineering*, vol. 22, no. 10, pp. 1345–1359, 2010.

[RIT 97] RITTER G., GALLEGOS M.T., "Outliers in statistical pattern recognition and an application to automatic chromosome classification", *Pattern Recognition Letters*, vol. 18, pp. 525–539, 1997.

[SCH 01] SCHÖLKOPF B., PLATT J.C., SHAWE-TAYLOR J., SMOLA A.J., WILLIAMSON R.C., "Estimating the support of a high-dimensional distribution", *Neural Computation*, vol. 13, no. 7, pp. 1443–1471, 2001.

[SIN 04] SINGH S., MARKOU M., "An approach to novelty detection applied to the classification of image regions", *IEEE Transactions on Knowledge and Data Engineering*, vol. 16, no. 4, pp. 396–407, 2004.

[SMO 97] SMOLARZ A., "Etude qualitative du modèle Auto-Binomial appliqué à la synthèse de texture", *Actes des XXIXèmes Journées de Statistique*, Carcassonne, France, pp. 712–715, 1997.

[TAR 95] TARASSENKO L., HAYTON P., CERNEAZ N., BRADY M., "Novelty detection for the identification of masses in mammograms", *Proceedings of the 4th IEE International Conference on Artificial Neural Networks*, Cambridge, UK, pp. 442–447, 1995.

[TAX 99] TAX D.M., DUIN R.P., "Support vector domain description", *Pattern Recognition Letters*, vol. 20, pp. 1991–1999, 1999.

[VAP 95] VAPNIK V.N., *The Nature of Statistical Learning Theory*, Springer-Verlag, New York, 1995.

[WID 10] WIDMER C., TOUSSAINT N., ALTUN Y., RATSCH G., "Inferring latent task structure for multitask learning by multiple kernel learning", *BMC Bioinformatics*, vol. 11, no. 8, p. S5, 2010.

[WU 09] WU R.-S., CHUNG W.-H., "Ensemble one-class support vector machines for content-based image retrieval", *Expert Systems with Applications*, vol. 36, no. 3, pp. 4451–4459, 2009.

[YAN 10] YANG H., KING I., LYU M.R., "Multi-task learning for one-class classification", *Proceedings of the 2010 International Joint Conference on Neural Networks (IJCNN)*, Barcelona, Spain, pp. 1–8, 2010.

[YOU 05] YOU S., KRAGE M., JALICS L., "Overview of remote diagnosis and maintenance for automotive systems", *Proceedings of the SAE World Congress*, Detroit, MI, 2005.

[YU 10] YU L., CLEARY D., OSBORNE M.D., Method and system for diagnosing faults in a particular device within a fleet of devices, US Patent 7826943, 2010.

[ZHE 08] ZHENG V.W., PAN S.J., YANG Q., PAN J.J., "Transferring multi-device localization models using latent multi-task learning", *Proceedings of the 23rd National Conference on Artificial Intelligence*, pp. 1427–1432, 2008.

Chapter 8

The APPRODYN Project: Dynamic Reliability Approaches to Modeling Critical Systems

8.1. Context and aims

8.1.1. *Context*

The safety of high criticity industrial systems relies heavily on relatively complex program systems (with regards to code size, architecture, number of variables examined, etc.). Due to the complexity of the interactions between physical processes and their control, the standard methods of safety assessment are limited. These methods involve constructing a structural function in the form of a temporally invariable Boolean equation (a combinatorial approach) of the variables representing the components' states. These approaches, however, are shown to be limited in the cases of "non-coherent systems".

Some systems cannot be represented by a temporally invariable Boolean function. Their structure function has been proved to be correctly represented by a language of events. This involves using a formalism of state-transition type to represent the behavior of the system allowing us to show the sequence of events rather than their combinations in the form of cuts or bonds. This is the first dynamic

---

Chapter written by Jean-François AUBRY, Genia BABYKINA, Nicolae BRINZEI, Slimane MEDJAHER, Anne BARROS, Christophe BERENGUER, Antoine GRALL, Yves LANGERON, Danh NGOC NGUYEN, Gilles DELEUZE, Benoîte DE SAPORTA, François DUFOUR and Huilong ZHANG.

aspect of the system's structure function. Beyond this first issue, the concept of 'dynamic reliability' covers a range of properties highlighted in the literature over recent decades that can be summarized as follows:

– the system's states are defined by a finite state machine (FSM). These states correspond to combinations of component states. Their number can be lower than the number of possible combinations of component states as the system can be stopped by a certain level of damage or a combination of component states that can be accessed by incompatible sequences of events;

– each of the system's states is also characterized by a set of continuous variables and integro-differential equations describing its evolution over time;

– the ageing of components is a function that generally depends on time, but can also depend on some continuous variables (e.g. temperature, pressure, speed, etc.). In addition, non-temporally dependent ageing laws can be considered (e.g. the accumulation of solicitation events);

– changes in the system's state are caused by different kinds of events: the failure or repair of a component, the crossing of a threshold associated with a continuous variable (e.g. an alarm involving the initiation of a security loop) or the crossing of a threshold associated with a numeral variable (e.g. an accumulation of events);

– different laws of probability can be associated with the failure or repair of a system's component depending on the system's state (e.g. the ageing of a component can depend on its request mode, its maintenance strategy, or the level of hazard associated with the system's present state, etc.).

We can only account for all of the aforementioned aspects in the same approach to a system's dependability assessment using simulations. This is because of the highly analytical complexity of such models.

8.1.2. *Objectives*

This project aims to test some dynamic reliability approaches that are used to support the probabilistic study of critical instrumentation and control systems' safety assessment in process manufacturing and energy industries. It focuses on potentially strong approaches that are not commonly used by safety and design engineers or operator manufacturers. It aims to evaluate the potential of these approaches by using a representative case study and comparing them with wider diffusion approaches, such as Stochastic Petri nets (SPNs).

8.2. Brief overview of the test case

8.2.1. *General remarks*

The project focuses on the water regulation system of a steam generator (SG) in a Pressurized Water nuclear Reactor (PWR). The feature of interest in this study is "unavailability". The system is designed to keep the water level in the steam generator around a reference point. This test case is designed to be representative of a real system and cover more difficulties than the basic test cases used in the literature on dynamic reliability. It can also be fairly easily adapted to other energy sources or situations in process manufacturing. A similar test case has been proposed and described by the *Nuclear Regulatory Commission* [MAN 08, NUR 07] comparing dynamic reliability approaches, *dynamic flowgraph methodology* and the Markov/*Cell-to-cell mapping technique*. However, as the reports that have been published do not provide sufficient information to constitute a reproducible model; hence there is the need to develop a complete test.

8.2.2. *Functional description*

We have modeled part of the secondary system in a pressurized water reactor, i.e. three systems, from the condenser, up from the steam collecting cylinder (VVP) and tubes that concentrate the steam coming from the steam generator. The rest of the secondary circuit is relevant to the test case but is represented in the form of disturbances $d(t)$. The systems are modeled as follows:

– extraction pumps (CEX);

– water flow control (ARE);

– turbo pump control (TPA); and

– part of the secondary circuit collecting the steam (VVP).

The reliability diagram is shown in Figure 8.1.

The VVP is a purely mechanical, passive subsystem. Its cylinder and tubes collect vapor from the SG and feed the turbopumps, driers, etc. A rupture in the VVP creates a single point of failure. Failures in the VVP input tube are indicative of the highest attainable level of reliability in the system. In the model, VVP failures also include other passive systems (tanks, reheaters and water control valve tubes).

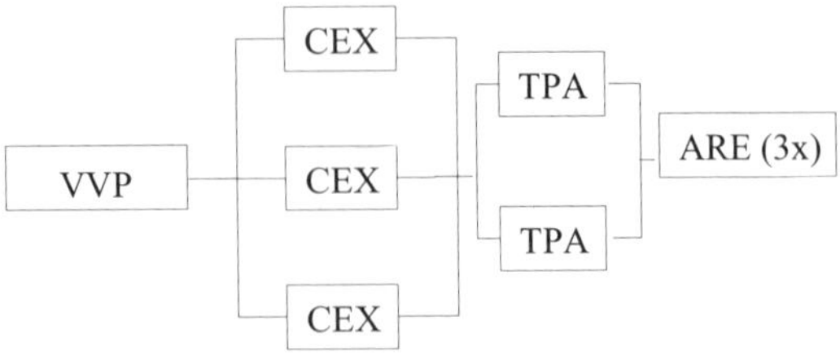

Figure 8.1. *Reliability diagram of the mechanical and electromechanical parts of the water feed system*

The three CEX pumps maintain the vacuum in the condenser and provide the influx of water. They are redundant: one of the pumps is on standby and only starts when one of the other pumps fails. The failing pump, once repaired, is left on standby.

The two TPA feed pumps ensure that the pressure in the steam generator is evenly distributed. Both function at the same time. As a result of a failure in one of the TPAs, the other goes into overdrive and takes up the load. In this model, the installation power automatically decreases when only a single TPA is functioning.

Lastly, there is the regulating water feed system itself (ARE), which regulates the water flow for each steam generator. It is made up of two parts: a small flow valve and a large flow valve. Regulation ensures that there are switchovers between the two valves. The valves' response time in terms of opening/closing allows us to follow variations in installation power. Two functioning profiles are considered and can be subject to exogenous disturbances other than those attributable to regulation (turbine stoppage, reactor stoppage, etc.).

8.2.3. *Modeling the process*

The process is represented by the physical importance variable, Y, which is the level of water inside the steam generator. Regulation relies on its measurement, y_m, to handle Qe, the flow of water injected into the steam generator.

In practice the system is nonlinear and several types of regulations can be used (Model Predictive Control, non-linear Proportional Integral Derivative, linear Proportional Integral Derivative, etc.). The variables are as follows:

– Y: the water level inside the steam generator;

–y_m: a measure of Y;

– *d*: disruption defined by the factory's operating mode. The turbine and condenser are not explicitly represented and some of these components are included in the disturbance *d*;

– *u* (*Qe*): command to the regulation system (flow of water injected into the steam generator); and

– $Y - y_m$: drift and noise on the sensors.

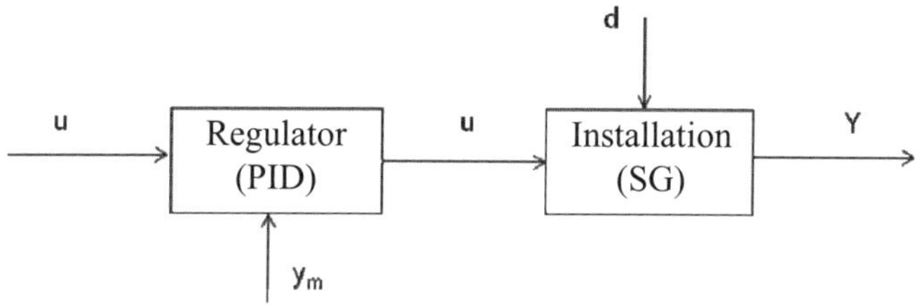

Figure 8.2. *Representation of the regulation of the water feed system that feeds the steam generator*

The installation section includes pipes, pumps, circuit valves and the SG model.

The behavioral model of the SG is obtained using a model published in 1998 and 2000 by EDF with the aim of benchmarking various command functions.

8.2.4. *Modeling command logic*

The command logic is made up of three components: sensors, actuators and industrial automata. The command controllers are implemented in the industrial automata. Two technological families of industrial automata can be used: conventional technology, which has limited autotest abilities and can be subject to undetected drifts; and programmed technology, which has more extensive autotesting abilities. In terms of APPRODYN, we will only examine conventional technology. Even without the risk of software failure, these automata can show errors in the logical conception or set up. Specification errors are not examined in this study.

The physical parameters required by the controller are: steam flow (*Qv*), water flow (*Qe*) and water/steam level (*NGe*). The faults affecting conventional instrumentation (request failure, blockage, etc.) are part of request faults and undetectable faults in the command logic model.

8.2.5. *Reliability data and state graphs*

Reliability data and state graphs have been created for each system modelized. We will study the simplest case, the steam collector (VVP). Many more data are needed for other systems: the worst case scenario is that of the ARE valves with four graphs for a dozen states and 30 parameters.

	Contribution to failure rate	**Pfd (request failure)**	**MTTR**	**Effect**	**Mode of failure**
Leaks	89 %	n/a	12	AAR (always detected)	Mode I: Mechanical origin
Cylinder rupture	11 %	n/a	168	AAR	Mode II: Mechanical origin

Table 8.1. *Example of reliability data (for the water input cylinder)*

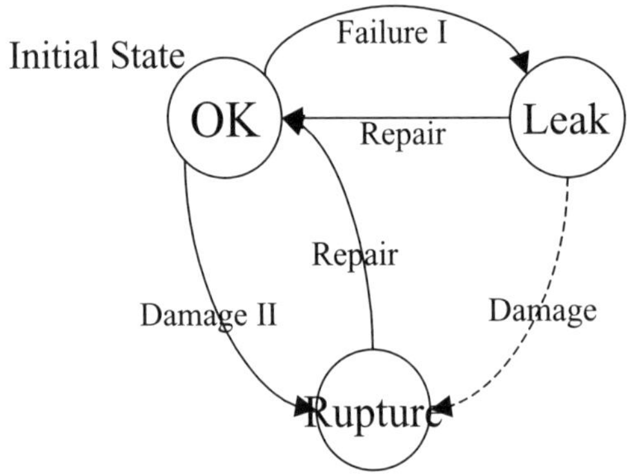

Figure 8.3. *Graph showing the different states of a water-feed cylinder*

8.2.6. *Ageing*

The ageing of materials can aggravate the risk of failure in repair or at request. Some ageing phenomena can occur through interaction with the procedure. These phenomena therefore need to be managed within a "dynamic reliability" framework. One means of representing these is to account for the cumulative effect of stops/restarts on the passive or electromechanical sections of the installation. The accumulation of overdrives is also an aggravating factor on the reliability of the TPA's turbine.

8.2.7. *Sensors*

This study has been designed to model a single type of sensor, the low range water level (*NGe)* measuring sensor. If the capability of a dynamic reliability approach to model this kind of sensor is proved in this case, it will almost certainly be applicable to other types of sensors. The sensors are redundant and the measures integrated via computational logic. Here, pairwise comparisons of each of the measures from each sensor are undertaken. The sensors are sensitive to mechanisms of degradation that are compensated for by regulation and are periodically tested. These optimization problems relating to periodic testing will not be examined here, although it may provide an interesting future area of research following dynamic reliability modeling.

8.3. Modeling using a stochastic hybrid automaton approach

8.3.1. *Main concepts and references*

We have chosen to use the FSM or finite-state automaton theoretical framework to conduct this study. This approach can result in a significant increase in the number of states but it has some advantages, as will be examined.

The FSM approach enables a formal construction of the system's model from embryonic automata representing the elementary component and combining them using the synchronization operation [CAS 08]. This feature presents an advantage compared to the Petri networks approach in which the construction of the model is intuitive and its properties (completeness, liveliness, boundedness, reachability, etc.) must be verified before use (particularly in simulation).

The second advantage, which is no less significant, is the FSM's natural ability to generate a complete set of the event sequences that are possible (its language) including critical sequences and, in the case of a stochastic approach, to evaluate the corresponding probabilities during simulation.

FSMs are used to represent the behavior of discrete events systems, i.e. systems whose discrete state only evolves following the occurrence of discrete events. In order to account for the stochastic nature of events (such as component failure), the concept of a stochastic automaton (of which the Markov chain is an example) needs to be defined. To account for dependency on certain events, derived from continuous phenomena and governed by differential algebraic equations, we need to define the concept of stochastic hybrid automaton (SHA). The stochastic hybrid automata have been formally defined and applied to academic case studies in the context of dynamic reliability [PER 09, PER 11]. This has enabled the modeling and probabilistic studies of systems with conflicts in the evolution of behavior [PER 10] and an assessment of the intensity of failure in complex systems in dynamic environments [BAB 11].

8.3.2. *What is a stochastic hybrid automaton?*

8.3.2.1. *Definition*

A SHA is an 11-tuple:

$$(\mathcal{X}, \mathcal{E}, \mathcal{A}, X, A, \mathcal{H}, \mathcal{F}, P, \chi_0, x_0, P_0),$$

where:

– $\mathcal{X}$ is a finite set of discrete states $\{\chi^1, \chi^2 \ldots, \chi^m\}$;

– $\mathcal{E}$ is a finite set of events $\{e_1,\ldots,e_r\}$, either deterministic or stochastic;

– X is a finite set of real variables evolving over time $\{x_1,\ldots,x_n\}$, where x indicates the vector of variables x_i, $x = [x_1 \ldots, x_m]^T$;

– $\mathcal{A}$ is a finite set of arcs of the form (χ, e, G, R, χ') where χ and χ' are the source and target states of the arc k, e_j is the event associated with the arc, G_k is the guard condition on X in state χ and R_k is the re-initialization function of X in state χ';

– A: $\mathcal{X} \times X \rightarrow (\mathbb{R}^+ \rightarrow \mathbb{R})$ is a function of "activities" allocating a function from $\mathbb{R}$+ into $\mathbb{R}$ to each element of $\mathcal{X} \times X$;

– $\mathcal{H}$ is a finite set of clocks;

– $\mathcal{F}$: $\rightarrow (\mathbb{R} \rightarrow [0,1])$ is an application that allocates a distribution function to each clock;

– $P = [p_i^l]$ is a matrix of probability distributions, where p_i^l is a distribution of state transition probabilities $p\left(\chi^i \middle| \chi^l, e\right)$. For example, if we have the same event e_q defining the transition of the discrete state χ^l towards the discrete states $\chi^1, \chi^2 \ldots, \chi^j$ (i.e. there are j conflicting transitions so the corresponding finite state automaton will not be deterministic), we can define the probability p_1^l of passing from state χ^l to the χ^1, the probability p_2^l of passing from state χ^l to state χ^2 and the probability p_j^l of passing from state χ^l to state χ^j, with $p^l_1 + p^l_2 + \ldots + p^l_j = 1$; and

– χ^0, x_0 and P_0 correspond to the initial discrete state, the initial value of the continuous variable vector in the initial state, and the initial distribution of transition probabilities, respectively.

The elements $\mathcal{X}$, $\mathcal{E}$ and $\mathcal{A}$ of the SHA define the discrete part of the finite-state automaton. In contrast, X, A, R and G define the continuous part. $\mathcal{H}$ corresponds to its temporized aspect and $\mathcal{F}$ and P express its stochastic aspect.

The automaton's functioning can be interpreted in the following manner: if the system is in state x^i, it is subject to a subset of events $\mathcal{E}$ allocated to different arcs exiting this state. At the occurrence of one of these events, e_q is allocated with arc k. If the guard condition G_k allocated to this arc is verified, the system goes into state x^j, the target state of the arc. The R_k function also associated with this arc defines the initial values of the system's continuous variables in state x^j. If e_q is associated to several arcs, the target state will result from taking a random variable drawn from the probability distribution p_i^l. In state x^i the evolution of real variables x in time is defined by function $f_i(\cdot)$.

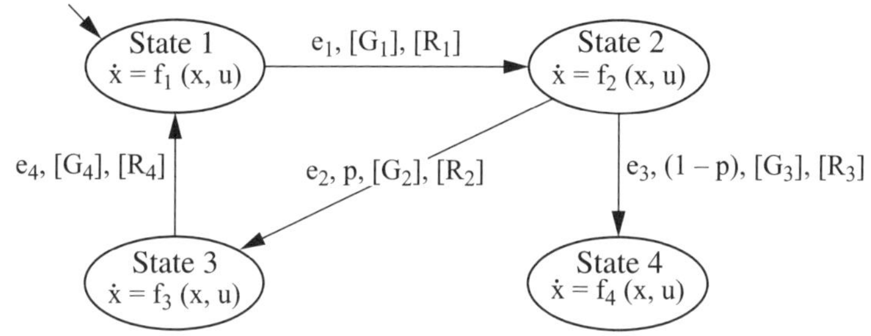

Figure 8.4. *Example of a stochastic hybrid automaton*

An example of a SHA with four discrete states is illustrated in Figure 8.4 ($\dot{x}$ corresponds to the derivative of variable x). Explicitly, the SHA is in discrete state 1 initially and the evolution of continuous variable x is given by function $f_1(\cdot)$. The occurrence of event e_1 and the verification of guard condition G_1 enables a transition towards discrete state 2. The re-initialization condition R_1 gives the initial value for continuous variable x in this discrete state. This variable evolves according to function $f_2(\cdot)$, and so forth. The durations of components' up and down periods are manifested by the clocks $\mathcal{H}$. These durations are random outputs of probability distribution functions $\mathcal{F}$.

The computer tool used to implement the SHA to assess the reliability of a dynamic hybrid system is the Scicos toolbox in Scilab. We have implemented the SHA on the basis of the hybrid automaton proposed by [NAJ 07]. The automaton defined by [NAJ 07] is the Scicos block illustrated in Figure 8.5 (for simplicity, the presentation of a single discrete state is shown). It is composed of I entry ports, where I corresponds to the number of discrete states (on the left of the block) and two exit ports (on the right of the block). The exit at the bottom of the block is a composite exit e of discrete events. It is activated when a transition occurs between discrete states. Each entry corresponds to a vector that contains the dynamics A of the system in state $\mathcal{X}^i$ (i.e. the evolution of continuous variables X), the initial values of X in the state (R), and the guard conditions G associated with the state's outgoing transition. The two exits correspond to a vector that indicates the numbers of the current discrete state $\mathcal{X}^i$, the previous discrete state $\mathcal{X}^{i-1}$ and the values of continuous state variables and their derivatives $\dot{x}$, respectively.

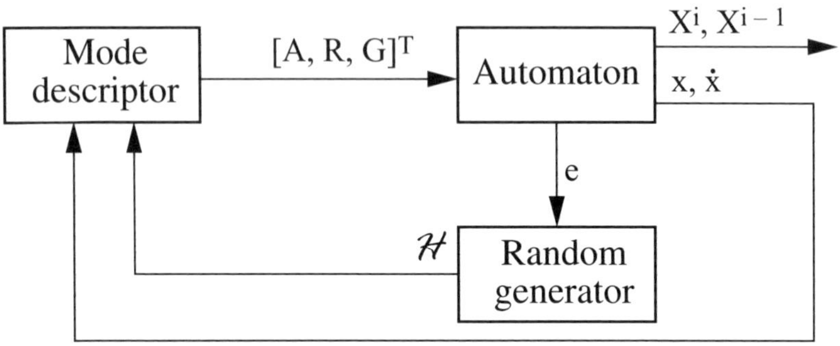

Figure 8.5. *Implementation of a SHA in Scilab/Scicos: illustration*

The random generator in Figure 8.5 corresponds to the stochastic temporized structure $\mathcal{H}$ in the given definition. The random generator carries out random samplings corresponding to random transitions and is activated on each change in discrete state by the discrete event exit of the automaton.

The mode descriptor of the SHA model given in Figure 8.5 corresponds to the system's different continuous dynamics. There can be as many continuous dynamics as there are discrete states.

8.3.3. *Structuring and synchronization approach*

Based on the approach developed for FSMs [CAS 08], this approach consists of defining an FSM to represent the behavior of each of the basic components and assembling those using a FSM synchronization operation. This results in an automaton representing the system whose properties are not in question, since the synchronization operation is formal. The advantage of this method is that it provides a formal model of the system that matches its specification as soon as the elementary automata do meet their specifications. Errors in the design of elementary automata or in their specifications can be demonstrated by the automaton's overall properties of accessibility, dead states, ability to restart, etc. For a SHA, we construct the underlying FSM using this method and complete it using stochastic events and differential algebraic equations associated with the states.

Another structured approach for constructing the underlying FSM consists of using some types of PNs, such as the hierarchical colored PNs defined by K. Jensen [JEN 09]. These PNs allow us to generate a complete reachability graph from the initial state. This graph is simply the FSM corresponding to the SHA.

8.3.4. *Modeling the case study*

In the framework of the case study, the following elementary automata and global automata (the result of synchronization) will be examined:

– VVP cylinder;

– three CEX pumps and the specification automaton ordering their simultaneous behavior, resulting in the global CEX automaton;

– two in-turbine TPAs and two non-turbine TPAs with two specification automata, and a specification automaton for simultaneous functioning of the two TPAs;

– small-flow ARE valve (ARE_{PD});

– heavy-flow ARE valve (ARE_{GD});

– control or global specification automaton.

The control automaton defines the power evolution and gives orders to start or stop and to open or close other components, thus controlling all other subsystems. The number of states and transitions for each of the basic automata and the global automata is given in Table 8.2.

Due to the complexity of the system considered in the case study and the large number of states in each subsystem due to this, synchronization is carried out in each subsystem (CEX, TPA, etc.). Relevant simultaneous functioning of the subsystems is achieved using synchronization variables that are shared by the subsystems, thus enabling interconnections. This approach is illustrated by an example of small-flow ARE automaton and the control automaton.

Sub-system	Number of states	Number of transitions
VVP cylinder	**3**	**4**
CEX (1, 2, 3)	15 (×3)	26 (×3)
CEX specification	46	114
CEX: global automaton	**121**	**225**
Non-turbine TPA (1, 2)	16 (×2)	27 (×2)
In turbine TPA (1, 2)	13 (×2)	22 (×2)
Specification for in-turbine and non-turbine TPA (1, 2)	*19 (×2)*	*30 (×2)*
TPA global specification	94	116
TPA: global automaton	**1,029**	**2,655**
Small-flow ARE (ARE_{PD})	**49**	**93**
Heavy-flow ARE (ARE_{GD})	**49**	**93**
Control machine	**44**	**60**

Table 8.2. *Basic and global automata for the case study*

A simplified diagram of the control automata showing part of the power increase period is given in Figure 8.6. The system's initiation implies a power (P) increase from 0 to 2% over time t (the timescale considered in this example is in hours) according to the equation $P = 0.2t$. This increase is carried out by safety systems. Once the power reaches 2%, the control automaton orders all of the subsystems (CEX, TPA and ARE) to start (for the purposes of simplification the starting failures are not fully shown in Figure 8.6). Power remains at 2% while not all of the subsystems are working correctly. This phase is followed by an increase in power. The evolution of power is not evenly distributed during the increase period, which implies the change in state equations (or the equation defining the evolution of power), for example $P = -6 + 0.8t$ or $P = -440 + 22.5t$. The transitions in bold (forcing power at 2%) represent the events modeled by synchronization variables, i.e. the external events common to several subsystems (the control automaton, ARE and TPAs in our example). The dashed transitions represent the events sent to the control automaton by the small-flow ARE automaton.

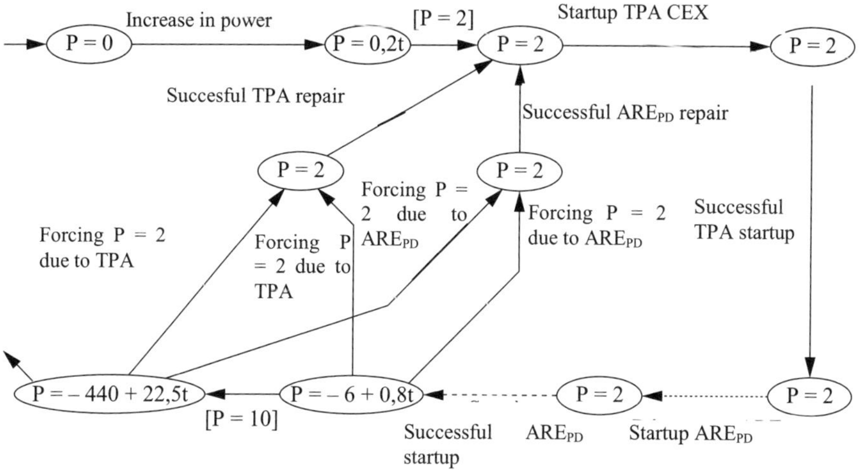

Figure 8.6. *A simplified diagram of the case study control automaton. The state equations are given. Transitions in the solid lines represent the synchronization variables, the long dashed lines show the events sent to the power automaton by the small-flow ARE automaton, and the short dashed lines show the events sent by the small-flow ARE automaton to the control automaton*

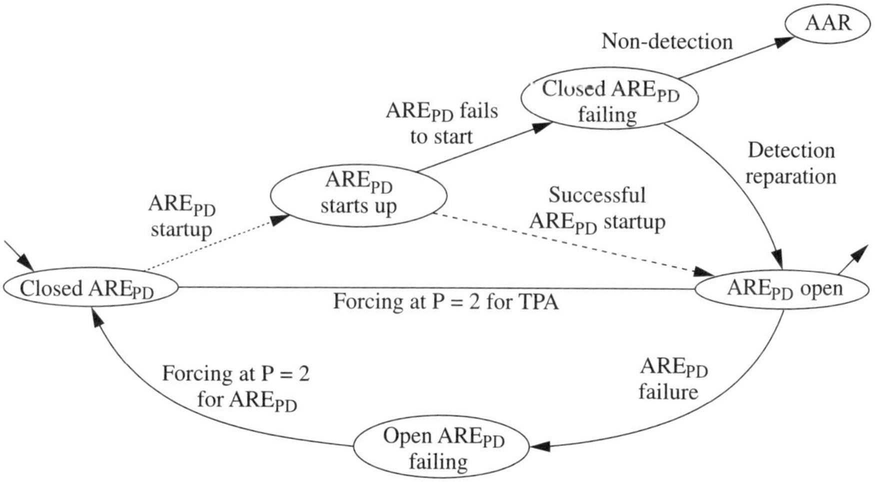

Figure 8.7. *A simplified diagram of the small-flow ARE automaton from the case study. The transitions in the solid lines represent the synchronization variables, the long dashes show the events sent to the power automaton by the small-flow ARE automaton and the short dashes show the events sent by the control automaton to the small-flow. ARE automaton AAR stands for automatic reactor stopping*

A simplified diagram of the small-flow ARE valve is given in Figure 8.7 (a single type of failure is considered in this figure for simplicity). During the power increase phase (ARE_{PD} open), the power equation is given by the control automaton $P = 0.2t$ or $P = -6 + 0.8t$ (depending on the global time, t). In this case, all of the subsystems can be forced to states corresponding to $P = 2.0\%$ (closing valves, stopping pumps, etc.) due to stochastic events (e.g. failure of TPAs or the ARE). Forcing is carried out using synchronization variables. The small-flow ARE automaton, in turn, sends information about opening failure/success to the control automaton.

Once 100% power is reached, the system operates steadily for a given time period. At the end of this period, a power decrease phase starts. The power decrease pattern correlates directly with its increase.

8.3.5. *Qualitative and quantitative results*

An example of a simulated trajectory is shown in Figure 8.8. This is a normal trajectory (without failure or forced power decrease) over a steady period of a month (the period running at full power). The water level (N_{ge}) and the entry flow (Q_e) for this trajectory are also given. There are considerable fluctuations of N_{ge} and Q_e during the phases of power increase and decrease. These fluctuations are due to the water level in the control system. During the steady phase the water level and flow remain stable.

We are also interested in a more detailed study of event sequences. To do so, the sequences likely to be produced have been identified and their frequency of occurrence has been empirically evaluated using 111 simulated independent trajectories. The results are given in Table 8.3. Only one trajectory results in an automatic reactor stopping (AAR). Indeed, an AAR is a rare event compared to the duration of operation considered (one month). Note that in the large majority of cases, the components operate without failure, with the exception of the TPA pumps. Only in approximately 20% of cases (the average over TPA1 and TPA2) are the TPA's trajectories normal. Nearly half of the simulated trajectories having a failure while operating in the in-turbine TPA. Between 10 and 15% (for TPA1 and TPA2 respectively) of trajectories contain a functioning failure of the non-turbine TPA. These failures are successfully repaired in the majority of cases. Note that the frequencies of failure in the in-turbine and non-turbine TPAs, when compared, correlate with the simulation parameters: the failure rate of the in-turbine part is four times higher than that of the non-turbine part.

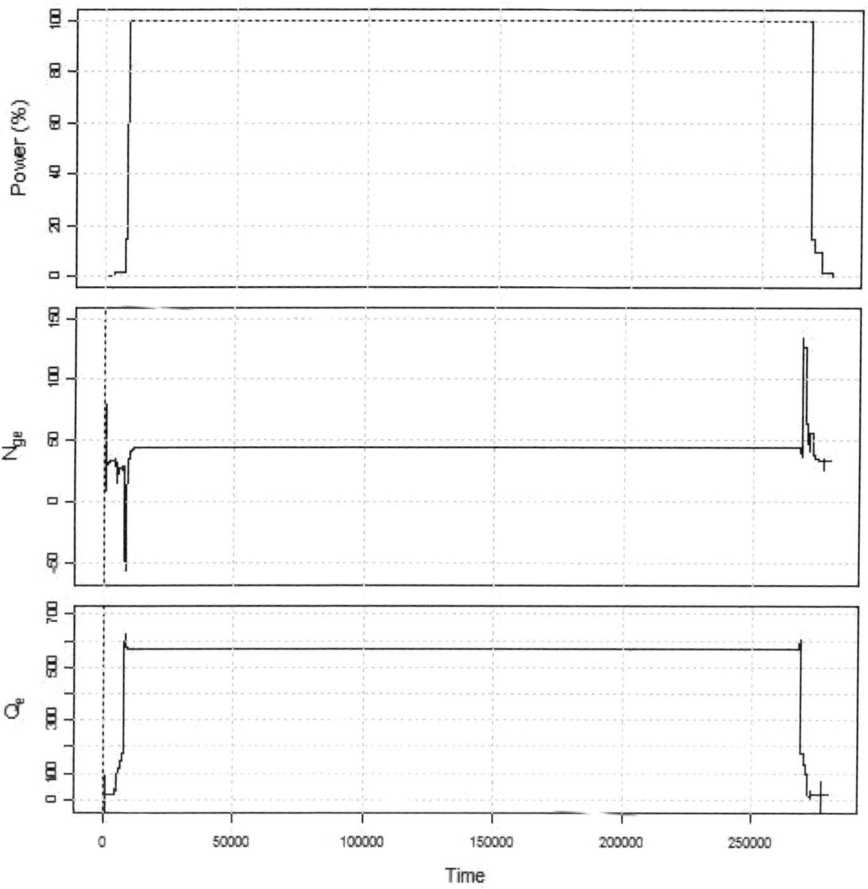

Figure 8.8. *Example of a normal trajectory (without failure or a forced decrease in power): evolution of power (duration of the stable period: one month), evolution of water level (N_{ge}) and of input flow (Q_e)*

The overall model obtained from the parallel composition of model components has the advantage of being comprehensive. However, its complexity in terms of the number of states and transitions involves time-consuming simulations. Optimization is therefore required to simulate more scenarios with a longer operational duration to evaluate the frequency of automatic reactor stops (AAR), thus enabling a detailed study of the critical event sequences causing these AARs.

Number (%) of trajectories		Type of trajectory	Detailed trajectory description
VVP Cylinder			
110 (99.1 %)		Normal trajectory	Functioning
1 (0.9 %)		Leak	Functioning → leak → AAR of the system
CEX			
109 (98.2 %)		Normal trajectory	Opening → functioning → closing
2 (1.8 %)		Functioning failure	Opening → functioning → failure of 1 out of 2 CEX → opening of the 3rd CEX → functioning → closing
TPA			
TPA 1	TPA 2		
31 (17.9 %)	28 (25.2 %)	Normal trajectory	Opening → functioning → closing
56 (50.5 %)	49 (44.2 %)	Functioning failure (in turbine)	Opening → functioning → turbine failure → successful repair → restart
12 (10.8 %)	17 (15.3 %)	Functioning failure (non-turbine)	Opening→ functioning → failure outside turbine → successful repair → restart
8 (7.2 %)	9 (8.1 %)	Failure to repair in turbine	Opening → functioning → turbine failure → repair→ failure to repair → successful repair → restart
3 (2.7 %)	8 (7.2 %)	Failure to repair non-turbine	Opening→ functioning → failure outside turbine → repair→ failure to repair → successful repair→ restart
1 (0.9 %)		Failure to start out of turbine	Opening → failure to open out of turbine → repair → restart
ARE			
Small flow	Heavy flow		
106 (95.5 %)	109 (98.2 %)	Normal trajectory	Opening → functioning → closing
4 (3.6 %)		Blocking	Opening → functioning → blocking → repair→ restart
1 (0.9 %)	2 (1.8 %)	Failure to open	Opening→ failure to open → successful repair → restart

Table 8.3. *Frequency of the occurrence of events sequences for each component*

8.3.6. *Conclusion and perspectives for the stochastic hybrid automaton approach*

Modeling using SHA allows us to evaluate the dependability parameters in a complex system, such as that considered in the case study. Beyond the empirical evaluation of the probability of dangerous states occurring, it allows us to analyze the complete set of sequences causing such problems and their respective probabilities. With the tools that are currently available, modeling remains complex and has a high dimension. This expansion in the model's size can be relativized, however, by several considerations: modern software tools are being developed and numerous approaches simplifying FSMs have been proposed (using the binary decision diagrams, see for example [HAM 05, POC 08]).

With the development of some packages to be included on the open source Scilab/Scicos platform, modeling can be significantly simplified. Furthermore, when SHA composition tools are formally developed, such as those found with FSM, only embryonic automata will need to be manually constructed and the rest will take place automatically without requiring graphic visualization of the entire automaton.

8.4. Modeling using piecewise deterministic Markov processes

8.4.1. *Principles and references*

In the field of dependability and safety, modeling is a crucial stage in analyzing the reliability of the physical process being considered. To do so, it is increasingly necessary to effectively and realistically account for various types of interactions that may exist between the system's different physical parameters and its overall behavior. In terms of dynamic reliability, one tendency that seems to emerge in the literature is a multi-model approach. The behavior of the physical process studied can therefore be described by different modes of operation, ranging from nominal to complete failure states with intermediate dysfunctional regimes. The layout of operational or accidental sequences comes from the occurrence of two types of events:

– the first type of event is directly related to a deterministic evolution of the system's physical parameters, such as crossing a threshold, for example; and

– the second type of event is stochastic and corresponds to the failure of certain components or external requests.

In this context, the system's continuous physical parameters (e.g. pressure in a pneumatic actuator, the temperature of an electronic component, etc.) define state variables and the evolution of these parameters is governed by the laws of physics, such as mechanical, thermodynamic or electromagnetic laws.

The current work is a continuation of research already undertaken by the INRIA-CQFD team that has sought to illustrate the implementation of a method combining the advantage of the modeling power of piecewise deterministic Markov processes

(PDMP) and the calculatory efficacy of Monte-Carlo simulation in order to consider problems of dynamic reliability. "Academic" [ZHA 09] and "industrial" [ZHA 08] size systems have previously been modeled and simulated using C++ or Matlab. To model this water level regulation system, however, we have chosen the Simulink and *Stateflow* in Mathwork software for three reasons:

– Several physical variables (*Nge*, *Ngl*, *Qe*, *Qv*, etc.) must be calculated at each stage in time. These variables evolve following nonlinear differential equations. A tool such as Simulink lends itself perfectly to this kind of problem. In addition, PID and MPC type controllers have already been implemented in Simulink in the form of a block system.

– The number of components that we need to model and their specific behaviors render the system highly complex. Each component can adopt a high number of states, whether nominal or damaged. The total number of possible combinations for all of the components is huge. A programming language, such as C++ (or Matlab), can solve this problem but as such it is very difficult to develop the code. Adding or suppressing a component or changing the system's configuration would be difficult.

– *Stateflow* increases the possibilities of Simulink due to its design environment for developing state machines and flux diagrams. It provides the language elements necessary to describe complex logic in a natural, legible and understandable form. It is object oriented, which allows us to model the VVP, CEX, TPA and ARE components separately and treat redundancy using a simple "cut and paste" technique.

Since *Stateflow* is an interactive graphic tool built into Simulink, it provides an effective environment for designing the SG simulator.

8.4.2. *What is a piecewise deterministic Markov process?*

A PDMP provides a general modeling framework for dealing with safety and dependability problems associated with physical systems. M is the finite set of possible modes for the system. For every m in M, E_m is an open subset of R^d. A piecewise deterministic Markov process is defined on the basis of three local characteristics (Φ, λ, Q) where:

– the flow Φ: $M \times R^d \times R \rightarrow R^d$ is continuous and for each $s,t \geq 0$, $\Phi(\bullet,t+s) = \Phi(\Phi(\bullet,s),t)$. This describes the deterministic trajectory of the process between jumps. For every (m,x) in $M \times E_m$, we apply:

$$t^*(m,x) = \inf\{\, t > 0 : \Phi(m,x,t) \in \partial E_m \}$$

which is the time required to reach the boundary of the domain;

– the jump intensity λ characterizes the frequency of jumps. For each (m,x) in $M \times E_m$, and $t \leq t^*(m,x)$, we have:

$$\Lambda(m,x,t) = \int_0^t \lambda(\phi(m,x,s))ds$$

– the Markov kernel Q represents the transition law and allows us to select the new position after each jump.

The trajectory $X_t=(m_t, x_t)$ of the process can therefore be defined as follows. We start from an initial point $X_0=(k_0, y_0)$ with $k_0 \in M$ and $y_0 \in E_{k0}$. The first jump time T_1 is determined by:

$$P_{(k0,y0)}(T_1 > t) = \begin{cases} e^{-\Lambda(k0,y0,t)}, & t < t^*(k_0, y_0) \\ 0, & t \geq t^*(k_0, y_0) \end{cases}$$

On the time interval $[0, T_1[$, the process follows the deterministic trajectory $m_t=k_0$ and $x_t=(\Phi(k_0, y_0, t)$. At the random time T_1, the process is subject to a jump. It therefore changes mode and is reinitialized at X_{T1}, a random variable that follows the law given by $Qk0(\Phi(k_0, y_0, T_1),\bullet)$. We then randomly take a new jump time $T_2 - T_1$, and on the interval $[T_1, T_2[$the process follows the trajectory $m_t=k_1$ and $x_t=(\Phi(k_1, y_1, t - T_1)$. In the same way, we iteratively construct the PDMP.

The SG system is different in that it is a regulated system. The flow Φ is the solution of a differential equation controlled by a PID controller. It does not have an analytical solution and is calculated at each time stage by Simulink.

The reasons for choosing PDMP modeling are twofold. First, it provides both a general and precise framework for modeling the problem. Second, this model offers the potential for optimal control in the future: optimal stopping, predictive maintenance, etc. [SAP 11a].

8.4.3. *Modeling the test case*

Let us consider scenario 1, represented by Figure 8.8. After a linear piecewise increase with a duration of 24 hours, the system reaches its stationary mode, 100% *Pn* and remains as such for 18 months followed by a fall lasting 24 hours. The aim is to simulate the system's behavior, which is subject to random faults. Simulation is stopped when the reactor stops (AAR).

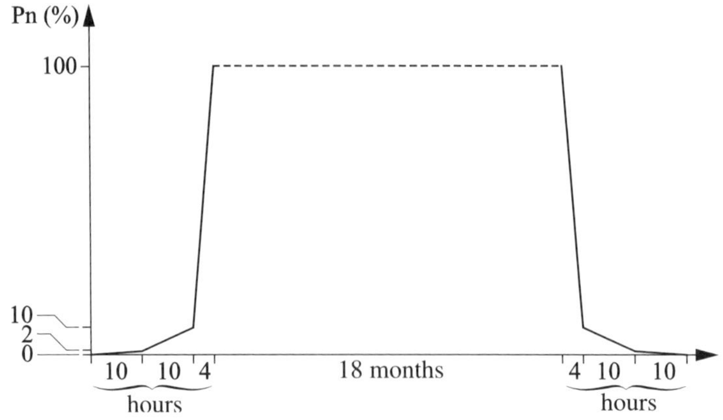

Figure 8.9. *Scenario n°1*

Figure 8.9 shows the results of the simulation. The Ramp block allows us to generate an increasing or decreasing ramp for variable *Pn*, which is one of the system's entry data. This ramp is saturated between [0,100] by the Saturation block. We will focus on four exit signals: *Nge*, *Pn* (water level and target), *Ngl* (broad water level) and *AAR* (signal that halts simulation). The SG generator is entirely modeled by the SG sub-block system shown in Figure 8.10, where we find:

– An installation *stateflow* block that regroups the VVP, CEX, TPA and ARE components, which we will examine later. This block is activated at each *time stage and each time Pn exceeds the thresholds of 2% and 15%.*

– A SG generator, modeled by a GNV subsystem with two inputs (*Qv*, *Qe*) and two outputs (*Nge*, *Ngl*). This system follows a system of nonlinear differential equations, since its coefficients depend on *Qv*. See [KOT 00] for a more detailed overview of this system.

– A PID controller, which takes the difference between the target *Pn* and *Nge* as its input data and the variable *Qe* as its output, for the water flow injected into the SG. Variable *Q*, represents disruptions from the ARE.

The main advantage of Simulink plus Stateflow modeling is that it takes place in the form of an interactive graph, thereby facilitating understanding of the model. If the system operates in nominal mode (e.g. increase, decrease, power mode of 100%), no component is faulty and the water level is regulated by the PID controller. If a component fails, it can provoke either an AAR or a minor fault. In the first case, simulation is stopped. In the second, the simulation continues and repairs are made. A drop and increase in ramp are programmed if necessary.

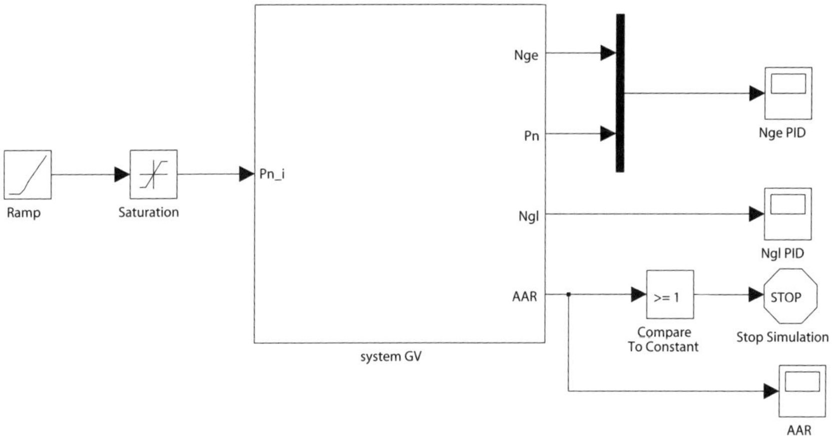

Figure 8.10. *Complete illustration of the simulator*

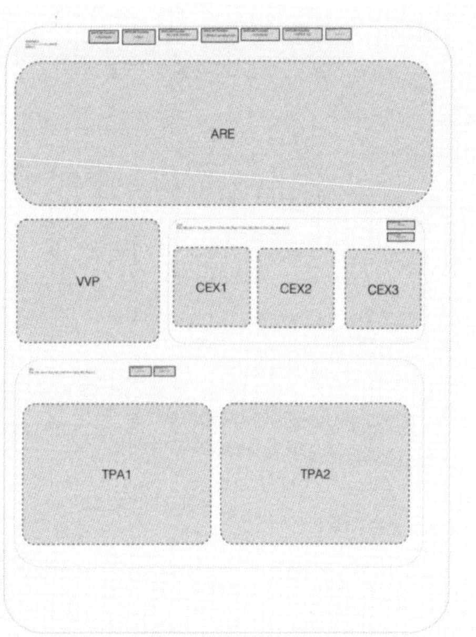

Figure 8.11. *Stateflow installation block*

The *Stateflow installation* block models the SG installation and includes the ARE, VVP, three CEX and two TPA components, as shown in Figure 8.11. We will now examine VVP and CEX in further detail.

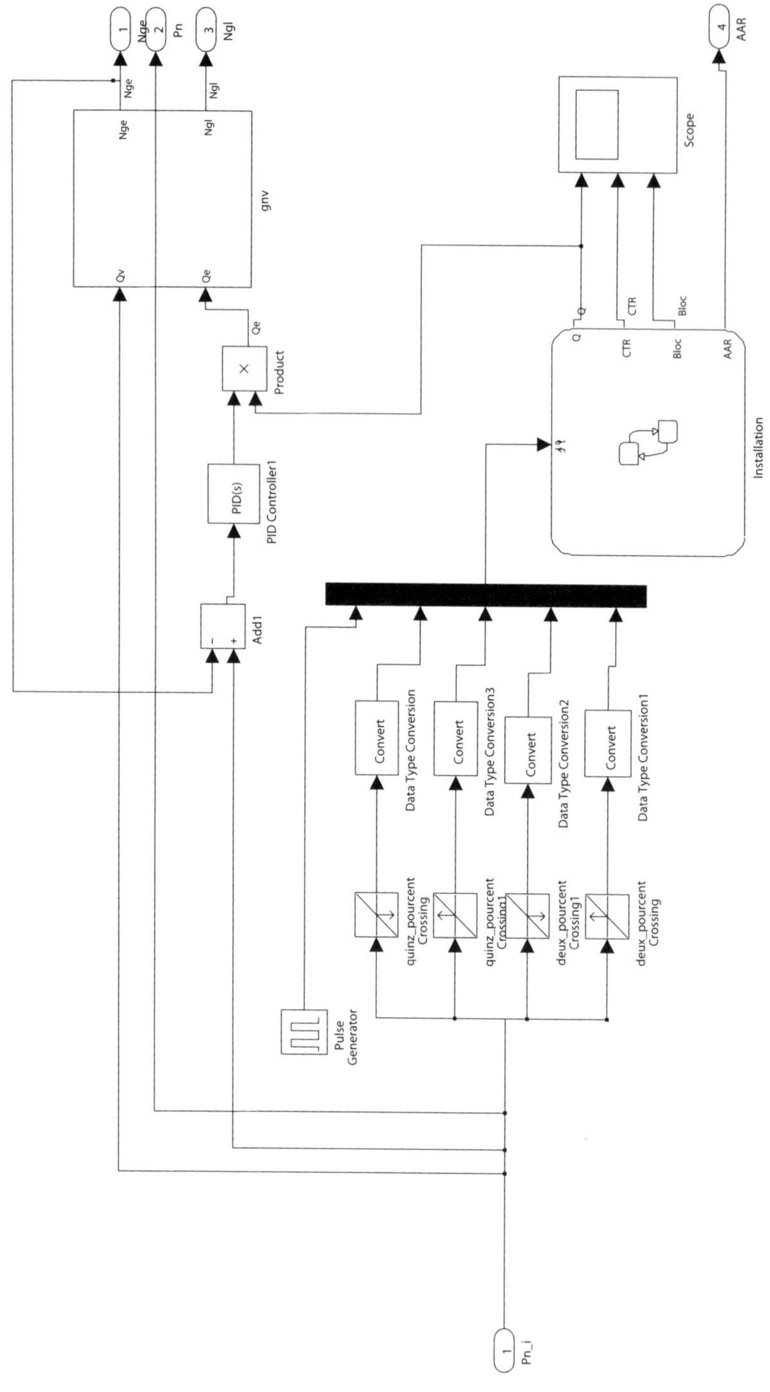

Figure 8.12. *Full illustration of the SG XE's steam generator*

8.4.4. *Modeling the VVP*

This component has three states (OK, leak and rupture). The time spent in each state follows an exponential law. When this component is activated, it is in the OK stage by default. A duration x is drawn from an exponential (E(4.05e-6/H)) and we also take a Bernoulli variable p (B(0.11)). When the time spent in this stage exceeds x, the transition after (x,sec) occurs. The component switches to the leak stage if p=1 and rupture if p=0. In this case, we send an AAR signal that halts the simulation. The codes 201 and 202 record the number of malfunctioning components and the type of malfunction.

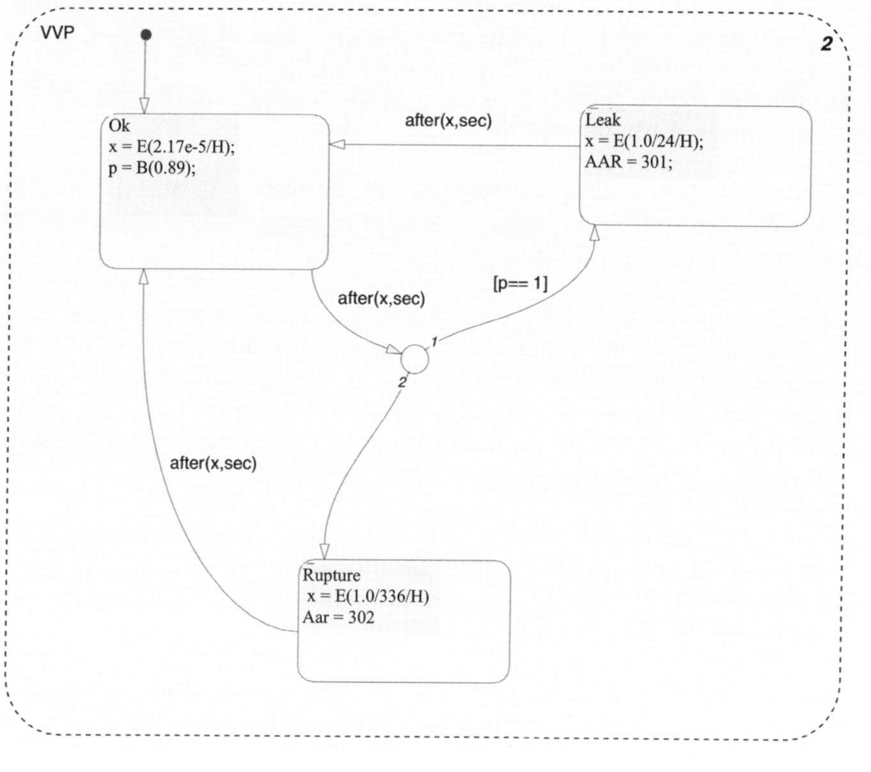

Figure 8.13. *VVP model*

8.4.5. *Modeling CEX*

The main function of CEX is similar but far more complex. Figure 8.14 illustrates the implementation details for CEX, which is in effect the CEX1 shown in

Figure 8.11. The two other CEXs are almost identical (cut and paste), with only the initial conditions being different. In the initial state, two CEXs are functioning and the third is held on *standby*.

8.4.6. *Qualitative and quantitative results*

The total duration of scenario 1 is 18 months. There are two types of regimens: transitory and stationary.

When the system is in transitory regimen, the water level *Nge* and the power *Pn* vary rapidly. To follow this command, we have to choose a sufficiently small period of time (0.6 seconds) during which the PID controller can work properly.

In the stationary regimen, all of the physical variables are constant and only component break-downs can alter the system's state.

Since the system is highly reliable, however, the nominal duration is often long and a short time stage is not appropriate because it significantly slows down the simulator.

There is a specific technique for solving this problem of inconsistent time stages. Two Simulink models have been created – one with and the other without the PID block. This allows us to choose two different discretization time stages. During the transitory regimen, we use a stage of 0.6 seconds and during the stationary regimen, a stage of 60 minutes is sufficient to simulate the failure of a component. A Matlab script manages the simulation.

To illustrate the results, we have taken a sequence of events without failure in full power mode over 10 days (rather than 18) months for illustrative purposes. Figure 8.15 shows that the PID controller has fulfilled its role as the regulator and the water level *Nge* coincides with the target *Pn*.

Out of the total number of 4,000 scenarios that were simulated, 2,190 AARs have been recorded with a probability of 54.75%, i.e. the system has a one in two chance of being subject to an AAR over a period of 18 months. Table 8.4 summarizes the number of AARs caused by each component in the system. Their percentage rate of occurrence in the 2,190 AARs is also shown.

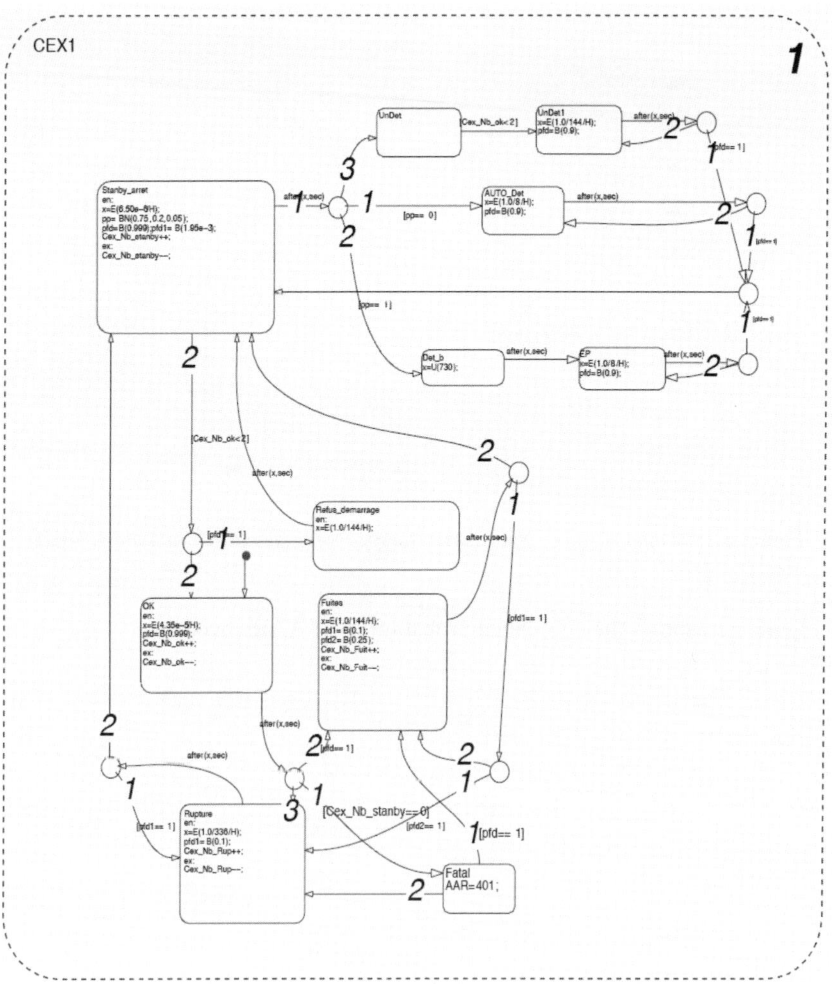

Figure 8.14. *Modeling CEX*

Subsystem	**Number of ARAs caused**	**Percentage**
VVP	792	36%
ARE	1,301	59%
CEX	50	2.28%
TPA	44	2.10%
Total number of AARs	2,190	100%

Table 8.4. *Number of AARs shown over 4,000 cases*

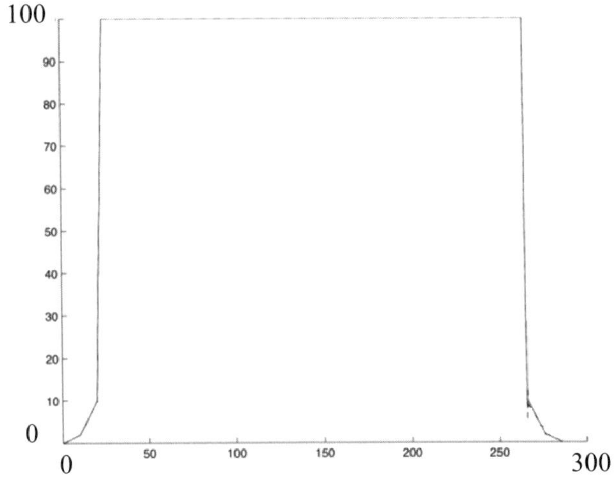

Figure 8.15. *Comparison of Nge and the target Pn for scenario 1 in nominal mode*

Figure 8.16 shows the cumulated probabilities of AARs over time.

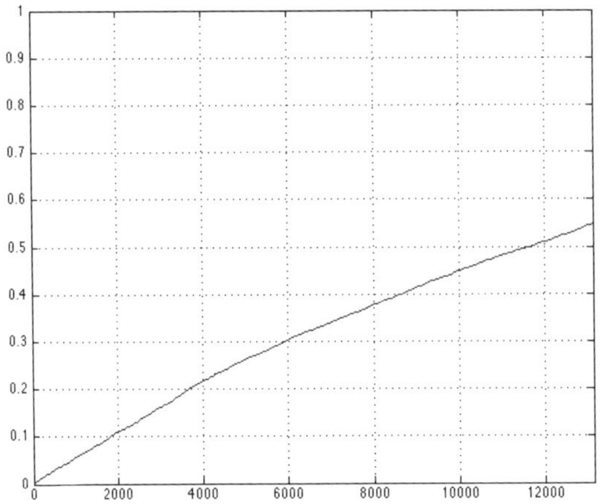

Figure 8.16. *Cumulated probability of AARs*

No AAR caused by the small-flow ARE valve was observed. This can be explained by the fact the time spent in transitory mode is too short (48 hours) in relation to the overall duration (18 months). This failure is therefore a rare occurrence. Also note that a high number of AARs is caused by the VVP cylinder

(36%). Its failure rate (2.17e-5) is comparable to that of CEX (4.35e-5) and TPA (5.9e-4), although these last components are redundant, which minimizes the AARs.

8.4.7. *Conclusion and perspectives for the piecewise deterministic Markov processes and simulation approach*

PDMP modeling lends itself well to the problem of dynamic reliability. The Simulink approach associated with *Stateflow* allows us to construct an interactive simulator. It therefore provides an interesting view on several aspects:

– graphic programming, because the source code strongly resembles a reliability diagram. In debugging mode, users can visualize states and transition stages by stage;

– progressive monitoring of the simulator. The VVP, CEX, TPA and ARE components are modeled separately first (assuming that the other components are 100% reliable) and then grouped together using a simple cut and paste approach. This allows us to easily add other components in the future. We can also approach this problem of component redundancy by pre-constructing a library of components; and

– limited number of components. This approach does not have the problem of combinatorial explosion. This is because the *stateflow* state machines are component oriented, i.e. the simulator calculates each component's state at each time stage. We can therefore think of the system's state as a vector whose size is equal to the component's number.

The main disadvantage of this approach is its execution time. For the test case considered, a scenario is simulated in around 30 seconds (on a portable MAC) and it takes eight hours for 1,000 Monte Carlo iterations. The test carried out in [ZHA 08] shows that a C++ simulator focusing on a problem of this size can undoubtedly function 10 or even 100 times faster, but at the cost of significantly slower programming and code generation, which is difficult to develop further. There are two solutions that can be applied to accelerate the simulation: parallel calculation and automatic generation of a code C. Both offer interesting perspectives for future developments. Other avenues are also open to exploration: the INRIA-CQFD team has proposed optimal control algorithms: optimal stopping, impulse control, etc. [SAP 10, SAP 11a, SAP 11b]. The Monte Carlo simulator is an essential part of all of these methods.

8.5. Modeling using stochastic Petri nets

8.5.1. *Principles and references*

Several classes of PNs have been developed to solve specific modeling problems [DAV 94, MUR 89]. PNs were originally used in deterministic processes, especially in industries using automation, control systems or in manufacturing. In recent decades, technological developments have meant that these same processes have become increasingly complex, allowing greater gains in terms of productivity and performance that are often costly in terms of maintenance. The aim is therefore to develop maintenance policies for system dependability so that maintenance operations are only carried out when needed, i.e. neither too late or too early. Among all possible methods allowing a combination of deterministic and stochastic behaviors due to potential failures and/or degradation modes, the class of stochastic PNs (SPNs) enables the implementation of a hierarchical modeling starting from each elementary component and moving towards the system level [ZIL 08].

8.5.2. *What is a stochastic Petri net?*

A SPN may be seen as state machine defined by the usual five-tuple (P, T, F, W, M_0) where:

– $P = \{p1, p2, ..., pm\}$ is a finite set of places;

– $T = \{t1, t2, ..., tn\}$ is a finite set of transitions;

– $F = (P \times T) \cup (T \times P)$ is a finite set of arcs linking places to transitions;

– $W : F = \{1, 2, 3...\}$ is a weighting function of each arc;

– $M_0 : P \rightarrow \{0, 1, 2, 3...\}$ is the initial network marking, i.e. the number of tokens in each place.

A transition is triggered by a probability law (exponential, Weibull, etc.) reflecting the component's failure, its repair time and the availability of a spare part, although this list is not exhaustive. Note that a deterministic transition can always be triggered in this stochastic framework with the use of a Dirac law. For a complex system, such as that studied within the APPRODYN project, the components interact with each other by exchanging value messages [DUT 97]. This flow of messages ensures cohesion and synchronization between all of the SPNs and can also carry quantitative information about the water level in the steam generator. In a colored PN, this kind of information is simply carried by the token itself.

For a better understanding of SPNs, we will consider the simple example of a one-component system. It is subject to preventive maintenance after a period of 1,000 hours and the maintenance lasts for two hours. During maintenance, the system is unavailable. The component failure rate is assumed to follow an exponential law. The corresponding SPNs are described in Figure 8.17.

Only two SPNs are used to model the system. The first describes the component behavior (faulty, nominal and in maintenance situations). The second explains the maintenance procedure. The SPNs are synchronized using the Boolean message M45, indicating the end of maintenance.

With large and complex systems, obtaining analytical solutions about the main dependability measures, such as availability and reliability, quickly becomes unavailable. For this reason, SPNs are animated using Monte Carlo simulations.

In summary, this method consists of simulating a significant number of independent scenarios, each describing the system's behavior for a fixed time period. For each scenario, the dates of occurrence and possible events, such as the appearance of a fault, are identified with random selections. Each scenario reproduces one of the system's possible evolutions, i.e. the sequence of events describing its behavior and the maintenance operations carried out. During the simulation of a given scenario, we can consolidate the number of occurrences of different events and the time spent in various possible states using dedicated measures. After simulating all scenarios, the measure values are the average of the total number of simulations carried out.

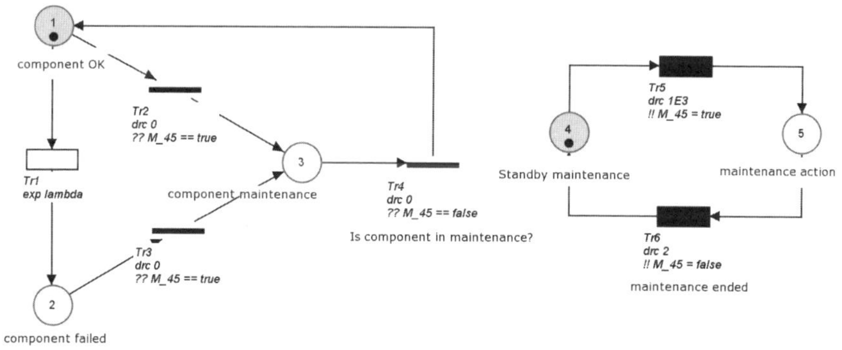

Figure 8.17. *SPN: an introductory example*

8.5.3. *Modeling framework*

In order to fully describe the system's behavior, three types of variables are used:

– *process variables* describing the physical variables involved in the system's dynamic;

– *information variables* describing the data or information evaluated, stored and exchanged between all components; and

– *component state variables* describing the system's structure, depending on its components and their corresponding modes.

The feedback control system is illustrated in Figure 8.18. The desired water flow *Qec* is calculated using a PID controller in its discretized form using the water level threshold *SetNge* and the measured water level value *Ngem*. Depending on the duty point (i.e. power, *P*), the ARE command logic switches between two small-flow and large-flow valves. The water level *Nge* is provided by a system of equations resulting from a representation in a state space that will not be detailed here. In this chapter, we will assume that the water level sensor does not fail and does not experience a deviation/drift, so that the measured value *Ngem* is always equal to *Nge*. For each time interval $[t, t + \Delta t)$, all of the variables are updated using a dedicated SPN. This update is carried out following a specific order respecting the system's dynamic and is only feasible by assigning a priority for each transition.

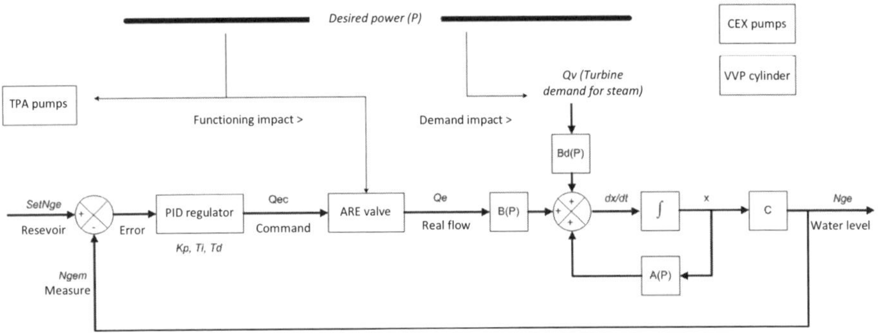

Figure 8.18. *Synthetic view of the model*

8.5.3.1. *Process variables and parameters*

The process variables are the variables described by the state space representation (which will not be examined here), such as the input water flow *Qe*, the steam flow *Qv* and the water level *Nge*. The process parameters *Tm*, *Tint*, *Tn*, *Fg*, *Th* and τ are the different time constants and must be updated according to the desired duty point (the power, P). In a first approach, this update takes place using a simple linear interpolation based on the characteristic points in the steam generator being studied. A complete description of these characteristic points as well as the state space representation is described in [KOT 00].

8.5.3.2. *Component state variables*

Each component state is represented using an integer variable. All of the variables of each component and functioning mode are summarized in Table 8.5.

Component	**State variable**	**Value**	**Description**
VVP Cylinder	VVP_Cylinder_State	=0 =1 =2	Working OK Breakdown mode I (leak) Breakdown mode II (break)
CEX_i=1,2,3	CEX_i_state	=0 =1 =2 =3 =4 =5 =6	Working OK Breakdown mode I (failure) Breakdown mode II Standby OK Standby (failure IIIa) Standby (failure IIIb) Undetected failure IIIc
TPA_i=1,2	TPA_i_state	=0 =1 =2	Working OK Standby OK Not OK
Outside_turbine_TPA_ i=1,2	Outside_turbine_TPA_ i_State	=0 =1 =2 =3 =4	Working OK Breakdown mode I Breakdown mode II Standby OK Failure to function
Turbine_TPA_i=1,2	Turbine_TPA_i_state	–0 =1 =2 =3 =4 =5 =6	Working OK Breakdown mode I Breakdown mode II Standby OK Failure IIIa Failure IIIb Failure IIIc
Small_flow_ ARE	ARE_pD_state	=0 =1 =2 =3 =41 =42 =43 =5	Working OK Breakdown mode I Breakdown mode II Breakdown mode III Breakdown mode IV a Breakdown mode IV b Breakdown mode IVc Breakdown mode V

Table 8.5. *Some components and their corresponding variables*

8.5.3.3. *Information variables*

Power P is the main information shared by all system components. P dictates their behavior and operating modes. For instance, starting/stopping orders are sent to the TPA and closing/opening orders are sent to valves. The water level setpoint *setNge* also depends on P, so that:

$$setNge = \begin{Bmatrix} 0.55\,P + 33\ if\ P \leq 20 \\ \text{otherwise} \end{Bmatrix}$$

8.5.3.4. *MOCA-RP*

MOCA-RP, or Petri net-based Monte Carlos is the tool used in APPRODYN. This software, currently maintained by SatoDev (www.satodev.fr), is designed to simulate the behavior of complex dynamic systems. MOCA-RP allows us to obtain statistical processing results concerning reliability, availability, productivity and other probabilistic parameters. It allows us to hierarchically split all kinds of case studies into elementary sub-problems and create libraries of reusable components. This tool therefore seems to be perfectly suited to the steam generation system in APPRODYN. The VVP SPN is described in Figure 8.19 as a guide.

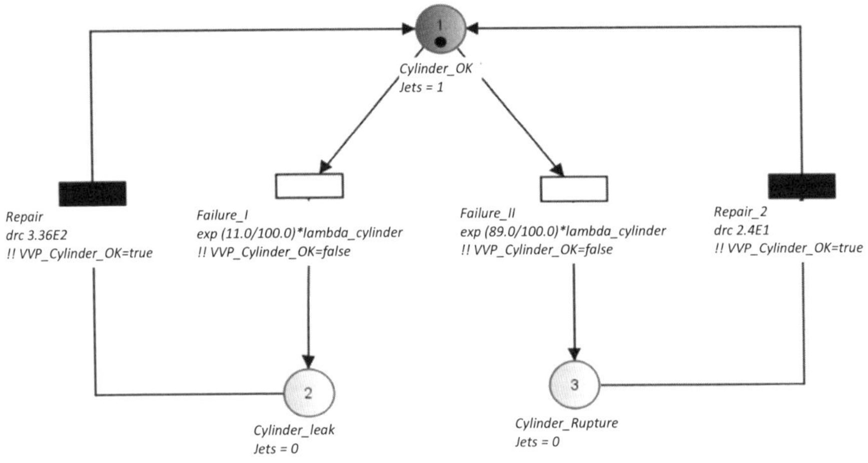

Figure 8.19. *VVP*

In summary, APPRODYN requires 45 SPNs including 228 places, 281 transitions, 664 arcs and 81 variables.

8.5.4. *Qualitative and quantitative results*

8.5.4.1. *Initial tests*

Only the feedback control law is tested with the operating profile illustrated in Figure 8.20. The first objective is to verify the discretized implementation of PID law inside a PN. The second objective is to test the robustness of this same law to power set variations in the range of 5–65%. We will not examine component failures here.

The control law is designed using the well-known Ziegler-Nichols method in the field of feedback control systems. This tuning is done with a dedicated software, such as Matlab Simulink. The results are described in Figure 8.21 (water flow) and Figure 8.22 (water level). The oscillations show that this tuning is not optimal and requires refinement.

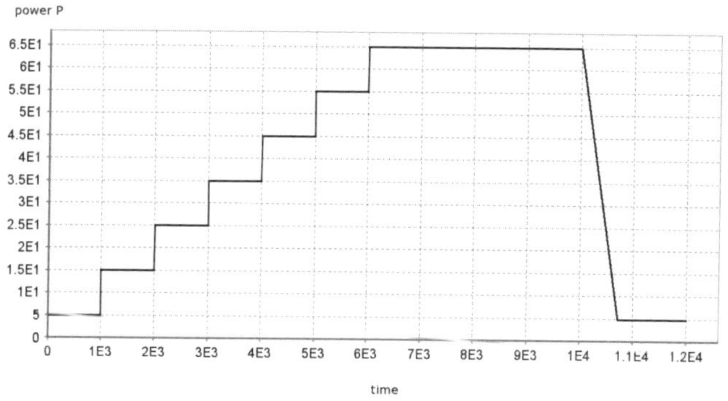

Figure 8.20. *Operating profile (PID test)*

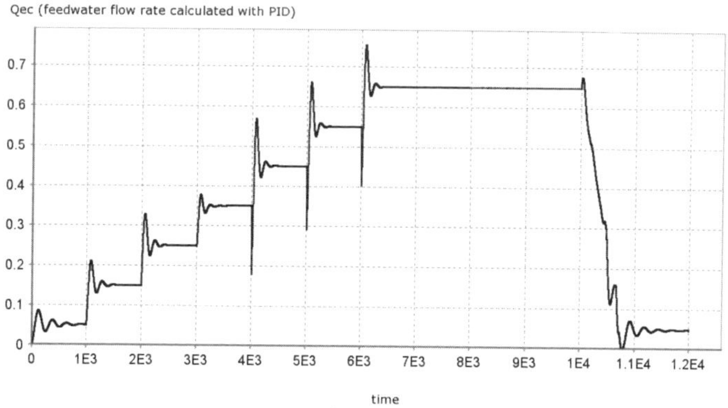

Figure 8.21. *Water flow, Qe*

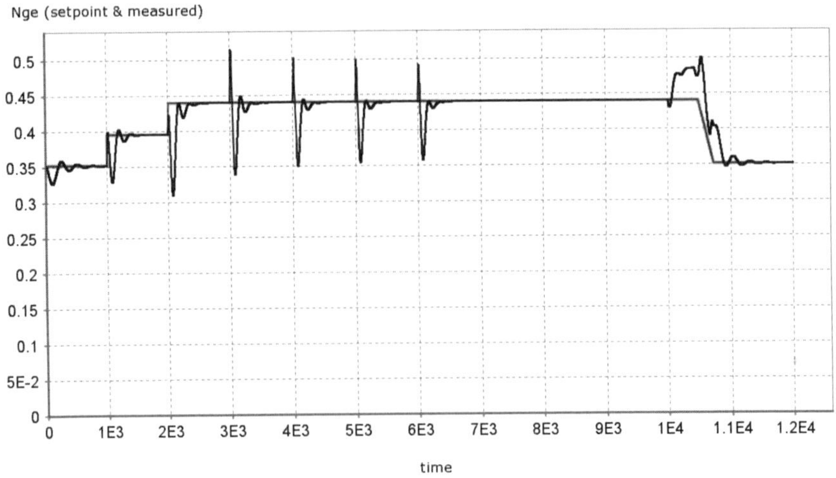

Figure 8.22. *Water level, Nge*

8.5.4.2. *Real-life tests*

Here, we aim to test the SPN using a realistic operating profile described in Figure 8.23 with an 18-month simulation period and a 0.1 hour test time.

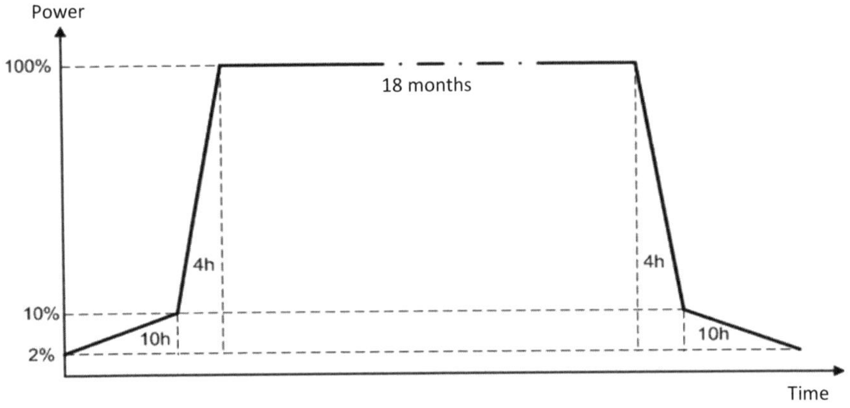

Figure 8.23. *Real operating profile (power setpoint)*

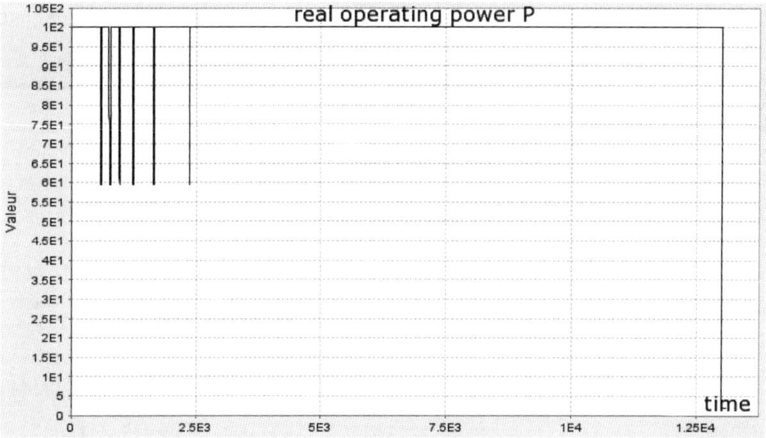

Figure 8.24. *Real operating profile (simulated)*

Figure 8.24 illustrates a real operating profile (one simulation run of 75 seconds) for the power, *P*, during the period of 18 months showing the increasing and decreasing phases. All component modes are accounted for. First, we can see that VVP and the set of CEX pumps do not fail during this period. If this happens, it would imply an ARA given the SG operating modes described in Chapter 1. Second, there are no type I, II or III failures of ARE valves because this would cause *P* to decrease towards 2%; the time required to carry out maintenance. Figures 8.25 and 8.26 describe the real water flow, *Qe*, as well as the corresponding water level, *Nge*.

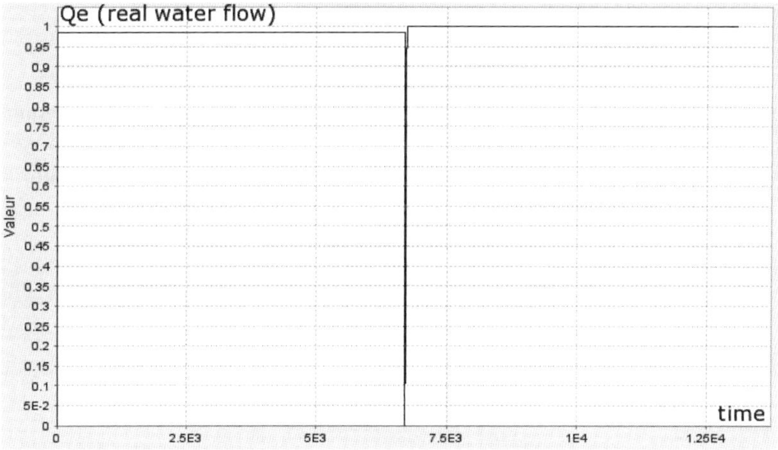

Figure 8.25. *Real water flow, Qe*

These two results clearly show that the control law is not sufficiently robust.

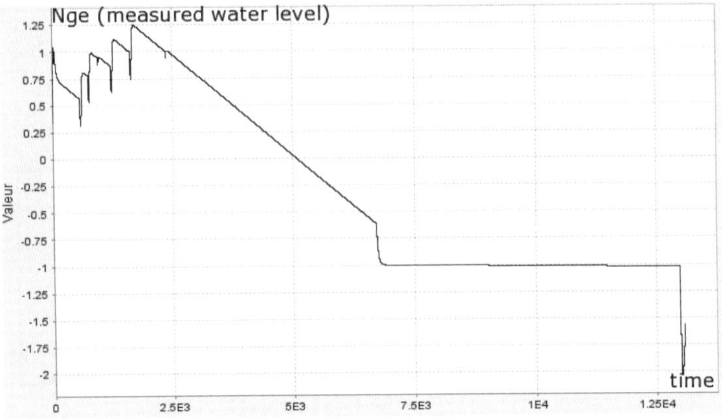

Figure 8.26. *Measured water level, Nge*

8.5.5. *SPN approach: conclusions and perspectives*

The initial results are not satisfactory. The main explanation is that a single adjustment of the PID law is carried out for the whole test case. It is therefore realistic to consider that this command law is not sufficiently robust to account for all the SG operational modes. This is the case because:

– the system is not an LTI system. The different matrices making up its state representation depend on the power *P* required (see [KOT 00]); and

– the control signal is naturally bounded.

Unfortunately, all dependability measures are strongly related to the controlled variable, i.e. the water level. The main objective is therefore not to overcome scientific challenges but to solve technical problems as a result of the tuning of a robust control law.

The positive aspect of this approach is the easy use of SPNs. The SPNs do not require any kind of computing knowledge. Common sense and modular programming are sufficient to achieve hierarchical modeling. SPNs aided with a user friendly graphical interface explain the reason why this approach is commonly used in industry and dependability engineering.

The remaining problems to be overcome to finalize the APPRODYN project beyond a robust PID law involve correctly managing the transition triggering priorities and the time during Monte Carlo simulations (time management is not the same for stochastic and deterministic events). Together these issues lead to interesting perspectives for the SPN approach in the context of APPRODYN.

8.6. Preliminary conclusion and perspectives

At the time of writing, the research project has not been fully completed and there are still some questions that remain (representing ageing and sensors and optimizing periodic tests). We can, however, attempt to make preliminary conclusions.

A test case, focusing on part of a secondary circuit in a pressurized water reactor, has been completed that is sufficiently broad to make further comparisons.

Project APPRODYN allows us to experiment with different dynamic reliability approaches within the context of probabilistic assessments, particularly in the case of energy production and process industries. We have explored the possibilities created by innovative approaches that are as yet rarely used, notably Hybrid Stochastic Automata and an approach combining Piecewise Deterministic Markov Process and simulation. We compared them with the more widely used approach, Stochastic Petri Network.

The project has demonstrated the potential of modeling fairly large hybrid systems, combining a combinatorial part with discrete events and an hybrid part with several approaches. The first results show that the significant expansion of the model's size can be relativized by accounting for the ability of modern computing tools and relevant simplification methods. Another area of progress concerns the development of Hybrid Stochastic Automata formal composition tools.

Lastly, we are currently in the process of defining a number of criteria that allow us to evaluate and compare probabilistic approaches to model hybrid systems (ease of programming, verification of the model, parallelization, ability to represent continuous variables and stochastic phenomena, etc.).

8.7. Bibliography

[AST 00] ASTROM K.-J., BELL R.-D., "Drum boiler dynamics", *Automatica*, vol. 36, pp. 363-378, 2000.

[BAB 11] BABYKINA G., BRÎNZEI N., AUBRY J.F., PÉREZ CASTAÑEDA G.-A., "Reliability assessment for complex systems operating in dynamic environment", *Annual Conference of the European Safety and Reliability Association, ESREL 2011*, Troyes, France, September 2011.

[BRI 11] BRISSAUD F., SMIDTS C., BARROS A., BÉRENGUER C., "Dynamic reliability of digital-based transmitters", *RESS*, vol. 96, 2011.

[CAS 08] CASSANDRAS C.G., LAFORTUNE S., *Introduction to Discrete Event Systems*, Springer Science, New York, 2008.

[DAV 93] DAVIS M.H.A., *Markov Models and Optimization*, Chapman & Hall, London, 1993.

[DAV 94] DAVID R., ALLA H., "Petri nets for modeling of dynamic systems", *A Survey Automatica*, vol. 30, no. 2, pp. 175-202, 1994.

[DUF 2002] DUFOUR F., DUTUIT Y., "Dynamic reliability: a new model", *$\lambda\mu$-13 and ESREL02*, vol. 1, pp. 350-353, 2002.

[DUT 97] DUTUIT Y., CHÂTELET E., SIGNORET J.P., THOMAS P., "Dependability modelling and evaluation using stochastic Petri nets: application to two test cases", *Reliability Engineering & System Safety*, vol. 55, no. 2, pp. 117-124, 1997.

[FAL 99] BENDOTTI P., FALINOWER C.M., "EDF benchmark for robust control techniques evaluations of proposed solutions", *14th World Congress IFAC*, Beijing, China, 1999.

[HAM 05] HAMIDI K., Contribution à un modèle d'évaluation quantitative des performances fiabilistes de fonctions électroniques et programmables dédiées à la sécurité, PhD thesis, Institut national polytechnique de Lorraine, Nancy, October, 27 2005.

[IRV 80] IRVING E., MIOSSEC C., TASSART J., "Towards efficient full automatic operation of the PWR steam generator with water level adaptive control", *Proceedings of the International Conference on Boiler Dynamics and Control in Nuclear Power Stations,* pp. 309-329, British Nuclear Energy Society, London, 1980.

[JEN 09] JENSEN K., KRISTENSEN L.M., *Coloured Petri Nets: Modeling and Validation of Concurrent Systems*, Springer-Verlag, Berlin-Heidelberg, 2009.

[KOT 96] KOTHARE M.V., METTLER B., MORARI M., BENDOTTI P., FALINOWER C.M., "Level control in the steam generator of a nuclear power plant. Decision and control", *Proceedings of the 35th IEEE*, pp. 4851-4856, December 1996.

[KOT 00] KOTHARE M.V., METTLER B., MORARI M., BENDOTTI P., FALINOWER C.M., "Level control in the steam generator of a nuclear power plant", KOTHARE M.V. *et al.*, *35th IEEE Transactions in Decision and Control*, January 2000.

[MAN 08] MANDELLI D., ALDEMIR T., KIRSCHENBAUM J., BUCCI P., MILLER D.W., STOVSKY M., EKICI E., ARNDT S.A., "A benchmark system for the reliability modeling of digital instrumentation and control systems", *International Probabilistic Safety Assessment and Management Conference PSAM 9*, Hong Kong, 2008.

[MUR 89] MURATA T., "Petri nets: properties, analysis and applications", *Proceedings of the IEEE*, vol. 77, no. 4, pp. 541-580, 1989.

[NAJ 07] NAJAFI M., NIKOUKHAH R., "Modeling hybrid automata in Scicos", *Multi-Conference on Systems and Control (MSC)*, Singapore, October 1-3, 2007.

[NUR 07] NUREG/CR-6942, *Dynamic Reliability Modeling of Digital Instrumentation and Control Systems for Nuclear Reactor Probabilistic Risk Assessments, US Nuclear Regulatory Commission*, Washington DC, 2007.

[PAR 85] PARRY A., PETRETOT J.F., VIVIER M.J., "Recent progress in SG level control in French PWR plants", *Proceedings of the International Conference on Boiler Dynamics and Control in Nuclear Power Stations*, 1985.

[PER 09] PEREZ CASTAÑEDA G.A., Evaluation par simulation de la sûreté de fonctionnement de systèmes en contexte dynamique hybride, PhD thesis. Institut national polytechnique de Lorraine, Nancy, March 30, 2009.

[PER 10] PÉREZ CASTAÑEDA G.A., AUBRY J.F., BRÎNZEI N., "Performance assessment of systems including conflict in the context of dynamic reliability", *International Journal of Adaptive and Innovative Systems*, , vol. 1, no. 3-4, pp. 233-247, 2010.

[PER 11] PÉREZ CASTAÑEDA G.A., AUBRY J.F., BRÎNZEI N., "Stochastic hybrid automata model for dynamic reliability assessment", *Proceedings of the Institution of Mechanical Engineers, Part O: Journal of Risk and Reliability*, vol. 225, no. 1, pp. 28-41, 2011.

[POC 08] POCK M., BELHADAOUI H., MALASSÉ O., WALTER W., "Efficient generation and representation of failure lists out of an information flow model for modelling safety critical systems", *Annual Conference of the European Safety and Reliability Association, ESREL 2008*, Valencia, Spain, 2008.

[SAP 10] DE SAPORTA B., DUFOUR F., GONZALEZ K., "Numerical method for optimal stopping of piecewise deterministic Markov processes", *Annals of Applied Probability*, vol. 20, no. 5, pp.1607-1637, 2010.

[SAP 11a] DE SAPORTA B., DUFOUR F., ZHANG H., ELEGBEDE C., "Optimal stopping for the predictive maintenance of a structure subject to corrosion", *Journal of Risk and Reliability*, 2011.

[SAP 11b] DE SAPORTA B., DUFOUR F., *Numerical Method for Impulse Control of Piecewise Deterministic Markov Processes*, Automatica, 2011.

[ZHA 08] ZHANG H., DUFOUR F., DUTUIT Y., ELEGBEDE C., "Application des processus déterministes par morceaux à un système de production pétrolière offshore", *Proceedings of λμ-16*, Avignon, France, 2008.

[ZHA 09] ZHANG H., DUFOUR F., DUTUITET Y., GONZALEZ K., "Piecewise deterministic Markov processes and dynamic reliability", *Journal of Risk and Reliability*, vol. 222, no. 4, pp. 545-551, 2009.

[ZIL 08] ZILLE V., BÉRENGUER C., GRALL A., DESPUJOLS A., LONCHAMPT J., "Multi-component systems modeling for quantifying complex maintenance strategies", *Proceedings of European Safety & Reliability Conference'08*, Valencia, Spain, pp. 3586-3591, 2008.

PART 3

Characterizing Background Noise, Identifying Characteristic Signatures in Test Cases and Detecting Noise in Reactors

A number of studies worldwide have been undertaken into fourth-generation reactors (Forum GEN IV). In this context, and given their results, work has focused on sodium-cooled fast reactors (SFR) with sodium as the caloporter fluid. Given the highly violent reaction between sodium and water, it is important to be able to detect even the slightest amount of water leaking into the sodium in a steam generator (SG) as early as possible. Research has therefore been carried out to apply continuous supervision techniques with a view to detecting any leak almost instantaneously. One of the means examined is that of passive acoustic detection of abnormal vibrations caused by a leak in the SG's external wall.

This section reexamines the acoustic signals recorded in 1994 in the SGs in the Scottish prototype fast reactor (PFR) with deliberate injections of gas into the liquid sodium. All of the GIS 3SGS partner laboratories have applied several techniques, particularly signal treatment and classification in order to:

– identify signatures (patterns) using acoustic signals collected during the reactor's operation and recognize them so that they can be used to detect the risk of boiling, leaks, shocks etc., as early as possible;

– determine the largest possible set of potential parameters characterizing signals;

– identify the most restricted subset possible that could distinguish the different classes (non-injection, injection of water, injection of argon);

– research particular signatures with time–frequency decomposition techniques (spectrogram, local Wigner-Ville transformation) and detect rupture (for faults in relation to a "normal" signal);

– detect changes in signals in order to recognize changes in operation using the matrix projection to determine model ruptures;

– perform analysis using the "components principle" method, which has a basic formulation and variables covering a number of situations;

– detect events that differ from measures considered to be standard.

Chapter 9

Aims, Context and Type of Signals Studied

Guaranteeing safe operation in fourth-generation liquid sodium cooled nuclear reactors involves monitoring steam generators (SG) in order to detect any water leak into the sodium. The sodium–water reaction is exothermic and produces sodium hydroxide (a source of corrosion) and hydrogen (an inflammable gas that causes an increase in sodium pressure).

Acoustic methods are potentially significant due to their very quick response time and have been the subject of both theoretical and experimental studies. Tests carried out in June 1994 in an evaporator module of the *Prototype Fast Reactor* (PFR) have consisted of injections of water, argon and hydrogen to categorize the noise created by the reaction between sodium and water.

A water or steam leak at high pressure (typically 180 bars in this type of apparatus) into the SG's sodium causes mechanical vibration in the fluid and the solid's structure. Coupling between the two media, which involves an exchange of mechanical energy, induces complex propagation of vibrations between the source and the external casing of the SG.

Passive acoustic detection consists of differentiating between the vibrations detected using accelerometers placed on the SG's outer wall when there is a leak

---

Chapter written by François BAQUÉ, Olivier DESCOMBIN, Olivier GASTALDI and Yves VANDENBOOMGAERDE.

from those corresponding to the normal state, which is considered background noise. This differentiation can be carried out using frequencies and/or amplitudes associated with this phenomenon.

Below is an illustration showing the transmission of elastic waves between the acoustic source and a sensor placed on the SG's outer wall.

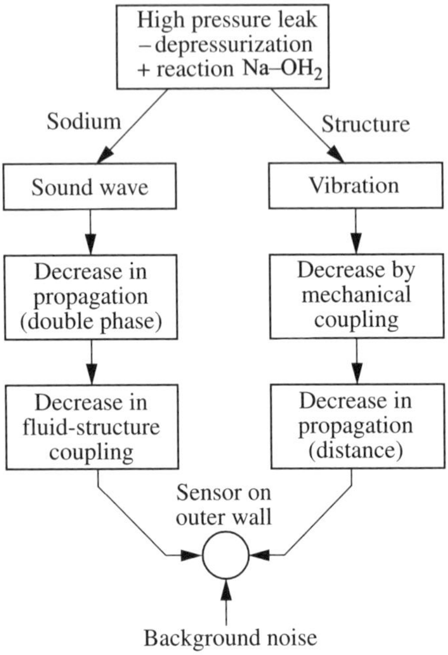

Figure 9.1. *Illustration of the transmission of elastic waves between the source (leak or injection) and a sensor placed on the SG's outer wall*

During the normal operation of a SG, different permanent and/or transitory phenomena create acoustic waves with an associated broad-band frequency spectrum. Below is a (non-exhaustive) list of these phenomena:

– water boiling;

– the flow of sodium;

– stress due to a thermal field;

– friction between structural elements (tube/tube support grid); and

– mechanical shocks of varying causes.

Motor pumps and, more generally, the other machines in the system also contribute to low-frequency background noise.

In the event of a water or steam leak, other phenomena related to the sodium–water reaction occur:

– vibrations in the structure caused by the high-speed ejection (through the cracking) of pressurized water or steam and its subsequent volumic expansion in the liquid sodium;

– the generation of pressure waves in the sodium due to turbulence caused by the jet bubbles (H_2O a H_2); and

– the emission of mechanical stress waves due to the rapid increase in temperature in the structures around the leak.

All of these phenomena result in the emission of acoustic waves with a broad-band frequency spectrum: the passive acoustic detection systems therefore function in a frequency domain that ranges from several dozen Hz to several hundred kHz, a band that allows the most precise distinction between the occurrences of leak or non-leak situations.

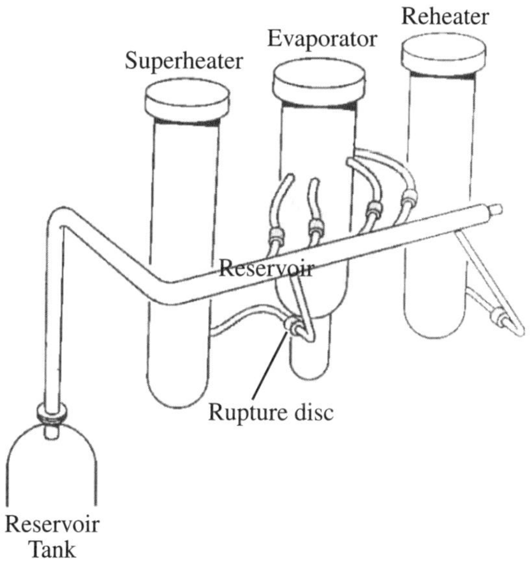

Figure 9.2. *A PFR steam generator*

The number of injections increased to more than 80. The signal was measured using six ENDEVCO 7704-17 accelerometers (with a resonance peak of around 45 kHz), each of them being fixed at the outer end of a wave guide that was welded onto the SG's external wall. The recording frequency band was [0–80] kHz.

The location of the six wave guides into the evaporator's outer wall is shown below. The location of these corresponds to WG 3, WG 5, WG 6, WG 7, WG8 and WG 10 sites.

Two injection systems have been used: one within the tube bundle; and the other in the space between the SG's casings.

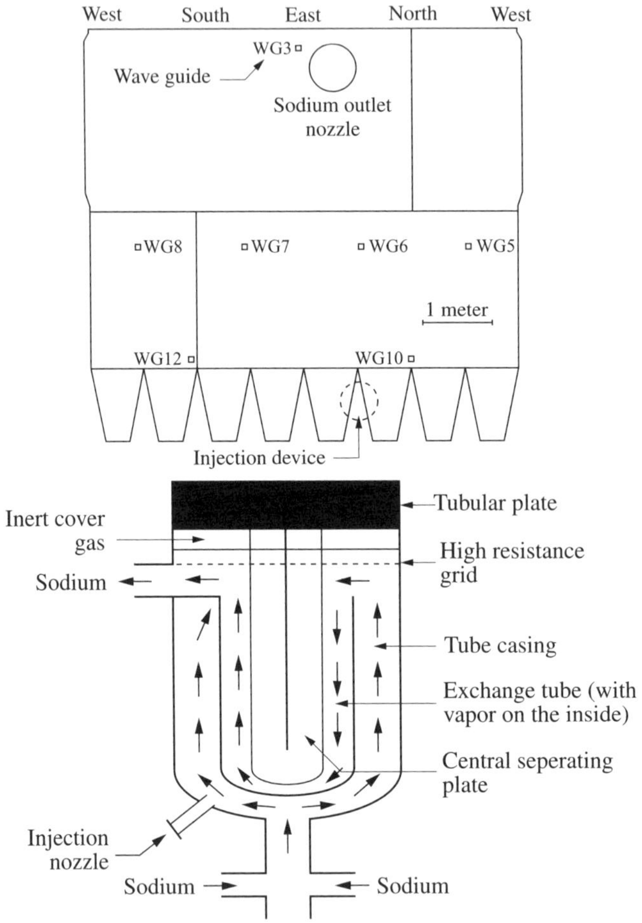

Figure 9.3. *Location of the six wave guides and the position of the injection device. The hydraulic sodium circuit in the PFR evaporator*

Figure 9.4 is a graphic representation of injections into a space with two dimensions of flow/pressure. Each point in the figure corresponds to an injection event in the SG. Note that the area representing a leak of water/steam into the sodium corresponds to high pressure and low flow conditions.

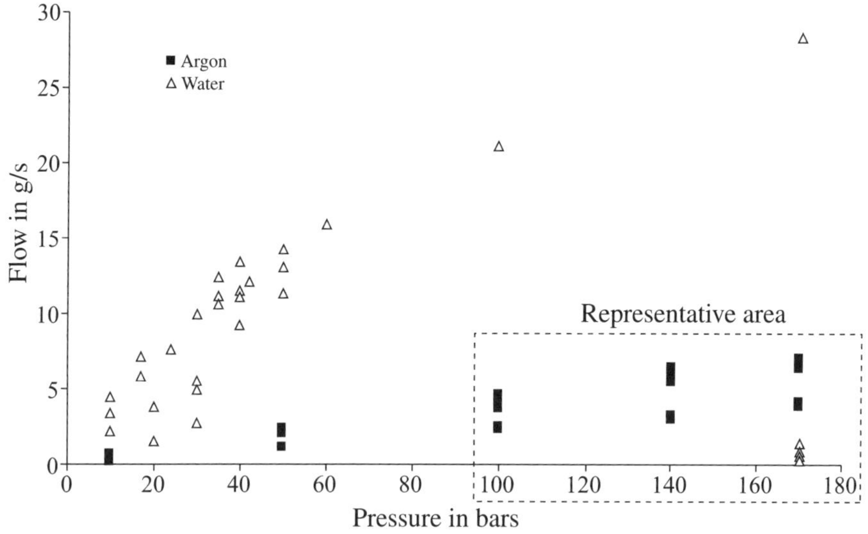

Figure 9.4. *Distribution of 80 injections into a two-dimensional flow/pressure space*

The numerical data used in the work associated with this current study are composed of 81 injections:

– 34 water injections; and

– 47 argon injections.

Figure 9.5 provides a graphic representation of the distribution of injections between argon and water and between injections within tube bundle and injections in the space between the SG's casings.

The signals recorded on magnetic tapes were digitalized by CEA in 1996 (sampling the signal at a frequency of 130 kH so as to have a useful bandwidth of up to 50 kHz). The systematic examination of signals' power spectral density (PSD) in the frequency band 0–60 kHz has shown that the majority of energy is concentrated in the band at around 0–5 kHz. The signals were under-sampled by a factor of 10, leading to the sampling frequency at Fe = 13 kHz. The analysis carried out in 1996 was conducted using the 0–6 kHz band.

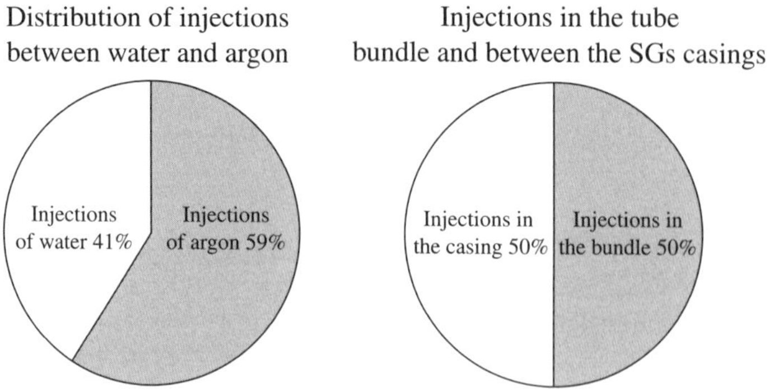

Figure 9.5. *Distribution of injections of argon according to their location*

An example of the distribution of PSD typically found with the injection of argon and water is shown in Figure 9.6.

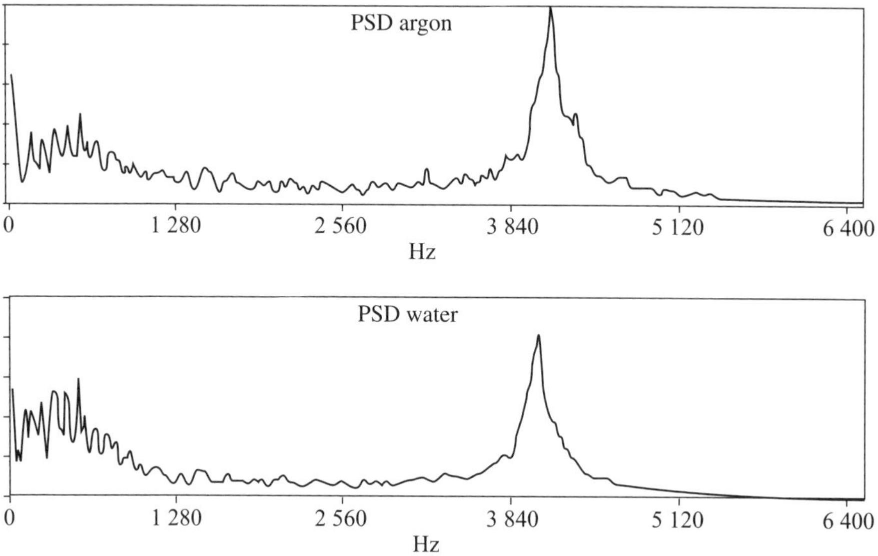

Figure 9.6. *Example of a PSD associated with injections of argon and water*

The 81 injection tests were used in 1996 to study a systematic audiophonic recording in order to label the database indicating the "clearness" (when the acoustic signal can easily be distinguished from background noise) associated with each injection noise. This subjective labeling can then be compared to analytical results. The signals were classed according to four categories:

– "clear": the injection is clearly audible;

– "slightly noticeable": the injection is not clearly distinguishable from the background noise;

– "inaudible": the injection is not clearly distinguishable from the background noise;

– "residual injection": the injection noise appears to be present from the beginning of the recording (between 0 and 10 sec).

Figure 9.7 provides a graphic representation of the distribution of injections according to the "clearness" criterion used in 1996.

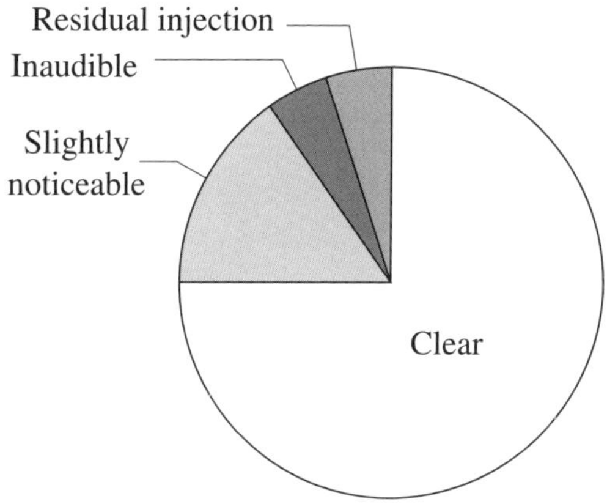

Figure 9.7. *Graphic representation of the distribution of injections according to the clearness criterion*

The injections represented by a vector of features (parameters) characteristic of the previously digitized signals have been classified using an unsupervised classification method.

This 1996 study has shown that the noise caused by water injections into the liquid sodium in a SG can be distinguished from the noises caused by argon injections: a recognition rate of 100% for argon and 70% for water has been obtained. In 1996 it was estimated that accounting for the granularity of signals significantly increases the recognition rate for water injections and decreases the rate of false alarms for argon injections.

It has been observed that no water injection carried out at pressures higher than 100 bars and with a flow of less than 10 g/s (representative area for SG water leaks in sodium reactors) was associated with any sound labeled as "clear" in 1996. This resulted in the analysis being re-examined in 2009 to improve signal recognition using "up to date" signal-processing tools.

Chapter 10

Detection/Classification of Argon and Water Injections into Sodium into an SG of a Fast Neutron Reactor

10.1. Context and aims

In the context of the research meant to develop a fourth-generation nuclear power plant, the use of liquid sodium as coolant is investigated. This solution would necessitate the development of specific monitoring tools. The reaction between (pressurized) water and sodium, a potential identified risk in steam generators (SGs), is a major issue in such a case.

To increase our knowledge about the acoustic response created by the reaction between sodium and water, we have studied signals measured during a specific experiment done at the shutdown of a PFR fast neutron reactor nuclear station in the UK in 1994. To simulate this chemical reaction, the AEA carried out tests using injections of argon and water into the liquid sodium in one of the SGs. This experiment aimed at testing the ability of the acoustic monitoring devices to detect a reaction between sodium and water (in case of a leak in one SG's tubes).

The study of these data, provided by the CEA [ORI 96, ORI 97a, ORI 97b], has a dual aim. On the one hand, it sought to assess the sensitivity of the acoustic monitoring system in the case of water–sodium reaction. On the other hand, it tried to compare the detectors' response to injections of water with those obtained using argon to determine the possibility, for monitoring system periodic controls, to simulate sodium–water reactions (consequences of which, are problematic for a new installation) using

---

Chapter written by Pierre BEAUSEROY, Edith GRALL-MAËS and Igor NIKIFOROV.

injections of argon into the sodium (as these injections do not produce heat emissions, their effect on the equipment is not problematic). As a result, the approach based on this substitution also allows us to test and characterize new monitoring systems designed for new generations of reactor.

The chapter first focuses on monitoring acoustic signals. Second, the problem of detecting and isolating injections of water and argon is introduced and the results of the accompanying feasibility study are presented. The third section introduces the characterization of the injection signals and the results of the obtained classifications. Finally, the concluding section summarizes the results of this study into detecting the reaction between sodium and water as well as the main differences between injections of water and argon observed with this data set.

10.2. Data

The monitoring device is composed of two passive acoustic sensors. Each record corresponds to a test and contains the 2 kHz sampled response of the two sensors and the command signal for injecting water or argon. The signals were filtered before analysis to reduce the number of bursts that disrupt them (Figure 10.1). Each test can contain from one to seven injections. Overall, the database allows the isolation of 43 argon injections and 30 water injections. The durations of injection and the fluid flow vary considerably among trials (from 10 to 600 s and 0.22 to 28.5 g/s).

Given the nature of the injection device [ORI 96, ORI 97a, ORI 97b], the delay between the injection command and the beginning of the fluid injection into sodium can vary greatly (ranging from several to 50 s). Therefore, it is very difficult to make any conclusion on detection delays because the injection starting time is known to have a very poor precision.

Detection of injections requires us to be able to classify the corresponding background noise at the nominal functioning stage. However, this can only be recorded during a fairly brief time lapse (a few seconds at most), which precedes the first injection in each test. As a result, little is known about it. The noise recorded after an injection is not usable as a noise reference because it is disrupted by the dispersal of residual bubbles. The brevity of the background noise recording is one of the main difficulties of this study.

A time-dependent spectral study has been carried out to analyze the temporal evolution of the spectral content of background noise (signal-off injection). Given the signal's slow evolution, this study has been carried out using spectrograms. It shows that, from a local point of view, the signal's frequential content varies over time, but on a larger scale it is independent of time (Figure 10.2). The signals are therefore assumed to be stationary on the points of interest (background noise, injection of water, injection of argon, etc.).

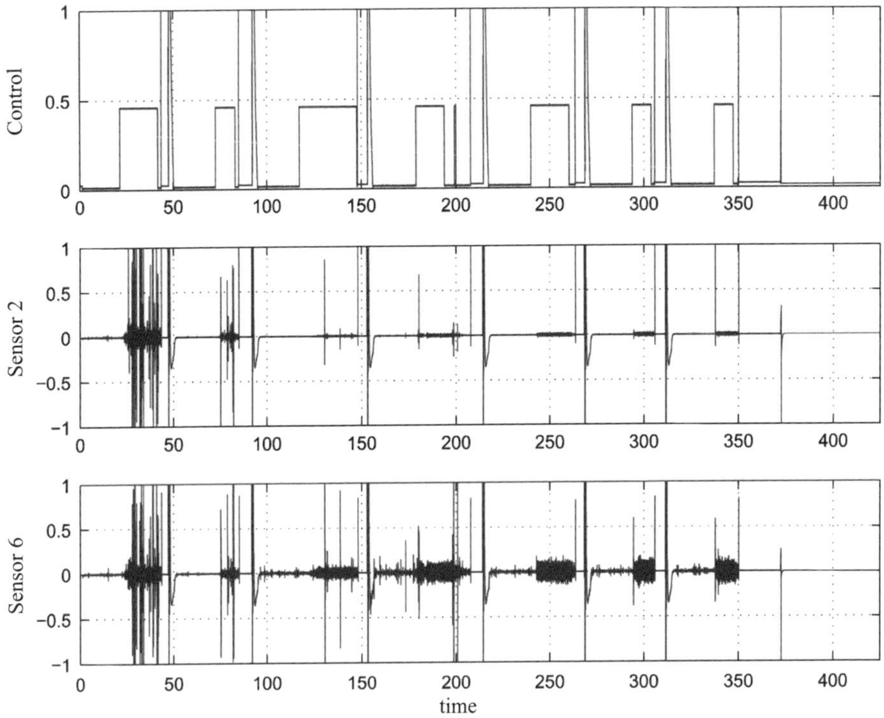

Figure 10.1. *Noise signals*

10.3. Online (sequential) detection-isolation

In this section, we examine the issue of online or sequential detection-isolation. This is followed by a brief analysis of existent methods of abrupt change detection-isolation in random processes. Finally, a recursive detection-location algorithm for the injection of argon or water into sodium in an SG and some experimental results concerning it are presented.

10.3.1. *Formulating the practical problem*

The problems requiring a solution are as follows:

– detection of argon or water injections into the sodium of a fast neutron reactor's SG;

– identification of the type of injection (argon or water);

with shortest possible detection-isolation delay under constraints on false alarms and false isolations.

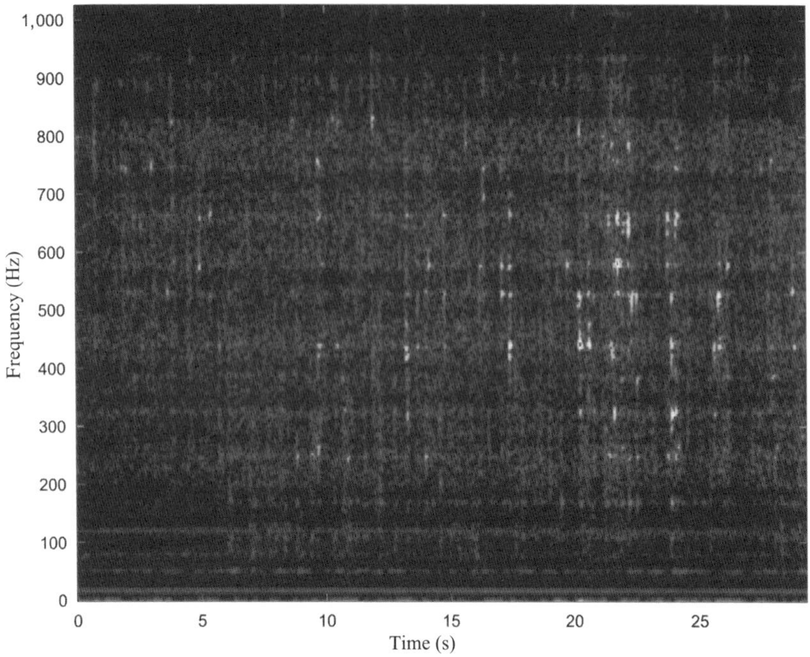

Figure 10.2. *Spectrograms of the signal pfr56*

We will present a model based on the hypothesis that the measures taken from each sensor $(y_k)_{k \geq 1}$ are defined by the following autoregressive-moving average (ARMA) equation (p, q):

$$y_k = \sum_{i=1}^{p} a_i y_{k-i} + \xi_k - \sum_{i=1}^{q} b_i \xi_{k-i}, \quad \xi_k \sim \mathcal{N}(0, \sigma^2)$$

A typical signal sample, prf44, after preprocessing by decimation and pre-filtering is shown in Figure 10.3. The point k_0 of an abrupt change or "rupture" in the model is marked by a vertical line. The change in the parameter vector is therefore:

$$\{a_i, b_i, \sigma^2\} = \begin{cases} \{a_i, b_i, \sigma^2\}_b & \text{if} \quad k < k_0 \\ \{a_i, b_i, \sigma^2\}_a & \text{if} \quad k \geq k_0 \text{ and injection of argon} \\ \{a_i, b_i, \sigma^2\}_e & \text{if} \quad k \geq k_0 \text{ and injection of water} \end{cases}$$

where $\{a_i, b_i, \sigma^2\}$ are the parameters of the ARMA model under different hypotheses.

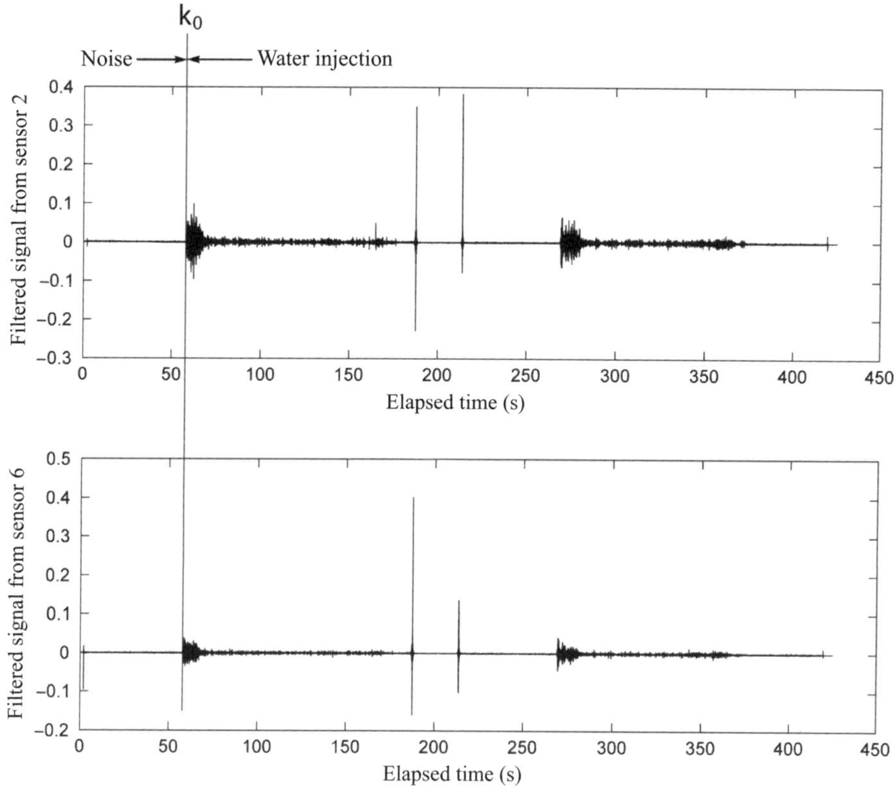

Figure 10.3. *Water sample pfr44: injection of water*

10.3.2. *Formulating the statistical problem*

Statistically speaking, the problem of sequential change detection-isolation can be represented as follows. Let us assume that there is a finite family of distributions $\mathcal{P} = \{P_i,\ i = 0, \ldots, K-1\}$ (with $K > 2$ hypotheses), whose densities are $\{f_i,\ i = 0, \ldots, K-1\}$, where $(Y_k)_{k \geq 1}$ is a sequentially observed independent random sequence:

$$Y_k \sim \left\{ \begin{array}{lll} P_0 & \text{if} & k < k_0 \\ P_l & \text{if} & k \geq k_0 \end{array} \right., \quad k_0 = 1, 2, \ldots$$

The change point k_0 and the type of change (identified by its label l) are unknown. The problem consists of *detecting* and *localizing* (identifying) the ruptures observed in the sequence $(Y_k)_{k \geq 1}$. The abrupt change detection-isolation algorithm calculates a *pair* (N, ν) using the $Y_1, Y_2, \ldots,$ where N is *the point of detection and isolation* of the rupture and $\nu \in (1, K-1)$ represents the *final decision*. The aim is to detect/identify a rupture with as short a delay and as few false alarms and isolations as possible.

10.3.3. *Non-recursive approach*

10.3.3.1. *Optimality criterion*

Let $P_{k_0}^l$ be the distribution of observations:

$$Y_1, Y_2, \ldots, Y_{k_0-1}, Y_{k_0}, Y_{k_0+1}, \ldots,$$

where $Y_1, \ldots, Y_{k_0-1} \sim P_0$ and $Y_{k_0}, Y_{k_0+1}, \ldots \sim P_l$. Examining the following criterion [NIK 95a, NIK 95b], we want the worst average detection delay:

$$\overline{\mathbb{E}}^*(N) = \sup_{k_0 \geq 1, 1 \leq l \leq K-1} \operatorname{esssup} \mathbb{E}_{k_0}^l (N - k_0 + 1 | N \geq k_0, Y_1, Y_2, \ldots, Y_{k_0-1}) \quad [10.1]$$

to be as low as possible in the following class:

$$\mathcal{K}_{T\beta} = \Big\{ (N, \nu) : \min_{1 \leq j \leq K-1} \mathbb{E}_0 \Big(\inf_{m \geq 1} \{N(m) : \nu(m) = j\} \Big) \geq T;$$
$$\max_{1 \leq l \leq K-1} \max_{1 \leq j \neq l \leq K-1} \mathbb{P}_1^l (\nu = j) \leq \beta_1 \Big\} \quad [10.2]$$

where $N(1), N(2), \ldots, N(m)$ is a series of false alarms, T is the minimal average time before a false alarm, and β_1 is the maximum probability of false isolation (exclusion).

10.3.3.2. *Non-recursive test*

The pair $(N_{\text{nr}}, \nu_{\text{nr}})$ (nr = non-recursive) is given by:

$$N_{\text{nr}} = \min\{N_{\text{nr}}^1, \ldots, N_{\text{nr}}^{K-1}\}, \quad \nu_{\text{nr}} = \arg\min\{N_{\text{nr}}^1, \ldots, N_{\text{nr}}^{K-1}\} \quad [10.3]$$

Using the following formula, we define N_{nr}^l :

$$N_{\text{nr}}^l = \inf \Big\{ n \geq 1 : \max_{1 \leq k \leq n} \min_{0 \leq j \neq l \leq K-1} [Z_{l,j}(n,k) - h_{l,j}] \geq 0 \Big\}, \quad [10.4]$$

$$Z_{l,j}(n,k) = \sum_{i=k}^{n} \log \frac{f_l(Y_i)}{f_j(Y_i)}$$

where $h_{l,j}$ are the detection-isolation thresholds.

10.3.3.3. *Non-recursive test performance*

The lower bound for the mean detection delay is defined by Theorem 10.1 (for further details, see [NIK 95a, NIK 95b]).

THEOREM 10.1.– *Defining the lower bound:*

$$n(T, \beta_1) = \inf_{(N,\nu)\in\mathcal{K}_{T\beta}} \left\{ \overline{\mathbb{E}}^*(N) \right\}$$

of the worst mean detection delay in the class $\mathcal{K}_{T\beta}$. The result is that:

$$n(T, \beta_1) \sim \max \left\{ \frac{\log T}{\rho_{\mathrm{d}}^*}, \frac{\log \left[n(T, \beta_1)\beta_1^{-1} \right]}{\rho_{\mathrm{I}}^*} \right\}$$

when $T \to \infty$ and $\beta_1 \to 0$ under the condition that $T\beta_1$ is approximately a constant, with:

$$\rho_{\mathrm{d}}^* = \min_{1\leq j\leq K-1} \rho_{j,0}, \quad \rho_{\mathrm{I}}^* = \min_{1\leq l\leq K-1} \min_{1\leq j\neq l\leq K-1} \rho_{l,j}$$

and

$$0 < \rho_{l,j} = \mathbb{E}_l \left[\log \frac{f_l(Y_k)}{f_j(Y_k)} \right] < \infty, \quad 0 \leq l \neq j \leq K-1.$$

THEOREM 10.2.– *The non-recursive algorithm [10.3]–[10.4] is asymptotically optimal in the class $\mathcal{K}_{T\beta}$.*

The optimality criterion [10.1]–[10.2] is generalized in the case of the dependent observations $(Y_k)_{k\geq 1}$ (e.g. in the case of ARMA models) in Lai's article [LAI 00].

The presently developed theory nevertheless has two disadvantages: its calculation charge (i.e. the number of "elementary" likelihood ratios at each given point) is significant and the optimality criterion does not take into account the probability of false isolation for a given change point ($k_0 > 1$). The simulations show that this probability strongly depends on the hypotheses' mutual "geometry" [NIK 00].

The sequential Bayesian detection-isolation is examined by [MAL 99, LAI 00]. The multi-hypothesis approach based on a sequential Shiryayev test has been proposed by Malladi and Speyer [MAL 99] adopting a dynamic programming approach.

Let us consider that $Q = \{q_1, \ldots, q_{K-1}\}$ is a *a priori* distribution of hypotheses after the abrupt change occurrence. Supposing that this distribution is independent of

k_0 and fixing an (*a priori*) distribution π of the change point k_0, Lai has proposed the following optimality criterion [LAI 00]:

$$\sum_{l=1}^{n}\sum_{k_0=1}^{\infty} q_l \pi(k_0)\mathrm{Pr}_{k_0}^{l}(k_0-1<N<\infty \cap \nu \neq l)$$
$$+\sum_{k_0=1}^{\infty}\pi(k_0)\mathrm{Pr}_0(N\leq k_0-1)\leq \alpha \qquad [10.5]$$

We can see here that $\mathrm{Pr}_0(N \leq k_0-1) = \mathrm{Pr}_{k_0}^{l}(N \leq k_0-1)$ because the event $\{N \leq k_0-1\}$ only depends on the observations $Y_1,\dots,Y_{k_0-1}$ and, therefore, the law $P_{k_0}^{l}$ gives the same distribution of $Y_1,\dots,Y_{k_0-1}$ as the law P_0. Lai [LAI 00] has established an asymptotic lower bound for an average "positive" delay to detection-isolation of each type of rupture $1 \leq l \leq K-1$ (using the Bayesian approach with $\alpha \to 0$):

$$\sum_{k_0=1}^{\infty}\pi(k_0)\mathbb{E}_{k_0}^{l}(N-k_0+1)^+ \geq \frac{|\log\alpha|}{\min_{0\leq j\neq l\leq n}\rho_{l,j}}(1+o(1)) \qquad [10.6]$$

Lai [LAI 00] has also introduced a second non-Bayesian approach based on a sequential window-limited test of size m, i.e. based on $Y_{t-m+1},\dots,Y_t$. For some kinds of application that are vital for security, we need to guarantee the probability of false alarms and isolations for a given time window (m_α) that is upper bounded by a given constant. The following lower bound has been set for the average "positive" delay for each type of abrupt change $1 \leq l \leq K-1$ (when $\alpha \to 0$):

$$\mathbb{E}_{k_0}^{l}(N-k_0+1)^+ \geq \left\{\frac{\mathrm{Pr}_0(N\geq k_0)}{\min_{0\leq j\neq l\leq n}\rho_{l,j}}+o(1)\right\}|\log\alpha| \qquad [10.7]$$

uniformly for $k_0 \geq 1$ under the following constraints:

$$\sup_{k_0\geq 1}\mathrm{Pr}_0(k_0-1<N\leq k_0-1+m_\alpha)\leq \alpha m_\alpha$$

and

$$\sup_{k_0\geq 1}\mathrm{Pr}_{k_0}^{l}(k_0-1<N\leq k_0-1+m_\alpha\cap\nu\neq l)\leq \alpha m_\alpha$$

for $1 \leq l \leq K-1$.

10.3.4. *Recursive approach*

Let us now examine another, recursive, approach, using a very low calculation charge that imposes constraints on the probability of false isolation when $k_0 > 1$ [NIK 00].

10.3.4.1. *Optimality criterion*

The first modification involves a new definition of the mean detection delay. In contrast to [10.1], it is defined now as:

$$\sup_{k_0 \geq 1} \mathbb{E}^l_{k_0}(N - k_0 + 1 \mid N \geq k_0)$$

instead of the previous definition:

$$\sup_{k_0 \geq 1} \operatorname{esssup} \mathbb{E}^l_{k_0}(N - k_0 + 1 | N \geq k_0, Y_1, Y_2, \ldots, Y_{k_0-1})$$

i.e. we can now estimate the mean detection delay using the following equation:

$$\overline{\mathbb{E}}(N) = \max_{1 \leq l \leq K-1} \overline{\mathbb{E}}_l(N), \;\; \overline{\mathbb{E}}_l(N) = \sup_{k_0 \geq 1} \mathbb{E}^l_{k_0}(N - k_0 + 1 \mid N \geq k_0) \quad [10.8]$$

The second modification is more important. Let us consider the following mode of observation: after a false alarm $N_r(m)$, we restart the algorithm (N_r, ν_r) at the point $n = N_r(m) + 1$. Therefore, we estimate the minimum mean time before the false alarm and the probability of false isolation using the following equations:

$$\min_{1 \leq j \leq K-1} \mathbb{E}_0 \left(\inf_{m \geq 1} \{N(m) : \nu(m) = j\} \right) = T, \; \max_{1 \leq l \leq K-1} \max_{1 \leq j \neq l \leq K-1} \sup_{k_0 \geq 1} \beta_{k_0}(j, l) = \beta, \quad [10.9]$$

where $\beta_{k_0}(j, l) = \mathbb{P}^l_{k_0}(\nu = j \neq l \mid N \geq k_0)$.

10.3.4.2. *Recursive test*

This recursive algorithm is given by the pair (N_r, ν_r), with:

$$N_r = \min\{N^1_r, \ldots, N^{K-1}_r\} \text{ and } \nu_r = \arg\min\{N^1_r, \ldots, N^{K-1}_r\} \quad [10.10]$$

The stopping time N^l_r is defined by the recursive formulas:

$$N^l = \inf \left\{ n \geq 1 : \min_{0 \leq j \neq l \leq K-1} [g_{l,0}(n) - g_{j,0}(n) - h_{l,j}] \geq 0 \right\} \quad [10.11]$$

$$g_{l,0}(n) = \left[g_{l,0}(n-1) + \log \frac{f_l(Y_n)}{f_0(Y_n)} \right]^+, \quad l = 1, \ldots, K-1$$

where $g_{0,0}(n) \equiv 0$, $x^+ = \max(0, x)$ and

$$h_{l,j} = \begin{cases} h_{\mathrm{d}} & \text{if} \quad l = 1, \ldots, K-1 \quad \text{and} \quad j = 0 \\ h_{\mathrm{l}} & \text{if} \quad j, l = 1, \ldots, K-1 \quad \text{and} \quad j \neq l \end{cases} \quad [10.12]$$

where h_{d} is the detection threshhold and h_{l} is the isolation threshold.

10.3.5. *Practical algorithm*

The aim is to identify the type of anomaly (the injection of water or argon) as soon as possible with few false alarms and false isolations.

After pretreatment, a recursive detection-isolation test [NIK 00] is carried out on each time step for each hypothesis and sensors. On the basis of residual errors $e_{k,b}$, $e_{k,e}$, and $e_{k,a}$ of ARMA, we calculate likelihood ratios to test the two alternative hypotheses $\mathcal{H}_e$ and $\mathcal{H}_a$ against the base hypothesis $\mathcal{H}_b$:

$$S_{k,e} = \frac{1}{2}\left[\log\frac{\sigma_{b,2}^2\sigma_{b,6}^2}{\sigma_{e,2}^2\sigma_{e,6}^2} + \frac{e_{k,b,2}^2}{\sigma_{b,2}^2} + \frac{e_{k,b,6}^2}{\sigma_{b,6}^2} - \frac{e_{k,e,2}^2}{\sigma_{e,2}^2} - \frac{e_{k,e,6}^2}{\sigma_{e,6}^2}\right] \quad [10.13]$$

$$S_{k,a} = \frac{1}{2}\left[\log\frac{\sigma_{b,2}^2\sigma_{b,6}^2}{\sigma_{a,2}^2\sigma_{a,6}^2} + \frac{e_{k,b,2}^2}{\sigma_{b,2}^2} + \frac{e_{k,b,6}^2}{\sigma_{b,6}^2} - \frac{e_{k,a,2}^2}{\sigma_{a,2}^2} - \frac{e_{k,a,6}^2}{\sigma_{a,6}^2}\right] \quad [10.14]$$

the detection functions:

$$G_{k,e} = \max\{0, G_{k-1,e} + S_{k,e}\} \text{ and } G_{k,a} = \max\{0, G_{k-1,a} + S_{k,a}\} \quad [10.15]$$

and isolation (classification) functions:

$$L_{k,e} = G_{k,e} - G_{k,a} \text{ and } L_{k,a} = G_{k,a} - G_{k,e} \quad [10.16]$$

The decision rule for the point k is the following:

– The hypothesis $\mathcal{H}_e$ is said to be accepted if the following conditions are satisfied:

$$G_{k,e} \geq h_d \quad \text{and} \quad L_{k,e} \geq h_l$$

– The hypothesis $\mathcal{H}_a$ is accepted if the following conditions are satisfied:

$$G_{k,a} \geq h_d \quad \text{and} \quad L_{k,a} \geq h_l$$

where the constants $h_d > 0$ and $h_l > 0$ are fixed beforehand.

10.3.6. *Experimental results*

The sequential detection-isolation procedure defined in the previous section has been applied to a number of samples from SG data. The parameters of the ARMA noise model $\{a_1, b_1, \sigma^2\}_b$ have been estimated using the first 40 s for each sample.

The first tests provide encouraging results. The behavior of the detection-isolation functions $G_{k,e}$, $G_{k,a}$, $L_{k,e}$, and $L_{k,e}$ is shown in Figure 10.4 in the case of an injection of argon with $k_0 = 60$ s.

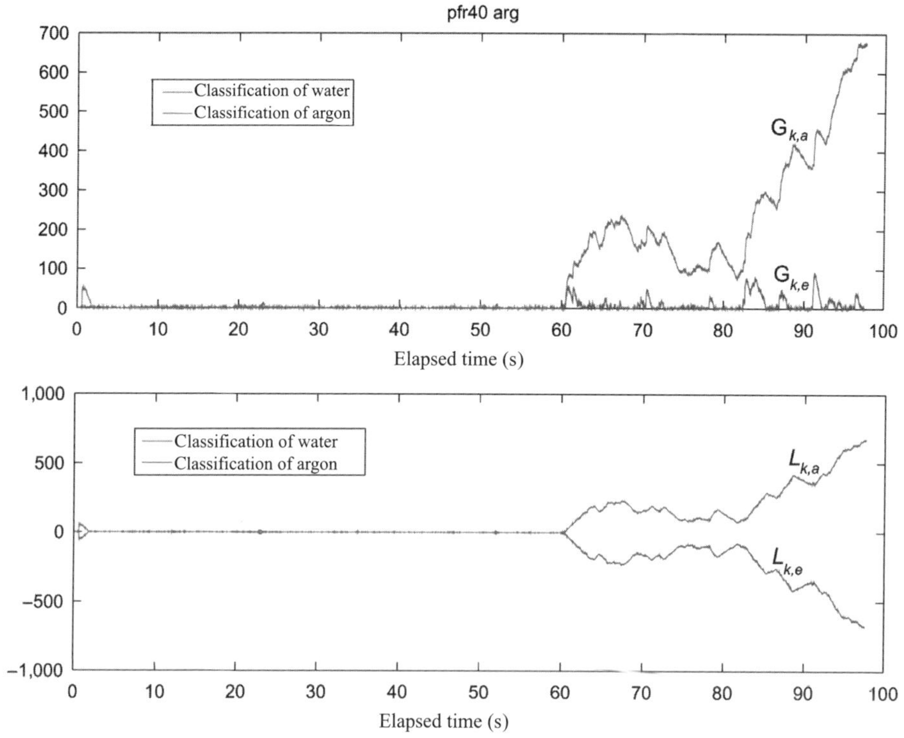

Figure 10.4. *Sample pfr40 arg: argon injections. The detection-isolation functions of the sequential test*

10.4. Offline classification (non-sequential)

10.4.1. *Characterization and approach used*

At the first part of the study, the detection capacity is not in doubt. However, the identification ability shows that the system's response to water injections is not identical to the response for argon injections. There is a wide range of options available to analyze these differences. Given that the signal stationarity hypothesis is fairly well confirmed, the results of an initial study [ORI 97b] have led us to analyze the spectral content of the injection signals.

10.4.2. *Initial characterization*

Figures 10.5 and 10.6 represent the mean normalized amplitude spectrums for the signals for argon and water injection for the two sensors. The mean normalized amplitude spectrum $SM((k/N)F_e)$ has been calculated as follows:

$$SM(\tfrac{k}{N}F_e) = \frac{1}{NB}\sum_{p=1}^{NB} SN_p(\tfrac{k}{N}F_e)$$

with

$$SN_p(\tfrac{k}{N}F_e) = \left(\frac{S_p(\frac{k}{N}F_e)}{\sum_k S_p(\frac{k}{N}F_e)}\right)^{\frac{1}{2}}$$

and

$$S_p(f) = \sum_{k=0}^{M_p/N-1}\left|\sum_{l=0}^{N-1} s_p\left(lT_e + kNT_e\right)e^{-2j\pi f(lT_e+kNT_e)}\right|^2$$

where *NB* is the number of injections, F_e the sampling frequency, $T_e = 1/F_e$, $s_p(t)$ is the p-th injection, M_p the number of samples from the signal s_p, and $j^2 = -1$. The difference between the mean spectrums is low. Comparison of the two spectrums taken from the two sensors shows a large amount of similarity and an almost identical structure (same modes). The SG seems to behave as a musical instrument that is affected by a variety of sounds. The reasoning modes and their harmonics are therefore affected in a similar way by injections of water and argon. The small differences observed can therefore be explained by the physical properties of the reactions between sodium and water and the reactions between sodium and argon that are not always completely identical.

Further examination of the spectrums shows that, on mean, the amplifiers' responses are greater at low frequencies with injections of water while high frequency components (from approximately 400 Hz) are more significant for injections of argon. Furthermore, we can see that interspectral disparity is significant and is more important than the difference between water and argon (Figure 10.7). This statement seems to contradict the conclusion in first part of the study and shows that characterization can be crucial for discriminating between the two types of injections.

To understand the previously observed discrimination, for each sensor we have separated the transients corresponding to the injections in 25 frequency ranges centered on each mode and calculated the energy in these previously defined energy ranges.

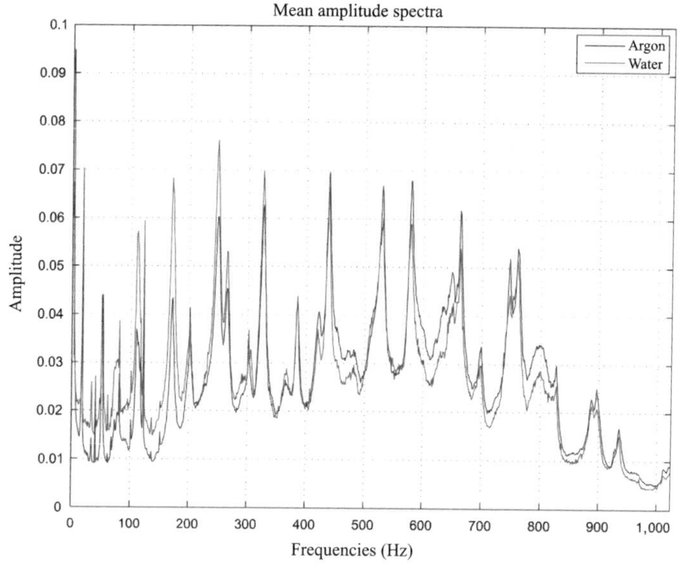

Figure 10.5. *Mean normalized amplitude spectrum of argon and water injections*

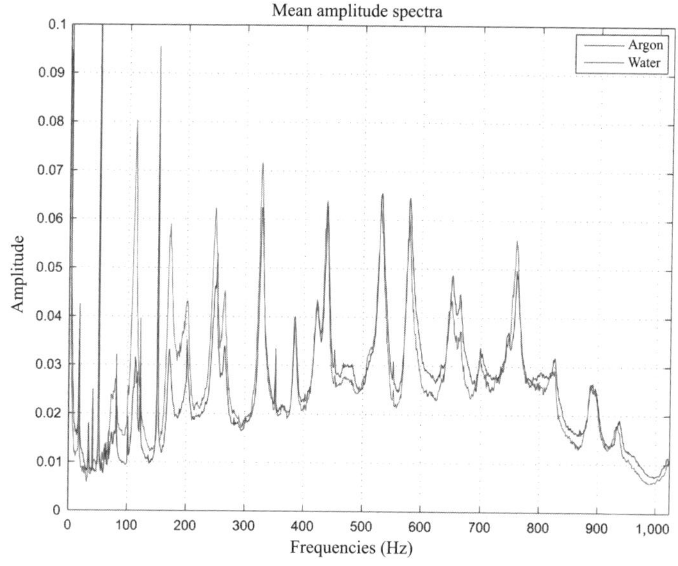

Figure 10.6. *Mean normalized spectrum of argon and water injections*

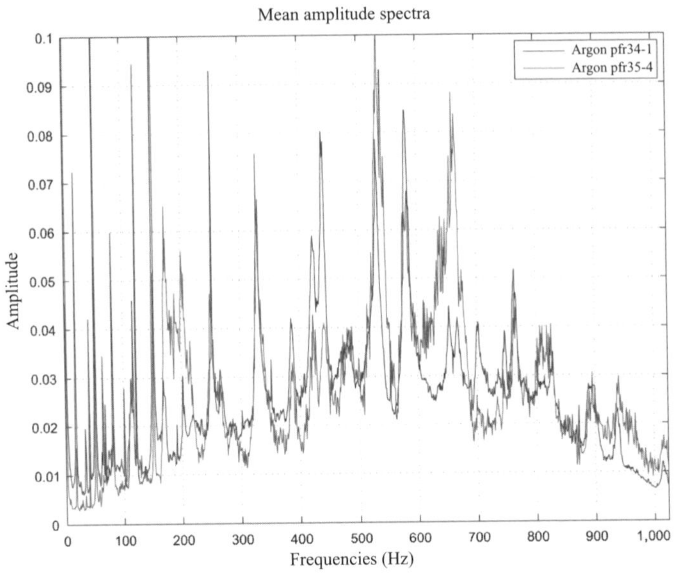

Figure 10.7. *Mean normalized amplitude spectrum for the two different argon injection signals*

To avoid considering the signal's amplitude that can vary significantly depending on the position of the leak and its size, the spectral amplitude of each mode is calculated after normalizing the amplitude spectrum according to its content in the frequency range 200–1,024 Hz. The restriction on the normalization calculation time for this range was set after reading the previous CEA reports. These testified to the appearance of components between 50 and 150 Hz that are likely to be in relation with the alternating current's frequency. By restricting the normalizing range, these components are excluded and do not affect the result.

10.4.3. *Effective features*

To identify more precisely the effective features, an analysis of the distribution of features was carried out in two stages. Since the two sensors are similar, we have considered to keep the same features for each of the sensors.

During the first stage, the feature values obtained from the 73 samples have been selected by observing marginal distributions. This suboptimal step was chosen due to the low number of examples in the database. It allowed us to identify nine features per sensor showing different marginal distributions for the two classes to be characterized (Figure 10.8).

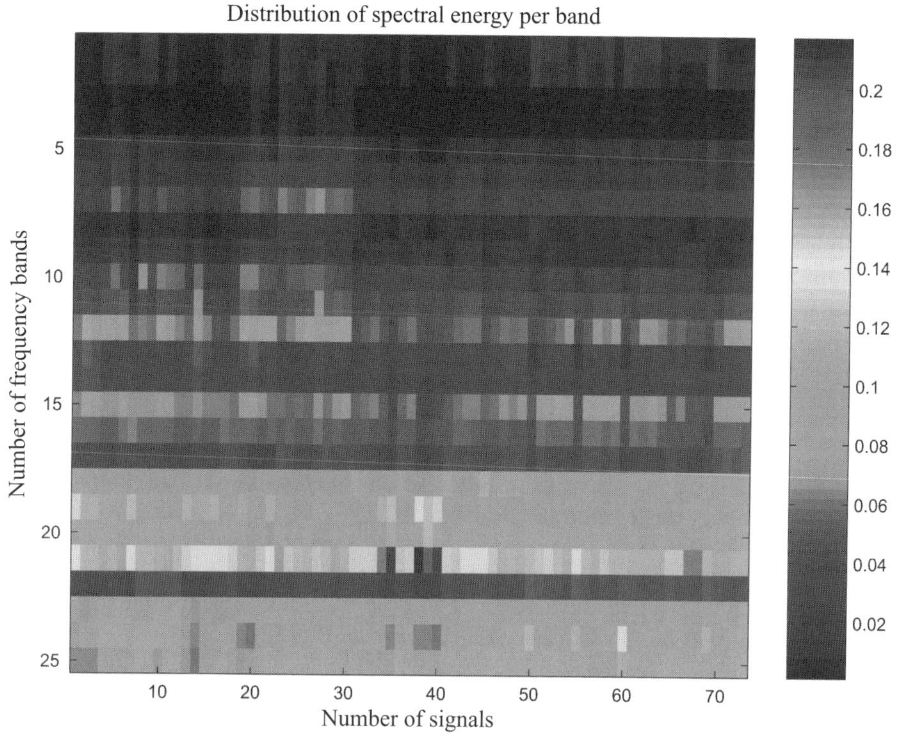

Figure 10.8. *Normalized amplitude in the 25 identified frequency ranges – 30 injections of water followed by 43 injections of water – sensor 2*

Features correspond to the following ranges: frequency range numbers 5(45, 64) 6(65, 89) 7(90, 119) 10(160, 184) 11(185, 224) 12(225, 256) 18(430, 489) 19(490, 549) 21(600, 679). These are ranges where a rupture can be seen in the figure between the 30 initial values (water) and the following 43 (argon) values.

The resulting representation space therefore has 18 spectral features (9×2 sensors) for 73 examples. No other feature has been added. We will see in the following section that this characterization allows us to obtain convincing classification results.

10.4.4. *Classification*

The aim is to identify the most effective characteristics among the nine selected features.

Given the relatively limited number of features (nine per sensor) and the fact that we have chosen to keep the same features for each sensor, it was possible to apply an exhaustive research method. The method consisted of creating all the possible combinations of the features (in the order of around $2^9 - 1$) making a decision rule, evaluating its performance, and identifying the best groups of features from these results.

Learning the decision rule was achieved using a two-class support vector machine (SVM) [CRI 00, BOS 92, COR 95]. This type of method consists of identifying a hyperplane separating the two classes. Therefore, we want to create a decision rule in the form $\mathcal{D}(x) = \text{sign}\,(g(\boldsymbol{x}))$ with $g(\boldsymbol{x}) = \langle \boldsymbol{x}, \mathbf{w} \rangle + b$, where $\mathbf{w}$ is a vector normal to the separating hyperplane, b is a threshold and $\boldsymbol{x}$ is an observation of $\mathbb{R}^d$. The risk r that must be minimized is given by:

$$r\,(\mathbf{w}, \mathcal{A}_n, C) = \frac{C}{n} \sum_{i=1}^{n} \xi_i + \frac{1}{2} \|\mathbf{w}\|^2$$

where ξ_i is a real positive whose value is in relation with the risk of a bad classification of $\boldsymbol{x}_i$, an element of the learning set $\mathcal{A}_n$.

The parameter C allows one to control the trade off between the training error and the model complexity. A small value for C will increase the number of training errors, while a large C will lead to a solution that is complex.

The optimization problem to determine the function $g(\boldsymbol{x})$ can be written in the form:

$$\begin{aligned} &\min_{\mathbf{w},\xi} \frac{C}{n} \sum_{i=1}^{n} \xi_i + \frac{1}{2} \|\mathbf{w}\|^2 \\ &\text{with } y_i\,(\langle \boldsymbol{x}_i, \mathbf{w} \rangle + b) \geq 1 - \xi_1 \quad \forall i = 1..n \\ &\qquad \xi_i \geq 0 \quad \forall i = 1..n \end{aligned}$$

where $y_i = -1$ if $\boldsymbol{x}_i \in \omega_0$ and $y_i = 1$ if $\boldsymbol{x}_i \in \omega_1$.

The solution to this problem is obtained by resolving the dual problem that can be expressed by the following quadratic form:

$$\begin{aligned} &\max_{\alpha} \sum_{i=1}^{n} \alpha_i - \frac{1}{2} \sum_{i,j=1}^{n} y_i y_j \alpha_i \alpha_j \langle \boldsymbol{x}_i, \boldsymbol{x}_j \rangle \\ &\qquad 0 \leq \alpha_i \leq \frac{C}{n} \quad \forall i = 1..n \\ &\qquad \sum_{i=1}^{n} \alpha_i y_i = 0 \end{aligned} \qquad [10.17]$$

and $\mathbf{w} = \sum_{i=1}^{n} y_i \alpha_i \boldsymbol{x}_i$.

The analytical form of the decision function g is given by:

$$g_{\mathcal{A}_n,C}(\boldsymbol{x}) = \sum_{i=1}^{n} y_i \alpha_i \langle \boldsymbol{x}_i, \boldsymbol{x} \rangle + b$$

These results can be generalized as nonlinear classification problems. The idea is to apply a linear discrimination method such as that described previously in a transformed space $\mathcal{X}_q$ with a dimension q. This transformed space is obtained by mapping the data from the initial space using a parameter function h: $\phi_h : \mathbb{R}^d \longrightarrow \mathcal{X}_q$ with $q > d$. Since the optimization problem and the decision function are expressed only according to the Lagrange multipliers α and the scalar products $\langle \boldsymbol{x}_k, \boldsymbol{x}_l \rangle$, the problem expressed in the transformed space only requires the definition of a scalar product. Every function K_h verifying the Mercer theorem can be used to calculate the scalar product in a transformed space without expressing the transformation function ϕ_h. The decision function in this case is written in the form:

$$g_{\mathcal{A}_n,h,C}(\boldsymbol{x}) = \sum_{i=1}^{n} y_i \alpha_i K_h(\boldsymbol{x}_i, \boldsymbol{x}) + b$$

where b is the bias and $\boldsymbol{x}_i \in \mathbb{R}_d$.

The kernel functions K_h generally depend on a parameter h that has significant influence on the class of detectors obtained for a set of values for the parameter C. Thus, it is required to determine C and h. In practice, the cross-validation method is commonly employed.

10.4.5. *Performance evaluation*

The performance of a learned rule is evaluated by estimating the probability of error. An empirical estimator is often used (ratio of the number of errors to the number of examples tested). To avoid a learning bias, the test data set is generally chosen independent of the learning data set. The most common performance evaluation method is V cross-validation [BRE 96]. This consists of randomly dividing the learning set $\mathcal{A}_n$ into V mutually exclusive subsets of approximately identical sizes: $\mathcal{A}_n = \mathcal{A}1 \cup \mathcal{A}2 \cup \ldots \cup \mathcal{A}V$. Typically, V = 5 or 10. At stage k, the decision rule is learned for the reduced learning set $\mathcal{A}_n - \mathcal{A}k$, and the performance criterion (the probability of error in our case) is estimated using $\mathcal{A}k$. Estimation of the global criterion is achieved by empirically averaging the estimated values obtained for each set $\mathcal{A}k$. This cross-validation method allows us to decrease the estimation bias in comparison with the simple validation method. A particular type of V cross-validation is the *leave-one-out* procedure for which $V = n$.

10.4.6. *Experimental results*

10.4.6.1. *Kernels*

The kernel functions K_h selected for the SVMs are Gaussian functions and have the form:

$$K_h(x_1, x_2) = \exp\left(-\frac{1}{2h^2}\|x_1 - x_2\|^2\right)$$

The decision function depends on two parameters (C, h) that are optimized on a set of possible values. Given the small number of examples, the performance has been evaluated by the *leave-one-out* procedure [DEV 82].

10.4.6.2. *Results*

The best result led to an error of classification which is equal to zero, that is $\hat{P}_e = 0$.

Several subspaces lead to this result. Among them, the spaces with the smallest dimension have three features per sensor (six in total). Table 10.1 shows the different frequency ranges for which this result has been obtained.

Space	Frequency ranges used
E_1	(185, 224) - (490, 549) - (600, 680)
E_2	(90, 119) - (185, 224) - (600, 680)
E_3	(90, 119) - (160, 184) - (430, 489)
E_4	(65, 89) - (185, 224) - (600, 680)
E_5	(45, 64) - (185, 224) - (600, 680)

Table 10.1. *Subsets of features leading an estimation of zero error*

These results confirm that despite the close similarity between the spectrums, the injections of argon and water can be clearly differentiated. Notably, they show that the joint distributions of features in the five subspaces differs significantly from one class to another.

A study based on the PCA (principal component analysis) has confirmed the discriminating nature of these subspaces (Figure 10.9).

We hypothesize that the detection methods have fulfilled their role and we therefore seek to differentiate between injections of water and argon. This problem can be formalized as a dual hypothesis test $\mathcal{H}_e$ against $\mathcal{H}_a$. Therefore, this consists of developing a decision rule that allows us to carry out this learning test using the available data.

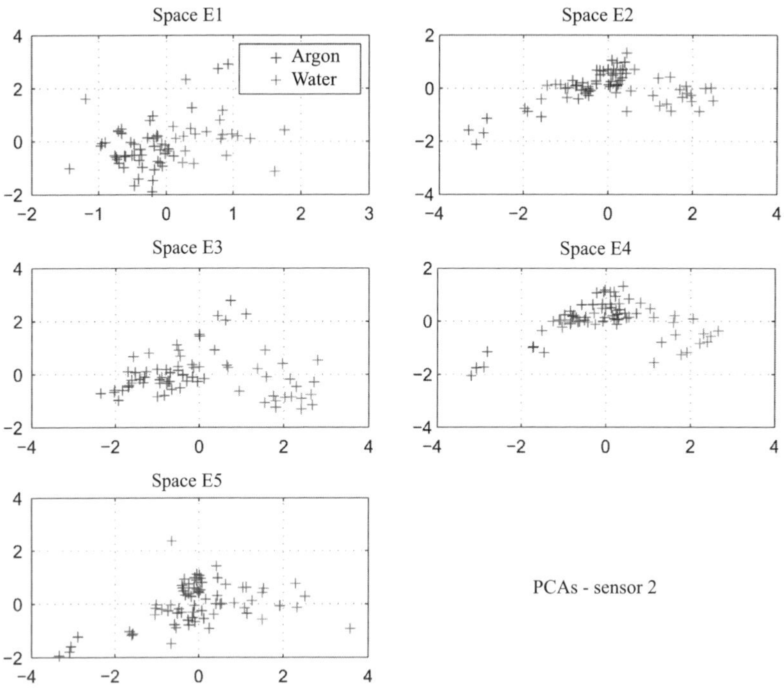

Figure 10.9. *Analysis of the main components in the five selected representation spaces – projection on the two initial inertia axes – sensor 2*

10.5. Results and comments

Several examples of detection-isolation results are shown, which clearly demonstrate the capacity of the tested device to correctly detect instationarities. Classification performance is indicated for several representation spaces and the sensitivity to learning parameters is discussed as well as the possible interpretation of features that allow us to distinguish between the classes "argon" and "water".

10.6. Conclusion

The research presented shows the possibility of detecting a reaction in sodium using acoustic sensors. The detection delay guaranteeing a high detection rate has not been evaluated, given the scarce amount of data and uncertainty on the time of injections (the command-injection delay is highly variable). They also show that acoustic emissions of water and argon injections are significantly different at some frequencies. These results show that argon is not a perfect substitute for water when testing the calibration of a water–sodium reaction detection system. It is necessary to better understand these differences and limit their impact. However, these results also show that the passive acoustic detection device is highly sensitive and that it carries important information on the reactor's operation.

10.7. Bibliography

[BEL 61] BELLMAN R., *Adaptive Control Processes: A Guided Tour*, Princeton University Press, Princeton, 1961.

[BLU 97] BLUM P.L.A., "Selection of relevant features and examples in machine learning", *Artificial Intelligence*, vol. 97, nos. 1–2, pp. 245–271, 1997.

[BOS 92] BOSER B., GUYON I., VAPNIK V., *Training Algorithm for Optimal Margin Classifiers*, Pittsburgh, PA, pp. 144–152, 1992.

[BRE 96] BREIMAN L., "Bagging predictors", *Machine Learning*, vol. 24, no. 2, pp. 123–140, 1996.

[COR 95] CORTES C., VAPNIK V., "Support-vector networks", *Machine Learning*, vol. 20, no. 3, pp. 273–297, 1995.

[CRI 00] CRISTIANINI N., SHAWE-TAYLOR J., *An Introduction to Support Vector Machines and Other Kernel-Based Learning Methods*, Cambridge University Press, Cambridge, 2000.

[DEV 82] DEVIJVER P.A., KITTLER J., *Pattern Recognition: A Statistical Approach*, Prentice-Hall, London, 1982.

[DUB 90] DUBUISSON B., *Diagnostic et reconnaissance des formes*, Hermès, Paris, 1990.

[LAI 98] LAI T.L., "Information bounds and quick detection of parameter changes in stochastic systems", *IEEE Transactions on Information Theory*, vol. 44, no. 7, pp. 2917–2929, 1998.

[LAI 00] LAI T.L., "Sequential multiple hypothesis testing and efficient fault detection-isolation in stochastic systems", *IEEE Transactions on Information Theory*, vol. 46, no. 2, pp. 595–608, 2000.

[MAL 99] MALLADI D.P., SPEYER J.L., "A generalized Shiryayev sequential probability ratio test for change detection and isolation", *IEEE Transactions on Automatic Control*, vol. 44, no. 8, pp. 1522–1534, 1999.

[NIK 95a] NIKIFOROV I., "A generalized change detection problem", *IEEE Transactions on Information Theory*, vol. 41, no. 1, pp. 171–187, January 1995.

[NIK 95b] NIKIFOROV I., "On two new criteria of optimality for the problem of sequential change diagnosis", *Proceedings of the American Control Conference*, Seattle, WA, pp. 97–101, 1995.

[NIK 00] NIKIFOROV I., "A simple recursive algorithm for diagnosis of abrupt changes in random signals", *IEEE Transactions on Information Theory*, vol. 46, no. 7, pp. 2740–2746, November 2000.

[ORI 96] ORIOL P.G.L., Analyse des enregistrements acoustiques des essais d'injections PFR en vue de la caractérisation de la réaction sodium-eau, CEA, Technical Report, June 1996.

[ORI 97a] ORIOL L., Analyse basse fréquence des essais PFR, CEA, Technical Report, December 1997.

[ORI 97b] ORIOL S.E.L., DEMARAIS R., Base de données numérique des enregistrements acoustiques des injections d'argon et d'eau faites dans un GV de PFR, CEA, Technical Report, December 1997.

[VAP 95] VAPNIK V., *The Nature of Statistical Learning Theory*, Springer Verlag, New York, 1995.

[VAP 98] VAPNIK V.N., *Statistical Learning Theory*, John Wiley & Sons, New York, 1998.

Chapter 11

A Dynamic Learning-based Approach to the Surveillance and Monitoring of Steam Generators in Prototype Fast Reactors

This research focuses on the surveillance and monitoring of evolving systems [ANG 04, LUG 11b, KAS 07, ANG 10, LUG 11a] using learning methods and dynamic classification. Like any evolving system, it changes from one mode to another suddenly (a jump) or progressively (a drift) over time. This evolution is the result of changes in the system due to a leak, damage to equipment or an adjustment, etc. When a static pattern-recognition method is used to construct models of classes for an evolving system, it allows us to classify new observations by comparing them to existing ones. It does not, however, take into account new characteristic information that is used to update models of classes (membership functions). As a result, a static classification system is not very well suited to representing the current characteristics of an evolving system. It is for this reason that this chapter applies the method that we propose for a steam generator in a prototype fast reactor. This method, based on the fuzzy K-nearest neighbors (FKNN) method [KEL 85] is semi-supervised and is called a semi-supervised dynamic fuzzy K-nearest neighbors (SS-DFKNN) method. This allows us to consider new information about an evolving system, detect unknown classes and adapt their characteristics. The SS-DFKNN method was developed to detect and monitor the evolution of dynamic classes online, adapt the latter and anticipate the current characteristics of a system.

Chapter written by Laurent HARTERT, Moamar SAYED-MOUCHAWEH and Danielle NUZILLARD.

11.1. Introduction

The monitoring and supervising of evolving systems requires continuous learning over time to account for evolutions and changes in their environment. Incremental learning methods are an effective solution for carrying out continuous learning. These methods allow the integration of information online and improve the estimation of class models. However, these methods adapt class models without challenging previous findings. They consider all patterns to be representative of classes. Indeed, some of these patterns become obsolete and are not designed to update class models. This therefore requires the use of a mechanism that allows us to ignore obsolete patterns and only keep patterns that are characteristic of classes after evolution. These methods are called dynamic learning and classification methods. The challenge with these methods is to select patterns that are representative of changes in class and avoid a catastrophic loss.

In the literature, class models are adapted by directly acting on the classifier's parameters by substituting or adding certain recent patterns that are representative of the learning set according to the state (stable, slow or rapid change) of the system [ANG 04, ANG 00, NAK 97]. This adaptation uses a sliding window, a selection criterion or a forgetting factor.

A sliding window, of either fixed or variable size, allows us to reduce or limit the increasing size of a database by accepting only the most recent *n* patterns [NAK 97]. The size of the sliding window must be well selected to obtain a compromise between a rapid adaptation and a sufficient number of representative patterns.

A selection criterion allows us to select patterns according to their age and usefulness [GIB 94]. The pattern's age cannot be the only selection criterion, since some patterns can be nonsensical or correspond to noise, etc. The usefulness of a pattern can be defined by its corresponding change in the system. For example, if one of the system's parameters evolves significantly following a change that is characteristic of the system, the displacement carried out by the patterns for this parameter corresponds to the significance of the evolution. Nevertheless, it is difficult to estimate the usefulness of patterns.

A forgetting factor is calculated on the patterns to detect those that are less characteristic of the system's functioning. This forgetting factor can occur at a constant or variable rate so that each pattern has a usefulness value of between 0 and 1.

Other methods adapt the parameters of their classifier as well as their structure and are known as evolving neural networks [AMA 06, COH 05, LEC 03]. In [ANG 04], the *evolving Takagi-Sugeno method* has been developed to account for the evolution of data. In the evolving Takagi-Sugeno method, a potential function is based on the distance between patterns. The potential of the first point is considered as being equal to one. It establishes the first neuron (or rule) that is considered a prototype (or center) in the first class. The new data may then have a potential that is close to or greater than this neuron. They therefore reinforce or confirm the information contained in the previous neurons. In [AMA 06, LEC 03], the neural network is based on the multi-prototype Gaussian modeling of non-convex classes. The activation function of each hidden neuron determines an observation's degree of belonging to a class prototype. During the initialization of the method, the first pattern allows us to create the first prototype, which constitutes the first class. The prototype is characterized by its center and its initial covariance matrix. Depending on the belonging values of the new acquisitions, the prototype (hidden neuron) can be adapted, eliminated or a new prototype can be created.

We have chosen to develop an approach based on the FKNN method [KEL 85]. This is well known and is often used for automatic learning applications. The method developed is known as the SS-DFKNN method that, as mentioned, allows us to take into account new information about an evolving system, detect unknown classes and estimate their characteristics. Semi-supervised methods are particularly well adapted to evolving systems where classes are not known in advance. As with FKNN, it uses the notion of distance between patterns to classify new data. Two indicators of evolution are calculated in relation to class parameters. These indicators are used in the detection and confirmation phase for class evolutions. During the adaptation phase of classes that have evolved, most informative patterns are selected and the classifier is updated. DFKNN allows us to respond to problems corresponding to dynamic classifiers such as cases of class drift, fusion, splitting and rotating.

11.2. Proposed method for the surveillance and monitoring of a steam generator

In this section, a SS-DFKNN approach that monitors the evolution of dynamic classes is examined. This method [HAR 10] has been developed to detect the evolution of dynamic classes online, adapt their characteristics and detect the appearance of new classes. This version is semi-supervised with the aim of:

– taking into consideration the initial patterns known in a system, such as the learning set X that represents the known classes;

– improve the characteristics of classes using new patterns; and

– detect new classes according to the new characteristics created by the evolution of a system.

The evolution of classes can even be considered in parts of the representation space where no pattern has been learned. In this version, a class that begins to move will retain its initial patterns while patterns corresponding to an evolution of this class will constitute a new class. It is this new class that will allow us to best estimate the system's current functioning mode. The main functioning phases for the SS-DFKNN method are shown in Figure 11.1.

11.2.1. *Learning and classification*

To start, labeled data are initially learned. For each class learned C_i, at least two patterns must be known to calculate its initial center of gravity ($\overline{x}_i^j$) as well as its standard deviation (σ_i^j) for each attribute j. These two values will be taken into account when calculating the method's two evolution indicators. For each time t, they are updated incrementally by:

$$\sigma_i^j(t) = \sqrt{\frac{N_i - 1}{N_i} \times \sigma_i^j(t-1)^2 + \frac{(x^j - \overline{x}_i^j(t-1))^2}{N_i + 1}} \qquad [11.1]$$

and:

$$\overline{x}_i^j(t) = \frac{\overline{x}_i^j(t-1) \times N_i}{N_i + 1} + \frac{x^j}{N_i + 1} \qquad [11.2]$$

where N_i is the number of patterns in C_i before the classification of x. $\sigma_i^j(t-1)^2$ and $\overline{x}_i^j(t-1)$ are the class's variance and center of gravity according to attribute j before the classification of x.

It is also important to specify that for each type of class, $\overline{x}_i^j(t)$ and $\sigma_i^j(t)$ can be calculated. In the case of classes of complex patterns, several Gaussian subclasses are created and merged in order to obtain other kinds of classes.

In the method's classification phase, each new pattern is classified sequentially according to its k nearest neighbors. It is therefore necessary to initially define parameter k.

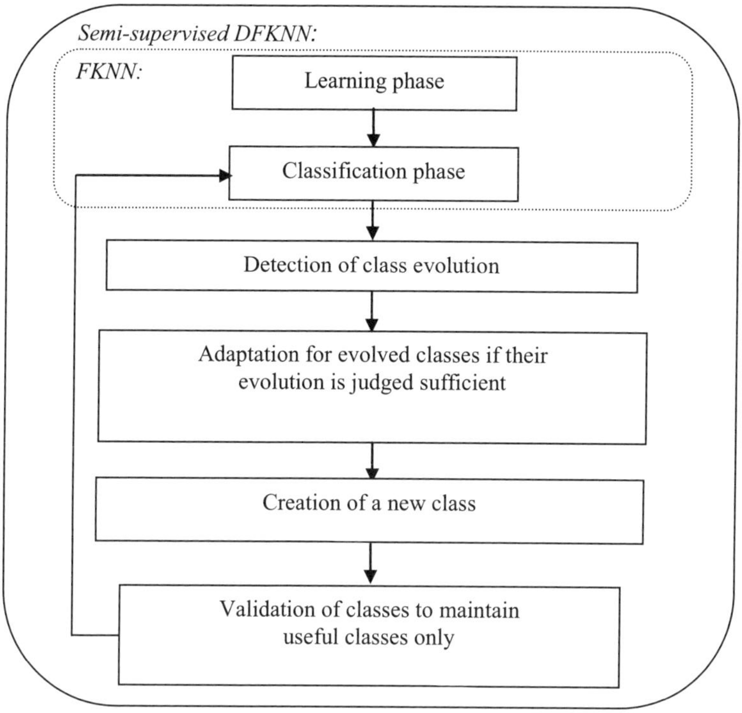

Figure 11.1. *The different phases of the SS-DFKNN method*

11.2.2. *Detecting the evolution of a class*

Two indicators of evolutions are used to detect changes in the characteristics of classes receiving the new pattern. The first indicator $ind1_j$ [11.3] allows us to obtain a compactness value for the class:

$$ind1_j = \frac{\sigma_i^j(t) \times 100}{\sigma_i^j(0)} - 100 \qquad [11.3]$$

$ind1_j$ is given as a percentage. $\sigma_i^j(t)$ represents the class's initial standard deviation. If at least one of the attributes j obtains a value of $ind1_j$ that is greater than the threshold th_1, then the class C_i has begun to change its characteristics. th_1 can be fixed as a small value such as five, which is effective for monitoring progressive evolutions in a class. The class can evolve suddenly, however, and a bigger value may then be necessary for th_1.

The second indicator $ind2_j$ represents the drift between the current point and its class' center of gravity in relation to its standard deviation:

$$ind2_j = \frac{\left| x^j - \overline{x}_i^j(t) \right| \times 100}{\sigma_i^j(t)} - 100 \qquad [11.4]$$

Here, $ind2_j$ is given as a percentage. If at least one attribute j obtains a value of $ind2_j$ that is greater than th_1 ($\max(ind2_j) \geq th1$) then the point has a weak membership value for class C_i. It is not sufficient, however, to have a single point that is a long way from the class in order to consider a change in the class's characteristics. The point can consist of noise and it is therefore necessary to define a value *NbMin* to represent the successive number of times that $ind2_j$ must exceed th_1 in order to confirm an evolution. If *NbMin* is fixed at a significant value, the delay in detecting the class's evolution may be too long. *NbMin* must be therefore defined as a compromise between the noise found in the representation space's patterns and the maximum possible delay in detecting evolution. As such, the evolution of a class is confirmed when *NbMin* successive values of the two indicators $ind1_j$ and $ind2_j$ are greater than th_1.

11.2.3. *Adapting a class after validating its evolution and creating a new class*

When the evolution of a class is confirmed, a new class is created by SS-DFKNN based only on the patterns representative of the evolution.

The different parts of this adaptation phase are:

– The creation of a new class, C', and the selection of patterns representing evolution. For the latter, the last pattern classified, x, is selected as well as its $k-1$ nearest neighbors. No new distance needs to be calculated because, during the classification of x, its nearest neighbors have already been identified. Subsequently, the points corresponding to the evolution are based on the most recent change.

– The last classified pattern x and its $k-1$ nearest neighbors that are already known are the only patterns kept to create C'.

– The selected k patterns are deleted from class C_i.

– The center of gravity $\overline{x}_i^j(t)$ and the current standard deviation $\sigma_i^j(t)$ of class C_i are updated.

– $\bar{x}_i^j(t)$ and $\sigma_i^j(t)$ are calculated for class *C'*. These values are quickly calculated because only *k* points are in the class.

– The number of classes is updated.

This phase allows us to follow the evolutions of a class online. If this latter is complex, a series of Gaussian subclasses could appear during evolution. The splitting and drift of classes is taken into account. If no evolution takes place, the patterns are classified normally.

If class *C* is considered to be useless, it may be necessary to delete this class *C* and solve the dataset's growing problem. Equally, two classes must sometimes be merged because they have ended with the same characteristics. The solutions to both of these problems are presented in the following section.

11.2.4. *Validating classes*

The classes' noise has already been considered with SS-DFKNN using the bias of the threshold *NbMin*. However, in some cases, it may be necessary to delete one or more classes:

– when a transitory class is created and kept, even after the final class destination is reached. Indeed, a transitory class does not represent a functioning mode;

– when a class that is considered to be noisy is created; and

– when a class containing little information is preserved for a given amount of time.

To consider these cases, suppressing the classes and updating the classifier, SS-DFKNN examines two validity criteria:

– an insufficient number n_1 of patterns is contained in the class; and

– no pattern is classified in the class since only n_2 patterns have been classified in other classes.

Each class that does not verify these validation criteria will be deleted. It is necessary to specify that this suppression of non-representative classes is not necessary for all applications. For example, for applications considering data from an important or critical nature, it may be preferable to preserve all of the characteristic data from all the classes.

Figure 11.2 shows an example of the suppression of a class. A class is initially known and its evolution occurs and leads the class towards its final destination. In this case, several patterns of the transitory class must be deleted. The parameters of the DFKNN method in this case are: $k = 5$, $th_1 = 5$, $NbMin = 5$, $n_1 = 10$, $n_2 = 20$ and $th_{Fusion} = 0.2$.

SS-DFKNN considers the first evolution of the class as a new class. A second class is then created and the transitory class that has therefore been considered non-representative is deleted.

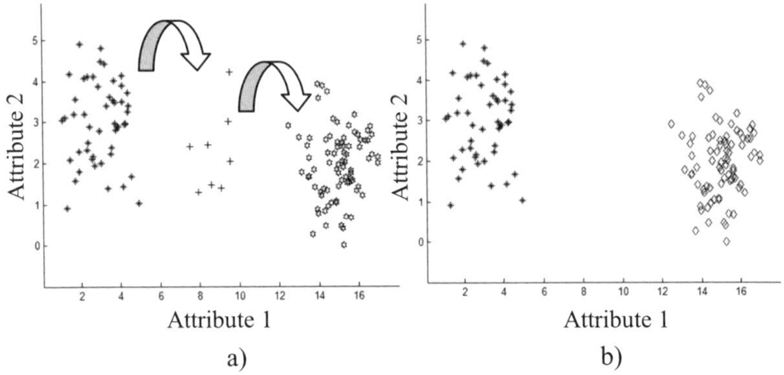

Figure 11.2. *Suppression of a class by SS-DFKNN. In a), the patterns of X are denoted by*, the transitory patterns are denoted by + and the patterns of the evolved class are represented by ✲. b) Classification results obtained by SS-DFKNN, the patterns of the transitory class are deleted*

In the case of merging classes, the measure of similarity proposed by Frigui [FRI 96] [11.5] has been integrated into SS-DFKNN in order to verify if after each classified pattern whether it is necessary to merge two classes which have a similarity which is greater or equal to th_{Fusion}:

$$\delta_{iz} = 1 - \frac{\sum_{x \in C_i \text{ and } x \in C_z} \left| \pi_i(x) - \pi_z(x) \right|}{\sum_{x \in C_i} \pi_i(x) + \sum_{x \in C_z} \pi_z(x)} \qquad [11.5]$$

where $\pi_i(x)$ and $\pi_z(x)$ are the membership values of x to C_i and C_z respectively. The nearer δ_{iz} is to one, the more similar the two classes are. The maximum value indicates that the two classes are overlap completely.

11.2.5. *Defining the parameters of the SS-DFKNN method*

As for all pattern recognition methods, the identified parameters influence the classifier's performance. We propose default values that are generally adapted to dynamic systems:

– k corresponds to the number of neighbors taken into account in k nearest neighbors methods to classify a point. k is the common parameter among k nearest neighbors methods and is the most important. It should be defined according to the size of the database, the noise present in the system's observations and the proximity of classes.

– th_1 is one of the most important parameters in SS-DFKNN. It is taken into account by the two indicators of evolution of SS-DFKNN. It detects evolutions of classes. A class that does not evolve will always have the same characteristics, even if noise appears. The characteristics of a class change, however, in the case of a sudden or progressive evolution. A value of th_1 that is equal to five is a good compromise, which allows us to adapt classes and their characteristics without needing to detect a significant evolution.

– *NbMin* allows us to validate an evolution. *NbMin* influences the results in terms of the class's adaptation time. It should be defined as at least equal to k ($k \geq 1$) with the aim of gaining sufficient representative patterns to fully estimate the characteristics of a new class. It should, also, not be defined by too significant a value as the detection of evolution could be not delayed. *NbMin* should be defined between k and $k + 5$ if k is small or large, respectively. These values have been determined by experimentation. If k and *NbMin* are small, the risk of false alarms becomes greater.

– th_{Fusion} is an optimization parameter used to merge classes. Even if no fusion appears, the simple appearance of a class indicates that there has been an evolution in the system. In this case, an alarm should be raised to call a human operator to verify the system's state. A value of th_{Fusion} that is near to 0.2 allows us to merge classes that have begun to present the same characteristics.

– n_1 is one of the parameters used in the class validation phase. It should be identified as being greater than k ($n_1 > k$), since a class will contain at least k patterns (at its creation). By default, a value of n_1 can be k*2.

– n_2 is one of the parameters used in the class validation phase. Its value should not be too low because after the creation of a class, it may be necessary to wait to classify more patterns in the class that has been created. If no pattern is classified in the new class after a significant number of points, however, the class is neither more nor less representative. This class can then be transitory or can correspond to noise. The value n_2 can be defined as 20 by default. This result has also been obtained by testing. This signifies that one pattern in 20 should be classified as being part of the new class with the aim of gradually confirming its usefulness. For the other classes, even if they do not receive additional patterns after a certain amount of time, they will not be suppressed because they have already confirmed their usefulness by having a sufficient number of patterns.

11.3. Results

This application concerns the development of a passive acoustic detection tool for water leaks in the steam generator (SG) of a prototype fast reactor. The aim of this approach is to detect water leaks of different flow and pressure at various locations with a very short delay in time.

The reaction produced by contact between the water and sodium used to cool the reactor's core causes an explosion. To simulate water leaks, argon was injected and the corresponding acoustic signals were recorded using sensors. This dataset has been taken from tests carried out by the Atomic Energy Authority on an SG in a prototype fast reactor. This dataset was then used by the *Commissariat à l'Energie Atomique* (CEA) – Atomic Energy Commission) to carry out tests. It is in this context that our research has been carried out.

11.3.1. *Data analysis*

An initial frequential and statistical analysis of the data was carried out to identify the informative and discriminative parameters required to define the representation space. The SS-DFKNN method was then applied using only the data known from a normal functioning class to detect new classes and their evolutions. Each signal had a sampling frequency equal to 2,048 Hz. Figure 11.3 shows an acoustic signal resulting from an injection of argon into the SG.

We have studied a significant number of statistical and frequential parameters. These included the mean root square value, Kurtosis, Skewness, the crest factor, the median value, coefficients a_i of the autoregressive dynamic model, etc. These parameters were calculated in a sliding window whose size was calculated by learning. This window should be relatively small in order to contain sufficient data without causing a significant delay in detection.

One of the current methods used by the SG detects the appearance of a problem in six seconds. One of our objectives was therefore to take less than six seconds to detect the argon injection in the sodium. The best window adapted to this application contained 8,192 data samples (4 s) and its sliding window contained 2,048 data samples (1 s). These 2,048 data samples corresponded to a drift of one second in the signal that allowed us to obtain a small delay in surveillance. The use of this sliding window therefore allowed us to follow the evolution of the SG's operation on a set of data online.

In terms of temporal parameters, we have sought to estimate the coefficients a_i of the autoregressive model. To do this, we needed to identify the maximal order p in the model. We began by seeking the order of the model for part of the statistical data using the Akaike information criterion and the method of principal component analysis. This criterion allowed us to fix order p of the model to 13. We were then able to estimate these 13 coefficients. Since our objective was not to predict an exact model of the system, however, we chose the most informative coefficients among the 13 examined. Principal component analysis showed that two parameters of the autoregressive model, coefficients a_3 and a_5, allow us to obtain inertia of 94.3%. These two autoregressive model parameters provided the lowest classification error for the two classes (injection and non-injection). This classification error was calculated using different combinations of parameters using the supervised fuzzy pattern logic (FPM) and support vector machines (SVM) methods. We also selected another parameter, the normalized average, which improved the classification result.

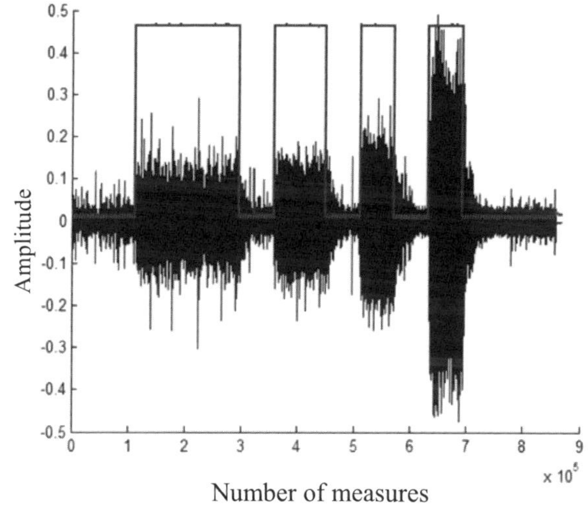

Figure 11.3. *The acoustic signal is composed of the injection and non-injection modes, with its injection command represented by the gray line*

As Figure 11.3 shows, the system's functioning modes evolved when several leaks occurred in succession so the accumulation of leaks therefore led to worsening of the problem. The SS-DFKNN method was used to monitor these evolutions and characterize each of the evolved classes. SS-DFKNN only needed to know information about one class in order to be initialized. The default classes were not all known *a priori*.

11.3.2. *Classification results*

The SS-DFKNN method used 53 points taken from the normal functioning (C_1) of the SG as a learning set. The patterns of a typical signal (Figure 11.3) are then classified. The patterns and classes corresponding to the anticipated classification result are shown in Figure 11.4. The classification result obtained is shown in Figure 11.5.

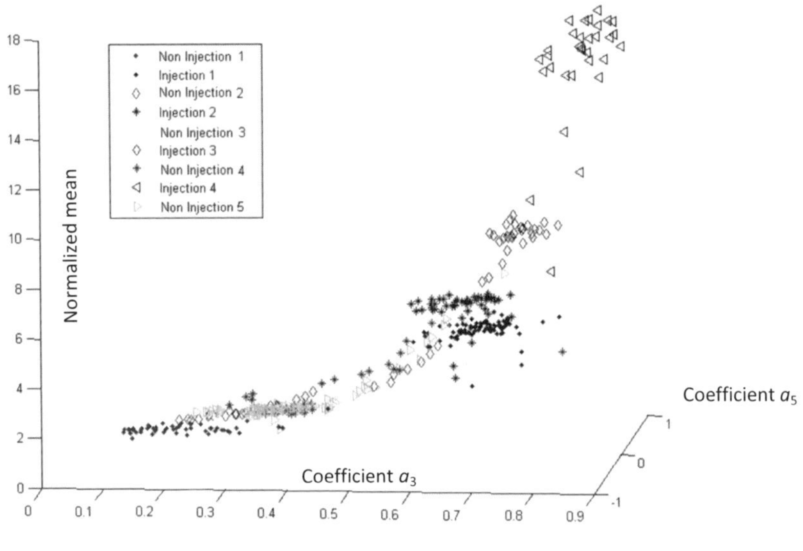

Figure 11.4. *Patterns corresponding to each injection and non-injection of the signal in Figure 11.3*

When the patterns of the signal in Figure 11.4 are classified, several evolutions of classes occur. Several classes are created and the resulting classes can help system experts characterize the system's different functioning modes. Some transition classes, notably between C_2 and C_4 (Figure 11.5), have been judged to be non-representative after some time and are therefore automatically discarded. These classes do not allow us to classify other signals' patterns. In terms of this application, the classes found allow us to identify the different functioning modes. The initial class C_1 has retained its characteristics and the classes obtained correspond to those estimated using the injection command (see Table 11.1).

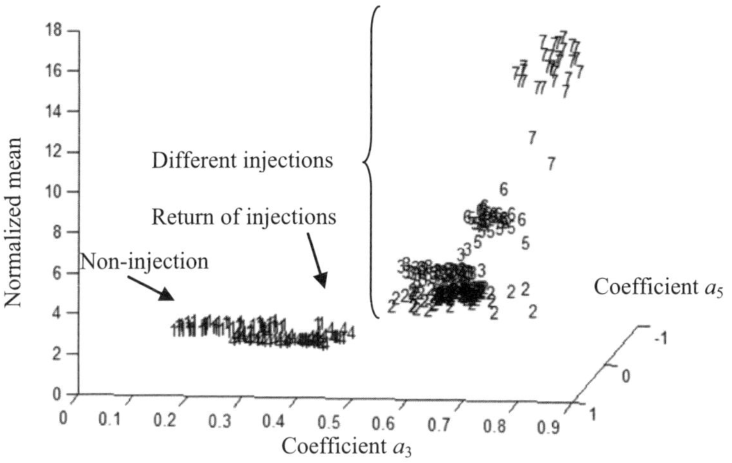

Figure 11.5. *Classification result obtained with semi-supervised DFKNN after classification of all patterns of the signal in Figure 11.3*

We can see in Table 11.1 that there is only a small delay in detecting the evolution of each class (a maximum of 2 s). In Figure 11.4, each injection is followed by several transitory points that are part of the repeat injection class. We can therefore say that once classification of the points in an injection class is carried out, the following points must be classified in the repeat injections class.

Number of injections or non-injections	**Corresponding class found**	**Delay in detecting the evolution (in numbers of windows)**
Non-injection	C_1	0
1st injection	C_2	1
Repeat injection	C_4	2
2nd injection	C_3	1
3rd injection	C_5–C_6	2
4th injection	C_7	2

Table 11.1. *Correspondence of the classes found with injections and non-injections. The delay in detecting the evolution of each class using the SS-DFKNN method is also indicated*

11.3.3. *Designing an automaton to improve classification rates*

To classify the points found in the transitory repeat injection zone (i.e. between injection and non-injection), we need to design an automaton in addition to the SS-DFKNN method. This automaton monitors the cycle of changes between the different modes of functioning (Figure 11.6).

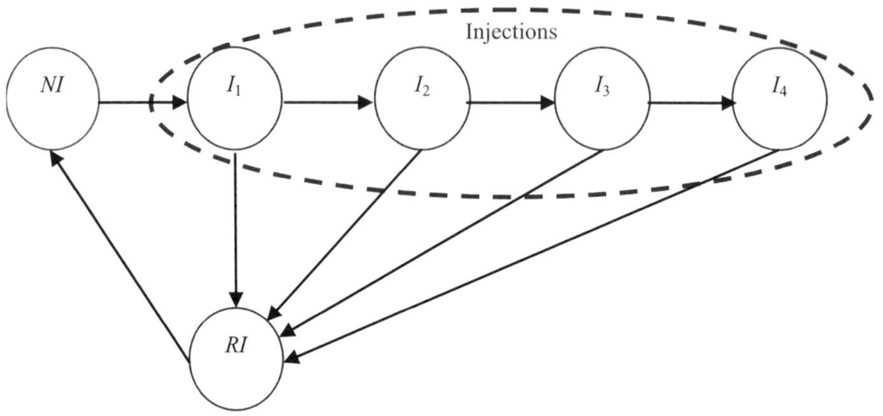

Figure 11.6. *Automaton used with SS-DFKNN to follow changes in functioning mode. NI corresponds to non-injection, I_k corresponds to the k^{nth} injection, and RI corresponds to the return of injections*

The automaton models designed determined the order of different transitions between normal functioning mode (non-injection) and faulty functioning mode (injection) over the course of the SG's operation. It also models the degrees of failure that are possible when an argon leak is found in the prototype fast reactor's SG. Note that the degree of intensity of injections follows the evolution of this machine, i.e. injection I_2 is more important than the injection I_1, I_3 is greater than I_2 and I_4 is greater than all of the previous injections. If an argon leak occurs in the SG and is significant (I_4), the system moves from normal functioning mode (*NI*) towards mode I_4, passing through the modes (I_1, I_2, I_3).

We have obtained the classification results shown in Table 11.2 by including the results of SS-DFKNN classification and the automaton that was discussed previously (see Figure 11.6). As a primary comparison, the classification results obtained by incremental fuzzy pattern matching (IFPM) [SAY 02], incremental support vector machines (ISVM) [CAU 01] and incremental k-nearest neighbors (IKNN) [ZHO 01] are also shown in Table 11.2. These results show the classification error rate, written as $err_{classif}$. The classifier's performance decreases significantly as the classes evolve over time. This criterion therefore shows the importance of updating the classifier's parameters in order to maintain its performance level. $err_{classif}$ is calculated by:

if $C_{est}(x_i) = C_{real}(x_i)$ then $e(C_{est}(x_i), C_{real}(x_i)) = 0$

otherwise $e(C_{est}(x_i), C_{real}(x_i)) = 1$

$$err_{classif} = \frac{\sum_{i=1}^{n} e(C_{est}(x_i), C_{real}(x_i))}{n} \quad [11.6]$$

where n is the number of patterns classified, e tells us whether a pattern is well classified, $C_{est}(x_i)$ corresponds to the estimated class for x_i and $C_{real}(x_i)$ corresponds to the real class of x_i.

Few patterns are misclassified using SS-DFKNN. The most significant result is that the first injection is perfectly detected. The patterns for this injection are well classified, at 98.9 % in less than 2 s. On the basis of this result, it is possible to detect the start of a problem in the SG very quickly. We can also see in Table 11.2 that the other injection and non-injection patterns are well classified, at 96.8%. This allows us to use this method to monitor the SG's functioning modes online.

	SS-DFKNN	**IFPM**	**ISVM**	**IKNN**
$err_{classif}$ between the 1st non-injection and the 1st injection	1.1%	100%	100%	100%
Overall $err_{classif}$	3.2%	55.9%	55.9%	55.9%

Table 11.2. *Classification results obtained by SS-DFKNN (k=5; th1=5; NbMin=6; th_{Fusion}=0,2; n_1=10; and n_2=20) in coordination with a machine by IFPM (h=5), ISVM (Gaussian kernel with variance=10 and regulation constant=5) and by IKNN (k=5)*

11.4. Conclusion and perspectives

The SS-DFKNN method has been developed in this chapter for the surveillance and monitoring of evolving systems. SS-DFKNN integrates two indicators of evolution that allow us to detect changes in classes' characteristics in order to correctly estimate the system's current mode of functioning. These evolved classes allow us to more precisely anticipate how functioning modes will change over time and monitor evolutions in complex classes (defined by several subclasses). SS-DFKNN only requires a few patterns in order to be initialized but the more representative the learning set is of classes' characteristics, the more evolutions will be correctly carried out. Classes' characteristics are therefore sequentially refined with the classification of new patterns.

SS-DFKNN has been applied to data from a fast neutron reactor's SG. The SS-DFKNN method is well adapted to systems whose abnormal mode of functioning cannot be known in advance. It allows us to examine the progressive evolution of each class and forget classes that are no longer useful. Nevertheless, this method requires us to define several parameters such as th_1, th_{Fusion}, $NbMin$, n_1 and n_2. It is for this reason that the development of a mechanism that enables the dynamic and adaptive identification of these parameters is an important aspect of this research

11.5. Bibliography

[AMA 06] AMADOU BOUBACAR H., Classification dynamique de données non stationnaire: Apprentissage séquentiel des classes évolutives, PhD thesis, Université des sciences et technologies de Lille, France, 2006.

[ANG 04] ANGELOV P.P., "A fuzzy controller with evolving structure", *Information Sciences*, vol. 161, no. 1-2, pp. 21-35, 2004.

[ANG 10] ANGELOV P.P., FILEV D., KASABOV N., *Evolving Intelligent Systems – Methodology and Applications*, John Wiley and Sons, New York, 2010.

[ANG 00] ANGSTENBERGER L., Dynamic fuzzy pattern recognition, Dissertation, Fakultät für Wirtschaftswissenschaften der Rheinisch-Westfälischen Technischen Hochschule, Aachen, Germany, 2000.

[CAU 01] CAUWENBERGHS G., POGGIO T., "Incremental and decremental support vector machine learning", *Advances in Neural Information Processing Systems*, MIT Press, Cambridge, MA, vol. 13, pp. 409-415, 2001.

[COH 05] COHEN L., AVRAHAMI G., LAST M., "Incremental Info-Fuzzy Algorithm for Real Time Data Mining of Non-Stationary Data Streams", *TDM Workshop*, Brighton, UK, 2005.

[FRI 96] FRIGUI H., KRISHNAPURAM R., "A robust algorithm for automatic extraction of an unknown number of clusters from noisy data", *Pattern Recognition Letters*, vol. 17, pp. 1223-1232, 1996.

[GIB 94] GIBB W.J., AUSLANDER D.M., GRIFFIN J.C., "Adaptive classification of myocardial electrogram waveforms", *IEEE Transactions on Biomedical Engineering*, vol. 41, pp. 804-808, 1994.

[HAR 10] HARTERT L., SAYED-MOUCHAWEH M., BILLAUDEL P., *A Semi-supervised Dynamic Version of Fuzzy K-Nearest Neighbours to Monitor Evolving Systems*, Springer-Verlag, Berlin-Heidelberg, 2010.

[KEL 85] KELLER J.M., GRAY M.R., GIVENS J.A., "A fuzzy k-nn neighbor algorithm", *IEEE Trans. Syst. Man Cybern.*, vol. SMC-15, no. 4, pp. 580-585, 1985.

[LEC 03] LECOEUCHE S., LURETTE C., "Auto-adaptive and dynamical clustering Neural network", *ICANN 2003*, Istanbul, Turkey, Proceedings, pp. 350-358, 2003.

[LUG 11a] LUGHOFER E., *Evolving Fuzzy Systems – Methodologies, Advanced Concepts and Applications*, Springer, Berlin-Heidelberg, 2011.

[LUG 11b] LUGHOFER E., ANGELOV P.P., "Handling drifts and shifts in on-line data streams with evolving fuzzy systems", *Applied Soft Computing*, vol. 11, no. 2, pp. 2057-2068, 2011.

[KAS 07] KASABOV N., *Evolving Connectionist Systems: The Knowledge Engineering Approach*, Second Edition, Springer Verlag, London, 2007.

[NAK 97] NAKHAEIZADEH G., TAYLOR C., KUNISCH G., "Dynamic supervised learning. Some basic issues and application aspects", *Classification and Knowledge Organization.* Springer Verlag, pp. 123-135, 1997.

[SAY 02] SAYED-MOUCHAWEH M., DEVILLEZ A., LECOLIER V.G., BILLAUDEL P., "Incremental learning in fuzzy pattern matching", *Fuzzy Sets and Systems*, vol. 132, no. 1, pp. 49-62, 2002.

[ZHO 01] ZHOU S., Incremental document classification in a knowledge management environment, Thesis, University of Toronto, 2001.

Chapter 12

SVM Time-Frequency Classification for the Detection of Injection States

12.1. Introduction

When the British fast neutron reactor *Prototype Fast Reactor* (PFR) ceased operation in 1994, a number of *end of life tests* were carried out. In particular, this included *leak detection system* (LDS) tests examining the acoustic detection of water leaks into the sodium of the steam generators (SGs). In this chapter, we study the use of the acoustic detection method on a series of recordings from LDS5 tests that involves injecting argon, hydrogen, and water into the PFR's third evaporator (EV3). The aim is to create an approach that provides an alternative to current chemical methods.

These recordings have been studied by the CEA (the French Atomic Energy agency). They present a significant low-frequency attenuation that is presumably caused by bubbles during the injection, although this statement cannot be proved due to electrical interference. Unsupervised classification was also carried out that proved the frequency band of 200 – 1,000 Hz to be optimal for detection but that did not enable the complete characterization of the sodium–water reaction. Therefore, other criteria are needed to make passive acoustic detection reliable.

In this chapter, we will study 32 series of acoustic recordings sampled at $f_s =$ 2,048 Hz taken using accelerometers placed at the end of the wave guides in the third secondary circuit's evaporator of the PFR. The first 15 series are injections of water while the remaining recordings are argon injections. Each series is composed of an injection command signal and two acoustic signals recorded using two sensors. The

---

Chapter written by Simon HENROT, El-Hadi DJERMOUNE and David BRIE.

aim of the study is to develop an algorithm that allows us to detect potential leaks in the SG in real time. Beyond this, we also want to provide additional information about the leak after detection.

12.2. Preliminary examination of the data

12.2.1. *Approach*

An initial examination of the recordings leads us to the immediate observation that an injection command causes a temporary increase in the signal's strength on each of the sensors. More precisely, a rising edge command results in a rising edge for the amplitude of the recorded signal while a falling edge command causes a weakening of the signal. These two phenomena do not occur immediately after the command, due to the injection protocol and the physical propogation of the fluid inside the SG.

However, this does not allow us to create a detection method based on the signal's absolute level for three reasons:

1) We want to mitigate the effect of potential variations in the sensors' amplification and join the recording instruments.

2) We do not have access to the signal-to-noise ratio in the context of an industrial installation.

3) No information is available on the temporal profile of a leak that is *a priori* distinct from an injection (which it resembles with a boost).

Therefore, we are interested in studying the existence of injections' spectral signatures.

12.2.2. *Spectral analysis of the data*

12.2.2.1. *Data format*

The data come from two distinct classes: class B that resembles a segment of data taken from the recording of an ambient sound and class I that corresponds to a segment from the recording of an injection. Class I is further divided into two subclasses, $I_{\mathrm{H_2O}}$ and I_{Ar}, which include the segments of injections of water and argon, respectively. $N = 32$ are the overall numbers of recordings. We proceed to the selection of N class B segments and N class I segments; $N_{\mathrm{H_2O}} = 15$ and $N_{\mathrm{Ar}} = 17$ that are the numbers of segments corresponding to the injection of water and argon, respectively. We have deliberately chosen a single segment per recording in order to limit data redundancy as far as possible.

12.2.2.2. *Welch method*

In this section, we examine the method used to calculate the power spectral density (PSD) X of a segment x with a size of $N_T = T \times f_s$, where T is the segment's duration in seconds. The spectrum is evaluated in the band [0, 1,024] Hz in order to satisfy the Nyquist–Shannon sampling theorem. The principle of the method is as follows:

1) x is divided into $M = [2N_T/(f_s + 1)]$ segments with a size of f_s with a 50% overlap.

2) We multiply each segment by a Hamming window.

$$g(n) = 0.54 - 0.46 \cos\left(\frac{2\pi n}{f_s - 1}\right) \quad [12.1]$$

3) We calculate the periodogram of each windowed segment using the fast Fourier transform (FFT).

4) The PSD is the average of the M obtained spectra.

The main objective of the method is to reduce the influence of noise by achieving a compromise between the variance of the periodogram and spectral resolution.

12.2.2.3. *Spectral study*

This involves the calculation of the average spectra of the classes B, $I_{\mathrm{H_2O}}$, and I_{Ar}. To do so, we calculate the PSD of each data segment using the Welch method. The spectra are nomalized to unit power. The results are shown in Figure 12.1. In each of the three classes, there are spectral peaks at various amplitudes centered on a set of frequencies E, which correspond to the structure's resonance frequencies. The main frequencies found are as follows:

$$E = \{112, 170, 248, 262, 326, 438, 530, 578, 648, 662, 750, 823, 886\}\,\mathrm{Hz} \quad [12.2]$$

From this frequency set, two distinct trends appear:

1) Class B is mainly characterized by low-frequency spectral lines, including the DC component and undesirable industrial lines at 17, 50 (electrical line), 122, 200, and 300 Hz. We can further divide the spectrum into two distinct bands $B_1 = [0, 350]$ Hz and $B_2 = [351, 1,024]$ Hz, with the maximal amplitude ratio for the two bands being greater than 30. There are also two low-amplitude spectral peaks centered on the last frequencies of E.

2) Classes $I_{\mathrm{H_2O}}$ and I_{Ar} have similar spectral patterns composed of spectral peaks, centered on the frequencies of E and similar amplitudes. Noisy spectral lines are dominated by main peaks, with associated amplitudes varying slightly for the two subclasses. The main difference between classes B and I occurs mainly in the $[0, 600]$ Hz band. As pointed out in [ORI 96], the separability of the two subclasses $I_{\mathrm{H_2O}}$ and I_{Ar} is best carried out in the $[70, 500]$ Hz band.

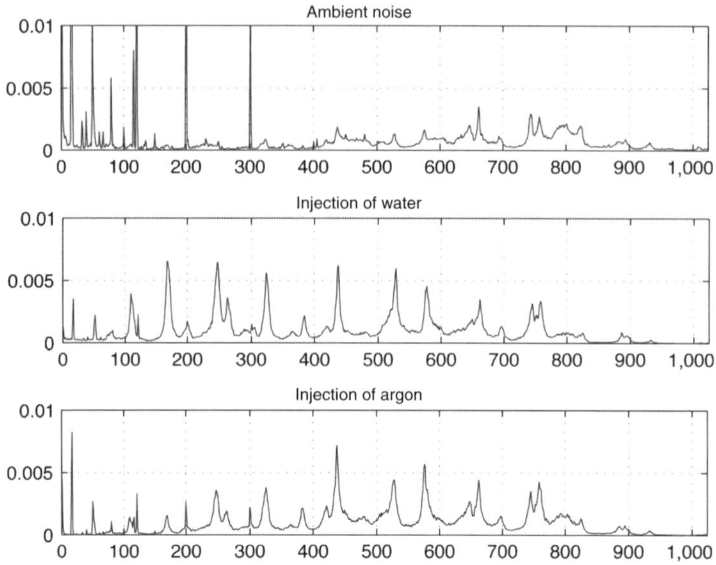

Figure 12.1. *Average PSD of the signal classes*

These results appear to validate the principle of a classification algorithm based on the spectral signature of an injection. To corroborate this observation, we need to measure the spectral similarity of the three signal classes. Since we already have three average PSDs, we treat them as three vectors and we calculate the correlation matrix R given by (with C being the covariance matrix):

$$R(i,j) = \frac{C(i,j)}{\sqrt{C(i,i)C(j,j)}} \qquad [12.3]$$

The calculation yields the following result:

$$\begin{array}{cccc} & B & I_{\mathrm{H_2O}} & I_{\mathrm{Ar}} \\ & \downarrow & \downarrow & \downarrow \end{array}$$

$$R = \begin{pmatrix} 1.00 & 0.12 & 0.37 \\ 0.12 & 1.00 & 0.79 \\ 0.37 & 0.79 & 1.00 \end{pmatrix} \begin{array}{l} \leftarrow B \\ \leftarrow I_{H2O} \\ \leftarrow I_{Ar} \end{array} \qquad [12.4]$$

The two injection subclasses are strongly correlated, although the correlation between injections and ambient noise is weaker. As a result, we will merge water and argon injection classes. It is important to note that the segments of the classes $I_{\mathrm{H_2O}}$ and I_{Ar} are also subject to the influence of ambient noise that provides a positive correlation.

12.2.3. *Class visualization*

In this section, we want to create a cluster of points that allows us to gain an insight into the separation problem. To do so, we use a *principal component analysis* (PCA) using the following approaches:

1) The previous segmentation is reordered in a $2N \times N_T$ matrix with the first N rows corresponding to class B and the next N rows corresponding to class I.

2) A $2N \times F_{\text{len}}$ matrix X is constructed by calculating the PSD of each row of x, where F_{len} is the number of samples in the frequency ($F_{\text{len}} = f_s/2 + 1$ if we keep only the frequencies between 0 and $f_s/2$).

3) We perform a singular value decomposition of X. The resulting principal vectors c_i's are listed in descending order of the corresponding singular values.

The data can then be projected onto graphs generated by two main vectors. Figure 12.2 provides some examples of these projections, showing distinct projected data clusters for classes B and I. The aim of the algorithm is to complete this discrimination using all the information available.

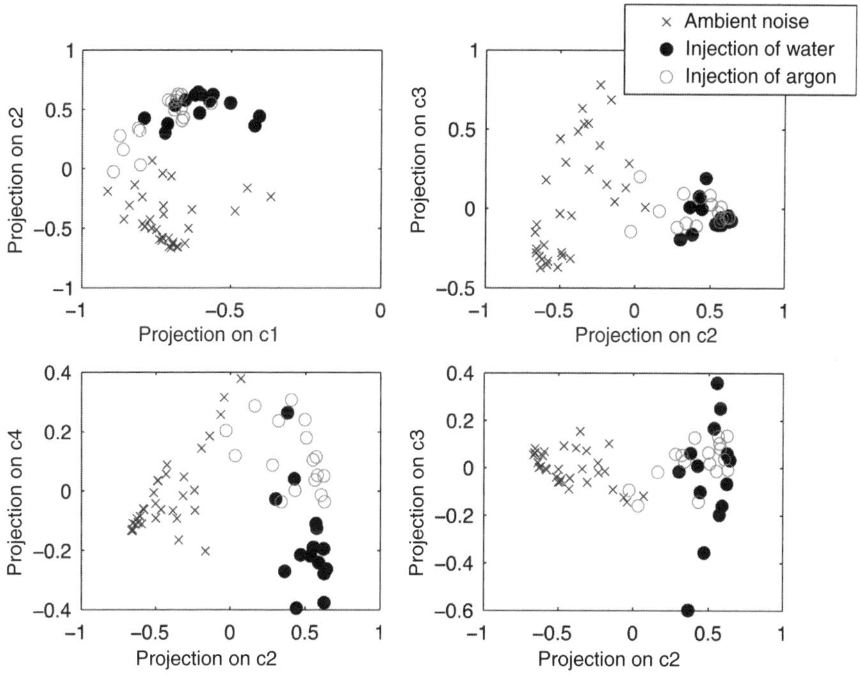

Figure 12.2. *PCA: projection of average PSDs on the graphs generated by a pair of principal vectors*

The rationale for injecting argon into the reactor during LDS5 tests was to simulate water–sodium reactions by producing a cluster of bubbles and the corresponding noise without creating hydrogen bubbles and heat that could damage the structure. A similar challenge will be faced when attempting to implement an LDS on a new SG with no water injections available. Hence, calibrating the algorithm will only be possible based on argon injections. The results of the preliminary analysis justify the assimilation of the spectral signatures for the two types of injection. Furthermore, they validate the principle of a class B/I discrimination algorithm based on their respective spectral signatures. Note that such a method does not depend on the absolute level of the signal.

12.3. Detection algorithm

The presence of distinct spectral characteristics for each signal classes allows us to create a leak detection method based on two key ideas. We therefore proceed to a continuous time-frequency analysis that is carried out on the signals received by the sensors using a sliding window. The resulting data segment constitutes the input of the decision algorithm. The aim of analyzing the signals in our database is to reconstruct a binary signal, which indicates the absence or presence of an injection on the basis of the signal recorded in one of the sensors that we will term the *class signal*. As seen previously, this kind of signal differs from the leak command signal due to the presence of delays and relaxation phases corresponding to injection tails.

Solving discrimination problems has been widely examined in the literature [HSU 03, FLE 09]. Here, we used a supervised classification approach known as *support vector machines* (SVM), or *wide margin separators*. The choice of this technique is motivated particularly by the wide range of data (the signals' spectra being composed of F_{len} = 1,025 samples) and its ease of implementation that is useful in the context of real-time detection. As we only consider classes B and I, we are working with dual-class SVMs (binary discrimination) because we are interested in a segment's adherence to the Class B (digitally treated by -1) or I ($+1$).

12.3.1. *SVM implementation*

12.3.1.1. *Principle*

There are several key ideas in SVM theory. The ultimate goal is to construct a function h that, for an input vector $X \in \mathbb{R}^N$, gives the class segment $y \in \{-1, 1\}$:

$$y = h(X) \qquad [12.5]$$

The first step consists of creating a *learning* set, i.e. a set of L-labeled data. Formally, this set takes the form:

$$(X_i, y_i), i \in \{1 \dots L\} \qquad [12.6]$$

Therefore, we examine $\mathbb{R}^N$ and seek to identify the *maximum margin* hyperplane separator between the two classes, the *margin* being the distance between the separation boundary and the nearest samples (*support vectors*). The SVM's *primal* formulation can be expressed by:

$$\min_{v,b,\xi} \frac{1}{2} v^T v + \lambda \sum_{i=1}^{L} \xi_i \qquad [12.7]$$

under the constraint:

$$\begin{cases} y_i(v^T \phi(X_i) + b) \geqslant 1 - \xi_i \\ \xi_i \geqslant 0 \end{cases} \qquad [12.8]$$

where v is the normal vector on the hyperplane and $b/\|v\|$ the original distance. To account for false classifications, we use the *soft margin* technique that consists of introducing the *slack variables* ξ_i, which allow us to minimize the constraints on the learning vectors. The constant λ controls the trade-off between the number of classification errors and the margin's width.

In general, the data are not linearly separable and the function ϕ transforms the representation space of the input vectors X_i into a higher, potentially infinite, dimensional space (*Kernel trick*). This technique allows us to carry out a linear discrimination in this space that is equivalent to a nonlinear discrimination in the original space. The aim is therefore to find the maximum-margin hyperplane in the higher-dimensional space whose existence and uniqueness is ensured by *Vapnik–Chervonenkis theory*. In practice, we do not directly know the transformation ϕ but construct the kernel function K defined by:

$$K(X_i, X_j) = \phi(X_i)^T \phi(X_j) \qquad [12.9]$$

Some common kernels used with SVMs are:

– linear: $K(X_i, X_j) = X_i^T X_j$;

– quadratic: $K(X_i, X_j) = (1 + X_i^T X_j)^2$;

– *radial basis function* (RBF): $K(X_i, X_j) = \exp(-\sigma \|X_i - X_j\|^2), \sigma > 0$;

– polynomial: $K(X_i, X_j) = (\gamma X_i^T X_j + r)^\eta, \gamma > 0$;

where σ, γ, r, and η are the kernel's parameters. Finally, we construct a validation set to which we apply the classifier in order to test the adequacy between the real classes and those determined by the method.

12.3.1.2. *Choice of attributes*

The term *attribute* refers to each component in the vector X. In our case, $F_{\text{len}} = 1,025$ are the attributes with each being the PSD's segments evaluated at a point of [0, 1,024] Hz. Another solution would consist of defining the frequency bands playing a specific role in the physical phenomena in order to reduce the number of attributes. This choice is justified by the following points:

1) The low computational complexity of the algorithm allows us to exploit all the frequency information.

2) The introduction of frequency bands playing a specific role may prevent the portability of the method to another SG (or may require an additional study).

3) Working in a high-dimensional space favors the existence of a linear classifier [HSU 03].

Conventionally, the attributes are scaled to account for their specific dynamic.

12.3.1.3. *Choice of kernel*

To select an SVM kernel, we construct a validation set similar to the learning set. We obtain a rate of 100% for the classification of validation segments with a linear kernel. As such, the use of another kernel does not allow us to improve the performance but worsens the computational complexity. This optimal result may be explained by the fact that we test the algorithm on "well-chosen" segments. More precisely, we can evaluate the performance of our detection algorithm on the transition zones between the classes B and I, which is examined in the following section.

12.3.2. *Algorithm calibration*

In this section, we want to fix two parameters involved in the implementation of the algorithm: the duration T of the segments used in the initial SVM learning phase (before use) and the duration w of the sliding window (the overlapping is fixed at 50%). We initially choose to evaluate the performance of the algorithm by using the following criteria:

1) the false alarm rate $f_a(T, w)$, i.e. the number of times where the algorithm detects an injection in the presence of ambient noise divided by the total number of decisions;

2) the rate of missed detections $n_d(T, w)$, i.e. the number of times the algorithm detects ambient noise in the presence of an injection divided by the total number of decisions.

The sum of these two criteria is the empirical probability of error $p_e(T, w)$:

$$p_e(T, w) = f_a(T, w) + n_d(T, w) \qquad [12.10]$$

The performance evaluation procedure is as follows:

1) We choose five recordings (PFR 35, 37, 39, 41, 43) whose structure B/I is relatively simple and carry out a complete segmentation of these signals using the cumulative sum (CUSUM) algorithm [PAG 54] based on changes in instantaneous power. Therefore, we have a "real" reference point for detecting these signals.

2) We create a mesh of values for the pair of variables (T, w), with each variable varying between 1 and 20 s with an interval of half a second. We set the limit value $T_{\max} = w_{\max} = 20$ s because a greater value will lead to an unacceptable detection delay (around $w/2$).

3) For each value of T, we generate an SVM learning base.

4) For each pair (T,w), we carry out a sliding window detection on each of the chosen recordings. The values of the classes are compared with the reference values and we then estimate the *empirical probability of error* for each of the recordings. We therefore calculate the quantities $f_a(T, w)$, $n_d(T, w)$, and $p_e(T, w)$.

The results are shown in Figures 12.3–12.5.

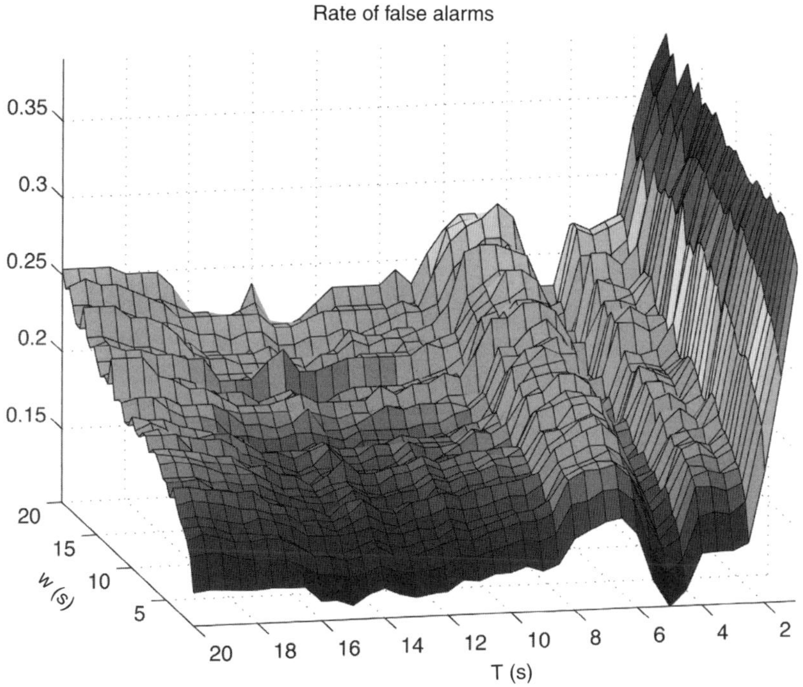

Figure 12.3. *Average rate of false alarms for T and w*

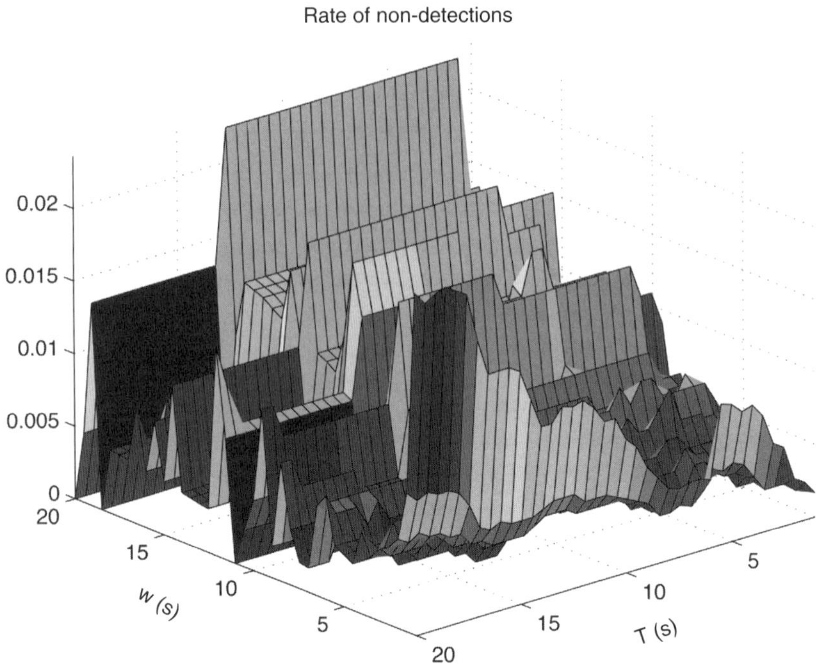

Figure 12.4. *Average rate of non-detections according to T and w*

An analysis of these results leads us to two immediate observations:

1) False alarms have a greater effect on the average error rate. In comparison, the rate of missed detections is relatively constant and low (at a maximum of 2.3%). This observation is important in terms of safety.

2) From Figures 12.3 and 12.5, we find the optimal amount of error $T = 5$ s. We therefore adopt this value to carry out learning.

To justify this decision, we will examine how the false alarm rate depends on the size of the sliding window w. There are two distinct criteria for this:

1) As seen previously, $f_a(w)$ measures the average rate of false alarms whose number of bad decisions commanding an injection is divided by the total number of decisions. This criterion does not provide any information on the distribution of false alarms; do they occur together or are they spread across several consecutive decisions?

2) $n_f(w)$ measures the number of times a false alarm begins (in the case of the appearance of several consecutive false alarms, this criterion is incremented only once). From a "users" perspective, the number of times the alarm sounds to indicate a leak is important.

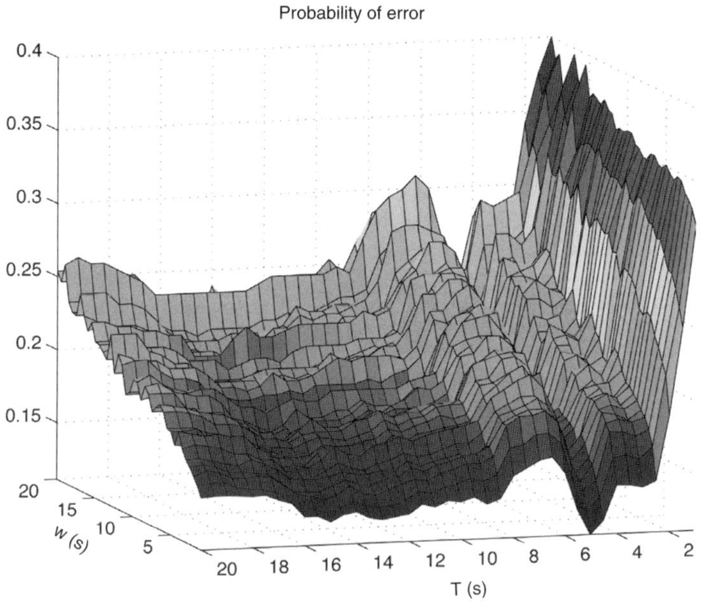

Figure 12.5. *Empirical probability of average error for T and w*

This study allows us to identify the temporal distribution of errors. The results are shown in Figure 12.6.

The study indicates the following tendency: as w increases, so does $f_a(w)$, but $n_f(w)$ decreases exponentiallly. These results can be explained as follows:

– sensitivity to noise decreases when w increases. Using a bigger window introduces an averaging effect on the more significant values, which, in turn, reduces the number of occasional false alarms.

– the detection delay increases when w increases. During the sliding window analysis of an injection, the transition between B and I only occurs when half the data comes from the wave pattern corresponding to the injection. This delay is therefore of the order of $w/2$. As a result, the rate of false alarms and non-detection increases slightly with w (due to a decision error before and after the injection, respectively).

We find that the value $w = 10$ s represents a good compromise between these two phenomena. We therefore choose the pair $(T, w) = (5, 10)$ by default; these values can be modified according to the performance requirements. Empirically, we see that the algorithm's reliability in terms of the number of false alarms increases when the sliding window is larger. The construction of an extended database could therefore allow us to improve the detection performance.

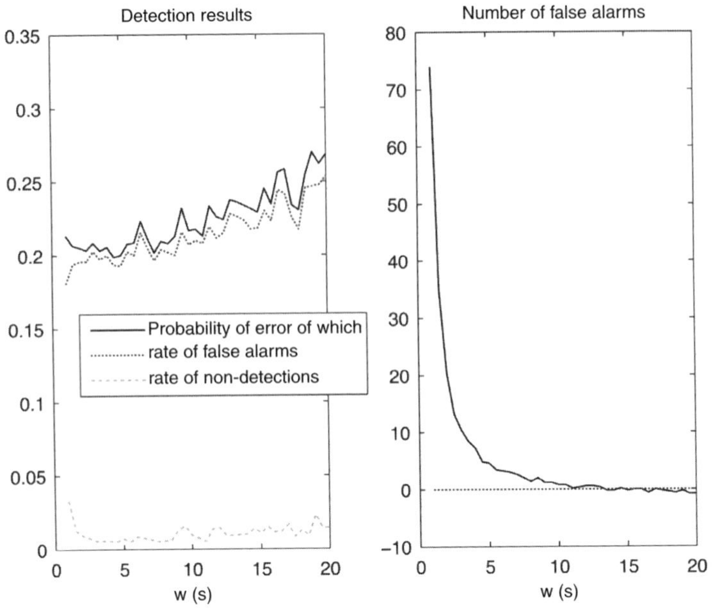

Figure 12.6. *Error and number of false alarms with w for $T = 5$ s*

12.4. Role of sensors

To reduce the number of false alarms, we use the logical operator AND to merge the decisions made for each of these sensors. We could also consider a three-level exit function that causes an intermediary alarm when different decisions are made on each sensor.

12.5. Experimental results

In this section, we present the detection results from recordings featured in the database.

Figures 12.7 and 12.8 show the detection results for injections of argon carried out on the two recordings pfr41 and pfr43 with $w = 20$ s. Graphically, the results correspond to the evolution of the signal's instantaneous power. The delay between the classification results and the command signal can be explained by the physical propagation of the fluid in the system and the width of the sliding window. We see that the algorithm correctly detects the four injections shown in the two recordings.

Figures 12.9 and 12.10 show the detection results obtained in the case of water injections on the recordings pfr37 and pfr60. In Figure 12.9, the detection signal does not show the same evolution as the injection signal, which can be explained by the proximity of injection peaks; the system does not return to ambient mode between injections. In Figure 12.9, the injection peaks are sufficiently spaced and the classification produces two distinct injection states.

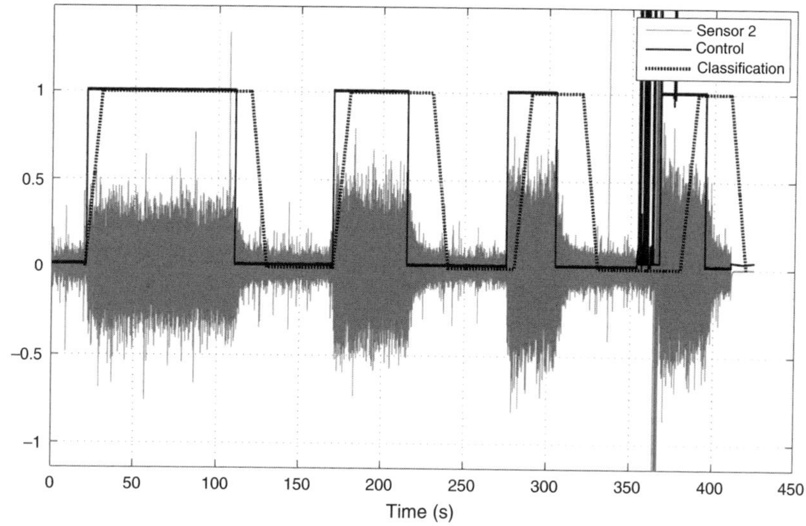

Figure 12.7. *Detection of an argon injection on the recording pfr41 with $w = 20$ s*

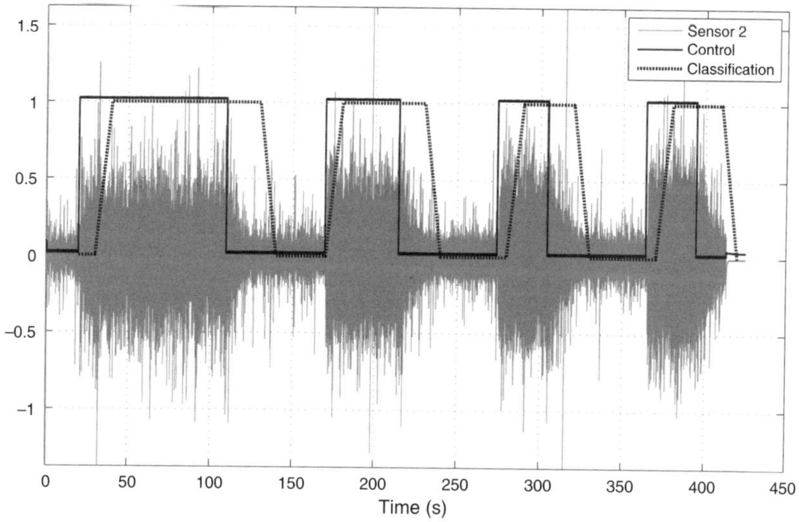

Figure 12.8. *Detection of an argon injection on the recording pfr43 with $w = 20$ s*

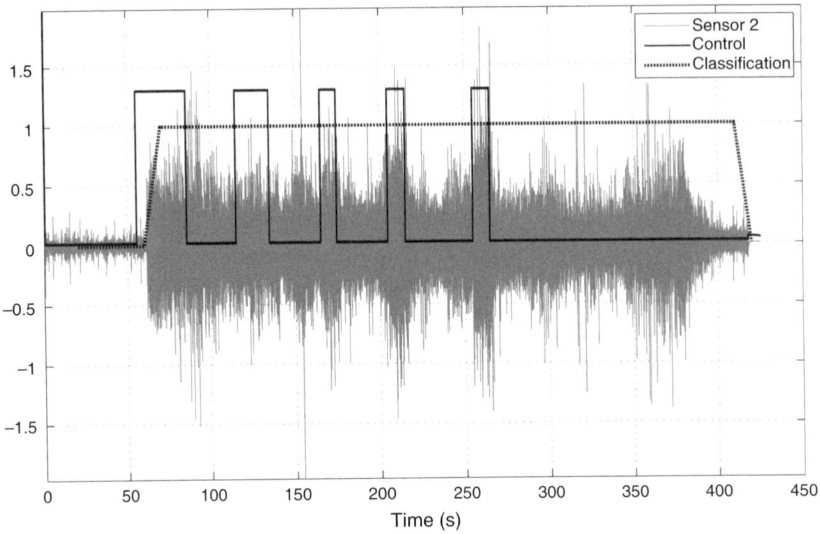

Figure 12.9. *Detection of an injection of water on the recording pfr37 with $w = 20$ s*

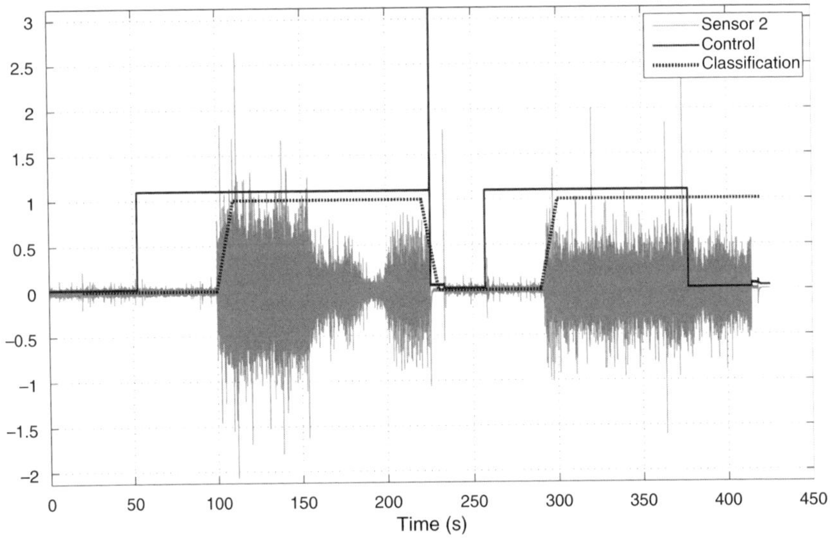

Figure 12.10. *Detection of an injection of water on the recording pfr60 with $w = 20$ s*

Figure 12.11 provides an example where the detection algorithm does not provide satisfying results. Here, the signal is corrupted by noise that hinders the classification of the different states. However, note that a series of punctual alerts is produced during the injection phase.

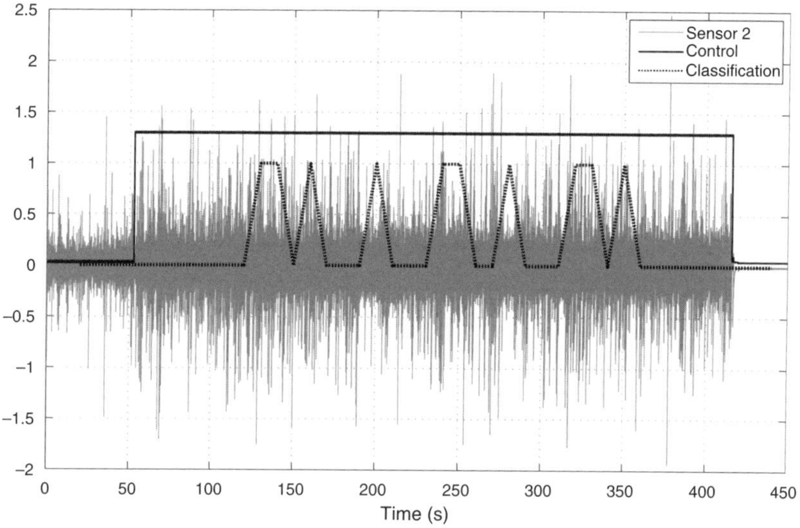

Figure 12.11. *Detection of argon injection on the "very noisy" recording pfr34 with $w = 20$ s*

12.6. Bibliography

[FLE 09] FLETCHER T., *Support Vector Machines Explained*, http://www.tristanfletcher.co.uk/SVM%20Explained.pdf, 2009.

[HSU 03] HSU C.-W., CHANG C.-C., LIN C.-J., A practical guide to support vector classification, Report, Department of Computer Science, National Taiwan University, 2003.

[ORI 96] ORIOL L., GRIVELET P., Analyse des enregistrements acoustiques des essais d'injections PFR en vue de la caractérisation de la réaction sodium-eau, NT SSAE/LSMI 96/0025, 1996.

[PAG 54] PAGE E., "Continuous inspection schemes", *Biometrika*, vol. 41, no. 1, pp. 100–115, June 1954.

Chapter 13

Time and Frequency Domain Approaches for the Characterization of Injection States

13.1. Introduction

13.1.1. *Framework of the study*

With the aim of characterizing injection states, we want to verify the following hypotheses:

– the frequential characteristics of signals recorded outside periods of injections and during the injections of water and argon are different; and

– the presence of an injection causes a modification of filtering characteristics between sources and the sensors. These modifications of characteristics differ from sensor to sensor.

Differentiation between "non-injection states" and 'water injections' must be independent of the water injection rate and any leak that occurs during the monitoring phase.

The aim is to develop methodologies that can be applied to other data or experimental situations without requiring too many additional studies. For this reason, the methods applied rely on a limited number of parameters whose significance is justified and automatic processing is necessary.

Chapter written by Jean-Philippe CASSAR and Komi Midzodzi PEKPE.

13.1.2. *Processing recordings*

The test recordings alternate between periods of non-injection and injection. Figure 13.5 gives a sample recording for the two signals taken from sensors and the injection command. For each recording, the place of injection is the same.

These recordings cannot be processed as a whole because they indicate different situations that must be distinguished from one another. It is therefore necessary to carry out initial processing to obtain recordings that can be linked to a well-defined experimental condition.

13.1.3. *Identifying the injection zones*

The injection zones are automatically extracted from the recordings using the injection command signal. On the basis of the information provided in the documents accompanying the data, each recording with an injection can be linked to experimental conditions, i.e. the position and rate of injection used.

13.1.4. *Extraction of "non-injection" zones*

Non-injection zones are extracted from the recording as long as possible before the first injection. The first four seconds at the start and the last four seconds of this period are eliminated in order to limit the interference that has occasionally been observed at the start of the recording or just after the injection. The relevance of the selected zone is then visually verified.

13.1.4.1. *Database*

The data extracted from the raw recordings allow us to construct a reference database that includes:

– 34 recordings with a water injection;

– 44 recordings with an argon injection; and

– 32 non-injection recordings (15 from recordings with a water injection and 17 with argon injections).

13.2. Analyzing the statistical properties of spectral power densities

Our hypothesis is that the signals have a random component and their different spectral components are represented using their power spectral density (PSD) [FLE 01, HOW 02].

The hypothesis that we want to verify is that the PSDs are different in different states, i.e. "non-injection", "water injection" and "argon injection". We therefore want to differentiate between:

– the states "with injection" and "without injection" to evaluate the capacity to detect a leak of water or argon; and

– "water injections" and "argon injections" to evaluate the capacity to differentiate between these two types of injection.

13.2.1. *Methodology*

13.2.1.1. *Calculating the power spectral densities*

PSDs are calculated using the longest period possible in the injection zone or the zone in the non-injection state concerned. We use the Welch method on the *Hamming* windows with a recovery rate of 1/3 between the windows. The resulting PSD is the average of the PSDs calculated on the set of windows. Its power is normalized to allow comparisons of PSDs independently of the signal's power and therefore size of the leak, if we hypothesize that these two sizes are interconnected.

13.2.1.2. *Modeling and recognition*

PSDs from the two sensors are concatenated into a single recording in order to integrate the potential effects of injections on the two sensors' PSDs into the analysis. The "form" of the recordings is characterized by how they are divided, which depends on applying the principal components taken from the principal component analysis (PCA) [SAP 90, KOU 95] to all of the available recordings. The number of principal components retained ensures that 95% of inertia in the recordings is accounted for.

The first approach relies on learning-based modeling using the set of recordings available to build models taken from unsupervised classification. In the recognition step, we search for the closest model to each new recording. This approach is applied in two steps in a hierarchical decision framework in order to differentiate between:

– step 1: states "with injection" and "non-injection"; and

– step 2: states with an "argon injection" and a "water injection" when the state 'with injection' is recognized.

The second approach concerns a new installation where no recordings with leaks are available. A model is constructed that corresponds to the "non-injection" recordings. A leak results in a significant gap between the current recording and the model.

Classification is carried out using a *K-mean* approach based on a Euclidian distance from the points to the center of classes in the principal component space. Initialization starts from a random point chosen in each class. The class model is therefore given by the center of the class. We can find this by assigning a new PSD (point) to the class with the nearest center.

To improve the difference between the forms of states with "water injections" and "argon injections", we only retain the frequencies where the average PSDs in these two classes show significant differences. Given that there are more than 30 tests, the distribution of the gap between averages can be assimilated to a Laplace-Gauss law. The frequencies retained are those where the standardized mean difference (SMD) between the two means [SAP 90] is greater than three (risk α=0.001). The frequencies retained are part of the "model" that will be used for recognition.

The performance of the two approaches proposed is estimated by *bootstrap* (random selections of sequences in the available recording).

13.2.2. *Results*

13.2.2.1. *Distinction between "injection" and "non-injection"*

Three principal components are retained from the PCA. Figure 13.1 shows the projection of recordings in the principal 2D plane. The classes' centers and the separator S_1 are automatically determined by supervised classification. The strong dispersion of the "non-injection" class leads this separator to classify 12 "non-injection" points in the "with injection" class. The choice of separator S2 classifies five "argon injection" as "non-injection" points. The fact that these points correspond to low injection rates may explain this classification error.

A comparison of PSDs corresponding to the centers of the two classes shows that the injections give a higher power in the 200–650 Hz zone and attenuation at low frequencies (< 200 Hz) in relation to the mean 'non-injection' PSD.

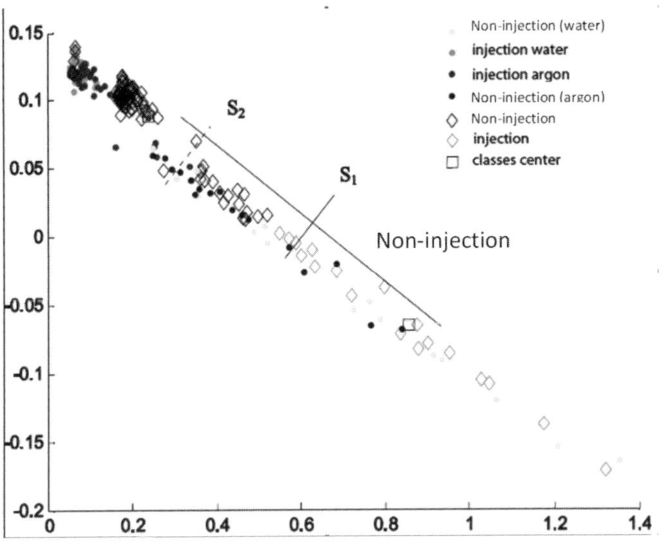

Figure 13.1. *Projections of recordings in the first principal plane (points) of classification using the k-mean approach (diamonds) and the center of classes (squares)*

13.2.2.2. *Water–argon distinction*

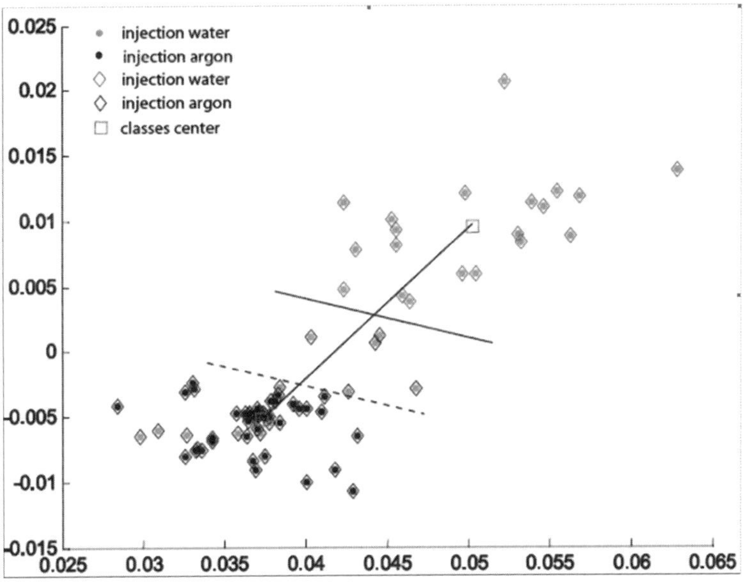

Figure 13.2. *Projections of recordings in the first principal plane (point) with classification using the k-mean method (diamonds) and the center of classes (squares)*

The frequency selection retains 194 frequencies out of 514. The eliminated zones lie mainly between 250 and 440 Hz, 500 and 580 Hz and 720 and 760 Hz.

PCA is applied to the recording at the selected frequencies. The six argon points that may potentially be recognized as "non-injection" are not used to obtain the model with the least amount of "interference" possible by the different points. Three principal components are retained.

Figure 13.2 shows the projection of recordings in the first principal plane. The separator S_1 of the unsupervised classification leads us to class 12 recordings corresponding to injections of water in the "argon injection" class. By weighting the points to account for the greatest dispersion of "water injection" recordings, this number can be reduced to eight. For five of these eight points, the injection rate is low.

13.2.2.3. *Validation and analysis of the results*

From the previous study, we have obtained a model that allows us to classify a new sample. This 'model' includes:

– the selected frequencies;

– the principal components for these frequencies;

– the coordinates of the centers of classes; and

– the position of the separator.

The recognition of membership of a class involves a comparison of the Euclidian distances in relation to the separator S_2 on the straight line linking the estimated centers of the two classes. Other distances are possible (e.g. Mahalanobis) by adding the variance–covariance matrices of the model's projections depending on the main components in each of the classes.

To evaluate the classification capacity of the model obtained, we carry out Monte-Carlo simulation. A random selection of samples with duration of two seconds (4,096 points) is carried out in the injection or non-injection zones. Each sample is then classified according to the principle explained above.

13.2.2.4. *Separation between "non-injection" and "with injection"*

Three thousand simulations were carried out to separate "non-injections" from "injections". The six "argon injection" tests classified as "non-injection" in the preliminary study were kept in the random selection.

We found that 4.87% (±1.1 % with a risk of 0.001) were badly classified and 2.87% (± 1.1 % with a risk of 0.001) "non-injection" points were classified as "with injection". The more significant dispersion of "non-injection" taken from recordings linked to injections of water (see Figure 13.1) means that these recordings provide an important contribution to these results. The six "argon injection" points kept in the test had 1.43% classification errors, i.e. 29% of the errors found.

In terms of detecting leaks, on the basis of these simulations, we can expect a limited "false alarm rate" linked to the capacity to detect water leaks: 30 tests out of 34 provided 100% detection, with this rate not falling below 82.3% for the four tests that did not provide 100% detection.

13.2.2.5. *Water and argon distinction*

Figure 13.3 provides a projection of the first principal plane of 1,000 random selections taken from recordings during injections of water and argon.

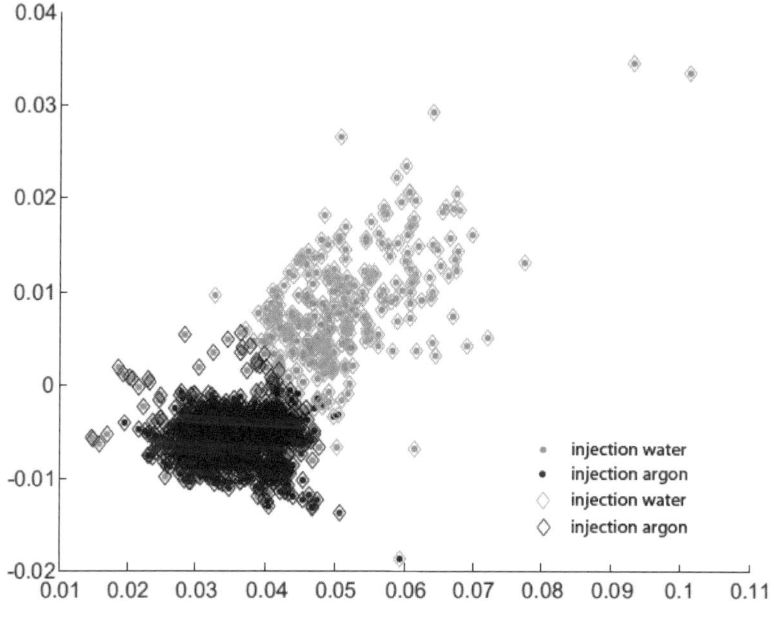

Figure 13.3. *Projections of randomly-selected recordings in the initial stage (points) and the affected classes (diamonds)*

Of the recordings made, 20.60% (±3.8% with a risk of 0.001) were incorrectly classified, of which 3.40% were "argon injections" that were labeled as "water injections". The classification errors therefore mainly involve "water injections", with an error rate of 43.00% (±7.3% with a risk of 0.001) in the 404 samples generated. This rate is greater than the rate of incorrectly classified tests in the preliminary study (24.00% of injections examined). The eight incorrectly classified points in the model's generation represent 24.50% of errors, which is more than half of the classification errors. For these eight injections, the rate of correct classification of these tests is zero or almost zero.

A low water injection flow rate and a localization in "EE" (the injection space between envelopes, see Chapter 9) lead to a significant increase in the number of points that are incorrectly classified. The positive aspect of this result, however, is that a low water injection flow rate is classified as an injection of argon. This result indicates that injections of argon are similar to small water leaks and allow us to test the performance of monitoring systems without the need for water injections.

13.2.3. *Exploring implementation in a new installation*

13.2.3.1. *Issues*

On the basis of the hypothesis that water injection tests cannot be carried out in new installations, we cannot generate a 'water injection' class that is adapted to the new installation. In this instance, the construction of a leak detection method is reliant on recordings taken in the installation, supposing that it is free of leaks.

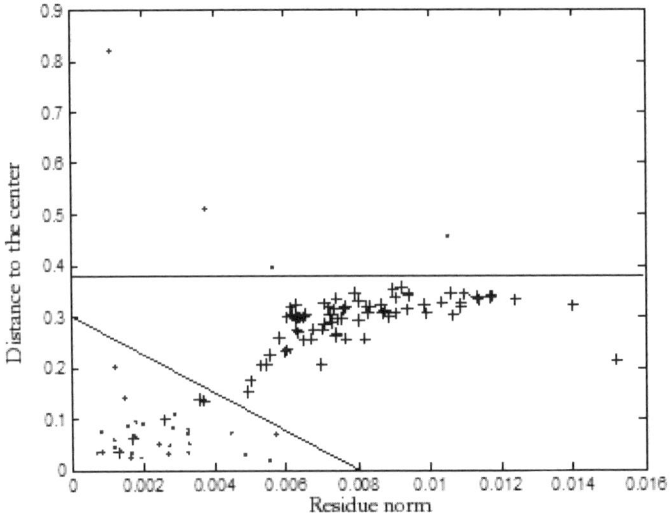

Figure 13.4. *Projection of residues – distances to the center:*
•: non-injections ; and +: injections

To implement a reliability study of this kind of strategy, a PCA model is created using the “non-injection” labeled recordings. We then verify its capacity, on the basis of proximity to the model, to distinguish the injections in this state. Proximity is determined according to two criteria:

– the Mahalanobis distance between the new recording and the center (also called a Hotteling T^2), which indicates the proximity of the principal component projection to that of the model; and

– the residue norm, which indicates the model’s capacity to represent the PSD obtained.

The newly recorded situation corresponds to the model and therefore to the ‘non-injection’ state if the two criteria are close to zero.

13.2.3.2. *Detecting leaks*

The PCA model uses four principal components that represent 95.6% of inertia.

In Figure 13.4, we can notice that:

– It is possible to establish a separator between “non-injection” and “injection”. Six points “with injections” have been classified as “non-injection”. They are the same as those previously obtained (see section 13.2.2.1), which correspond to argon injections. Five of them have a low output.

– Four “non-injection” points diverge strongly from the model: three correspond to tests carried out on water and one is at the start of a test carried out on argon. A visual examination of the signals and tests preceding the four incorrect points does not allow us to determine the cause of this incorrect classification.

– All of the recordings corresponding to injections of water are correctly classified as being ‘with injection’.

– The treatment of the aspect of residue–distance provides a better separation of the ‘non-injection’ and ‘with injection’ points than if one of the two criteria had been used alone.

13.2.3.3. *Conclusions on the PSD approach*

The results obtained show that an approach using the DSP of recordings has the capacity to distinguish:

– between ‘non-injection’ and ‘with injection’ states; and

– between, “with injection”, the ‘water injection’ and ‘argon injection’ states during unsupervised study.

Recognition is carried out using a group of model classes composed of the (three) main components retained, the centers of classes and consideration of the Euclidian distance for each class. The results show:

– the ability to detect water or argon injections with a false alarm rate of 2.87% (detection of ‘injection’ when the system is deemed to be in a “non-injection” state), mostly caused by recordings of water, and a 100% detection rate for water injections; and

– the “water”–“argon” distinction gives better performance for larger water injections.

The same results are obtained with the study that only uses data available in a new installation. An examination of the poorly classified points shows that:

– a small “argon” injection may not be detected while a small ‘water’ injection is often well-classified as an injection; and

– small “water” injections may be classified as “argon”.

This observation confirms the possibility of using tests on argon to verify the ability of the leak-monitoring system to detect small water leaks.

13.3. Analysis of the filtering characteristics

We hypothesize that a leak modifies the filtering characteristics between sources and sensors and that these modifications are different for each of the two sensors. An autoregressive (AR) model is proposed to represent the filtering characteristics between the source and sensor. This model avoids the use of non-linear optimization techniques, which are often blocked in local optima. If these hypotheses are verified, the detection of leaks relies on comparing the AR models associated with each sensor.

13.3.1. *Estimating filtering characteristics using an AR model*

AR modeling expresses output at the moment k using previous p samples [BRO 98]:

$$y_k = -a_1 y_{k-1} - a_2 y_{k-2} - \ldots - a_p y_{k-p} + e_k$$

where a_i $(i = 1,...p)$ are the coefficients of the AR model and e_k is a white noise.

The estimation of the AR model's coefficients is carried out using the following Yule-Walker equation:

$$\begin{pmatrix} Rss(0) & Rss(-1) & \dots & Rss(1-p) \\ Rss(1) & Rss(0) & \dots & Rss(2-p) \\ \vdots & \vdots & \vdots & \vdots \\ Rss(p-1) & Rss(p-2) & \dots & Rss(0) \end{pmatrix} \begin{pmatrix} a_1 \\ a_2 \\ \vdots \\ a_p \end{pmatrix} = \begin{pmatrix} -Rss(1) \\ -Rss(2) \\ \vdots \\ -Rss(p) \end{pmatrix}$$

where $Rss(m)$ is the autocorrelation function obtained empirically using N samples:

$$Rss(m) = \sum_{n=0}^{N-1} y_k y_{k-m}, \quad m = 1, 2, ..., p$$

The p coefficients of the AR model are estimated using the least squares method with the Yule-Walker equation.

We have two sensors numbered 2 and 6. Coefficients $\hat{a}_i^j$ $(j \in \{2,6\},\ i = 1,\dots,p)$ are estimations of the models' coefficients of p order associated with each sensor.

13.3.2. *Comparing filtering characteristics*

Modifying the filtering characteristics changes the value of the filtering coefficients estimated using the signals recorded by the sensors. In the case of a stable AR filter, the coefficients are impulse responses that decrease when the filter is stable. The modifications therefore have a strong influence on the first coefficients of the AR model. It is for this reason that we have chosen to study the first coefficient of each of the two associated filters.

The SMD allows us to account for the random nature of this gap in the analysis of modifications. It is calculated using the following formula:

$$e_k = \frac{\bar{\hat{a}}_{1,k}^{2} - \bar{\hat{a}}_{1,k}^{6}}{\sqrt{\dfrac{\hat{\sigma}^2_{a^2_{1,k}} + \hat{\sigma}^2_{a^6_{1,k}}}{N}}}$$

where the integer N represents the number of points in a sliding window used for calculation:

– $\bar{\hat{a}}_{1,k}^{2}$ (respectively $\bar{\hat{a}}_{1,k}^{6}$) is the average estimated at moment k of the first AR model coefficient of the measures taken from sensor 2 (respectively sensor 6);

– $\hat{\sigma}_{a_{1,k}^{2}}^{2}$ (respectively $\hat{\sigma}_{a_{1,k}^{6}}^{2}$) is the variance estimated at moment k for the first AR model coefficient of the measures taken from sensor 2 (respectively sensor 6).

This SMD is calculated online. The study of these variations according to whether or not there are any leaks should allow us to validate the proposed hypotheses.

13.3.2.1. *Detecting leaks using SMD analysis*

The SMD is studied in three functioning zones:

– the functioning zone before the first injection;

– the injection zone; and

– the functioning zone after injection.

This breakdown into three zones instead of two (injection and non-injection zones) comes from the observation of the influence of leaks remaining after they have been stopped.

13.3.2.2. *"Non-injection" zone after the first injection*

Calculation of the SMDs has shown that the extensions of their values in this zone are essentially included in the interval [1:17].

13.3.2.3. *Injection zones*

In the injection zones, the SMD values noticeably decrease as leaks occur, as shown by Figures 13.5 (argon injection) and 13.6 (water injection). This SMD is less than one in all the water or argon injections.

The value of the gap depends on the leakage flow rate. For argon injections, for example, small values in the flow rate have a small decrease in the SMD which does not decrease past one. However, for larger flow rate values, its value decreases noticeably and passes below -4.

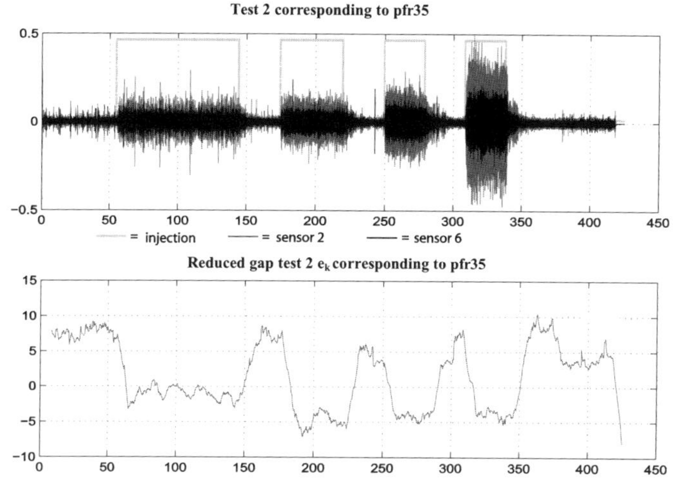

Figure 13.5. *Variation in the SMD during argon injections*

13.3.2.4. *Non-injection zones after injection*

Once injections stop, the SMD increases, but it does not occur in the interval [1:16]. The SMDs calculated in the non-injection zones after one or more injections change and often have values that are lower than those in the zone after the first injection.

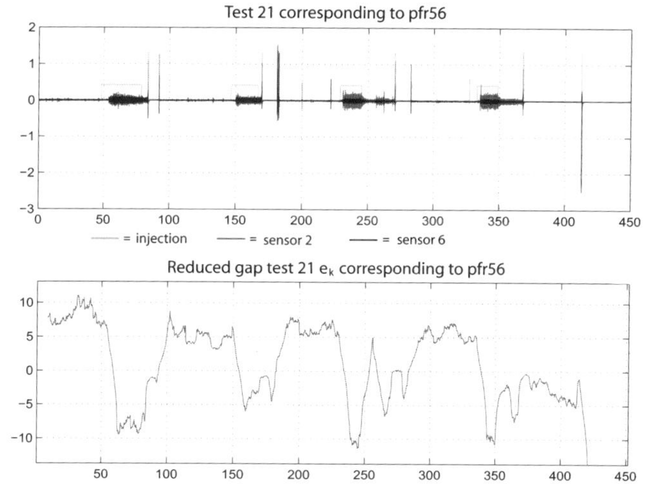

Figure 13.6. *Variations in the SMD with injections of water*

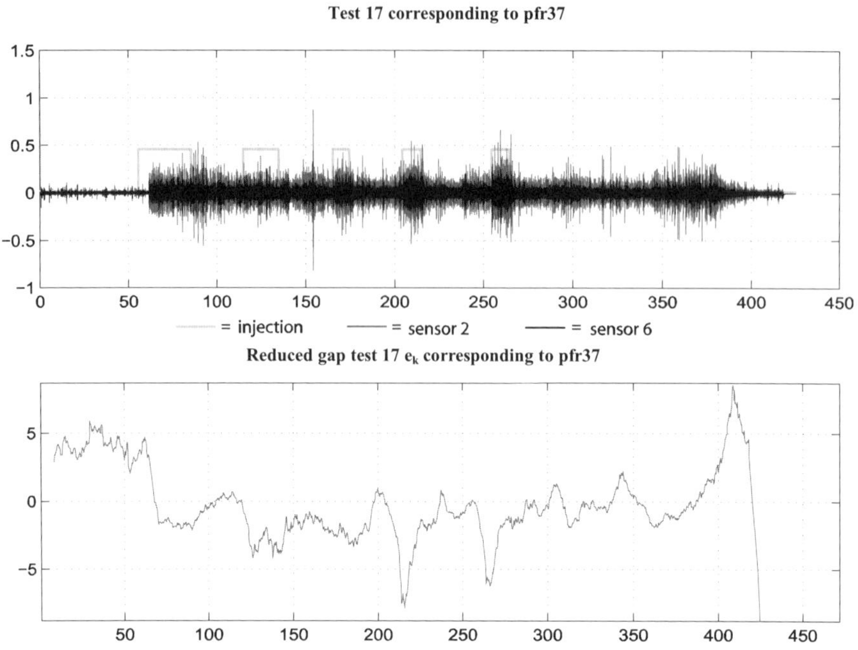

Figure 13.7. *Variation in the SMD depending on the injection*

Figure 13.7 illustrates the effect of previous injections on the average value of the SMD. While the effect of anterior injections does not disappear, the SMD increases, though it does not reach the value of the first non-injection zone.

Note, however, that there is the continued effect of previous injections for some short periods between leaks. This effect is strong when the amount of injection water is significant. We should highlight that this effect does not prevent the detection of new leaks because there is still a noticeable decrease in the SMD with each new injection.

13.3.3. *A leak detection algorithm*

On the basis of previous observations, we have proposed a leak detection algorithm based on the online estimation of the SMD.

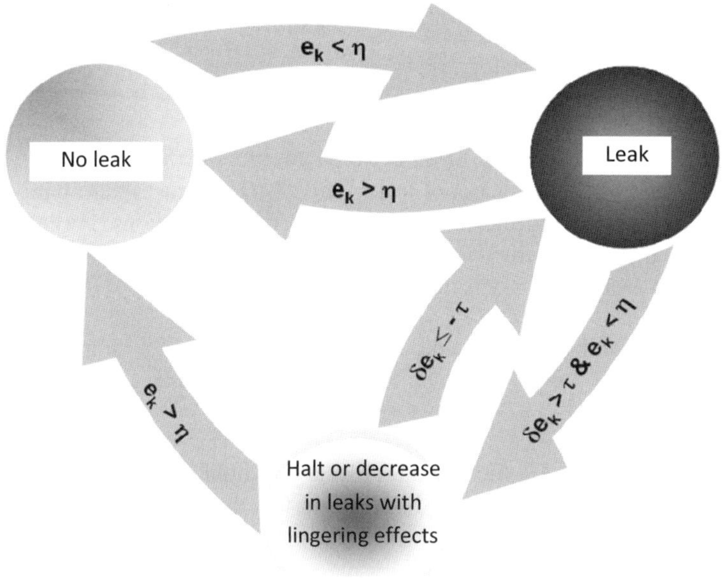

Figure 13.8. *The installation's different modes of functioning*

The aim is to be able to identify three modes:

– a safe operating mode without a leak;

– a failing operating mode with leaks; and

– a mode with a decrease in undesired effects corresponding to stopping a previous leak with continuing influence or decreasing the leak flow rate.

Figure 13.8 illustrates the algorithm proposed for the detection of leaks and supposes that initially the installation is in a functioning state. We suppose here that there is no initial leak. Two thresholds η and τ and an instantaneous rate of the SMD δe_k are used in the algorithm. The threshold η is a low limit of the SMD in safety mode and τ the instantaneous rate beyond which the SMD changing is supposed significant.

13.3.4. *Conclusions on the autoregressive signal modeling-based approach*

The study of variations in the SMD between the initial coefficients of the AR models associated with each sensor depending, on whether there are leaks or not, has verified that:

– the presence of a leak modifies the filtering characteristics between the source and sensors; and

– this modification is different for each sensor.

This justifies the use of this SMD between the first coefficients of an AR model for sensors 2 and 6 as an indicator of leaks. This SMD decreases during injections of water and argon, but the decrease is greater with water injections. Furthermore, the SMD value increases when injections cease. An algorithm based on these observations is proposed for detecting leaks.

This method can be applied to other installations and used with multiple sensors. A multidimensional study would be necessary in this case, with the use of the generalized SMD.

A study of an operating mode without leakage would improve leak detection by determining the dispersion of the SMD in this functioning mode. This could thus enable us to better estimate the threshold for leak detection.

13.4. Conclusion on frequential and temporal approaches

The approaches examined in this chapter fall into two categories:

– a method based on detecting changes in state; and

– a recognition method for a given sample without taking any previous decisions into account.

It is possible to envisage these two approaches being integrated for system state recognition. This should occur in models' distance spaces (including residues). This approach has the benefit of being able to classify use of the same type of information in contrast with approaches that construct recognition models based on a set of raw information that may vary.

In the hypothesis examining implementation of a new installation, the signals available are those of "non-injection" and potentially, if argon injection systems are considered, argon injection signals. In this case, two models are constructed (and can be updated) on the basis of these signals that will be used for detecting:

– the presence of an injection (use of the "non-injection" model); and

– a leak other than argon (use of the "argon" model with a leak).

The reliability of this approach was evaluated using the PSD approach with the resulting performance being comparable to that using signals linked to water injections.

13.5. Bibliography

[BRO 98] BROCKWELL P., DAVIS R., *Time Series: Theory and Methods*, Springer, New York, 1998.

[FLE 01] FLEURY G., *Analyse Spectrale,* Ellipses, Paris, 2001.

[FUC 94] FUCHS C., BENJAMINI Y., "Multivariate profile charts for statistical process control", *Technometrics*, vol. 36, pp. 182-195, 1994.

[HOW 02] HOWARD R.M., *Principles of Random Signal Analysis and Low Noise Electronic Design: The Power Spectral Density and its Applications*, Wiley, New York, 2002.

[KOU 95] KOURTI T., MACGREGOR J.-F., "Process analysis, monitoring and diagnosis, using multivariate projection methods", *Chemometrics and Intelligent Laboratory Systems*, vol. 28, no. 1, pp. 3-21, 1995.

[SAP 90] SAPORTA G., *Probabilités, Analyse de Données et Statistique,* Technip, Paris, 1990.

PART 4

Human, Organizational and Environmental Factors in Risk Analysis

The analysis of accidents and incidents in complex systems shows how organizational, human, technical and environmental dimensions are independent. The different components of complex systems are sufficiently secure that a single failure will lead to a serious incident. Accidents not incidents are the result of a combination of more or less serious factors influencing these four dimensions. It is therefore necessary to have access to tools and methods that allow us to analyze these four families of factors. These tools must allow us to identify conjunctions between serious factors in technical and human systems that could weaken the "defense" system.

The MARATHON project aims to deal with this issue. Its methodological framework allows us to examine the system on three levels: organizational, technical and actions (i.e. human operators). The SOMAIR project (the role of the "human–machine system" interface in integrated risk analysis) focuses on actions by examining the influences of these levels.

Chapter 14

Risk Analysis and Management in Systems Integrating Technical, Human, Organizational and Environmental Aspects

14.1. Aims of the project

In its strategic risk analysis guidelines following the analysis of accident scenarios (such as Chernobyl, for instance), EDF identifies the need to look beyond technical aspects in risk analysis and consider organizational and human aspects. In this context, CRAN (the Research Center for Automatic Control of Nancy)[1] and EDF[2] have initiated research [LEG 09a] that has led to the development of an integrated risk analysis approach (IRA) based on a unified knowledge model using Bayesian networks. This model allows us to integrate technical information about systems – which are themselves composed of subsystems, each with a large number of interconnected views – and to include human aspects such as maintenance and control actions, since these are carried out in an organizational context that influences efficiency and safety (i.e. an evaluation of the effectiveness of maintenance barriers). In addition to this, the technical system is often subject to environmental hazards and regulatory constraints (hazards that develop over time) that require parameters affecting operational performance and demands to be

---

Chapter written by Geoffrey FALLET-FIDRY, Carole DUVAL, Christophe SIMON, Eric LEVRAT, Philippe WEBER and Benoît IUNG.
1 www.cran.uhp-nancy.fr.
2 www.edf.fr.

considered. This has led to the integration of information related to these constraints in the IRA.

The MARATHON project, carried out with the cooperation of EDF and CRAN within the GIS 3SGS framework, has furthered research in this field with the aim of extending the IRA methodology to other areas of application. It aimed to highlight its limitations and provide potential solutions to them but its primary focus was on tackling epistemic and random issues of uncertainty inherent to the model created using this methodology. The aim of this study into uncertainty was to quantify its influence in quantitative risk assessment knowing that this takes:

– a random (probabilistic) form for technical or environmental aspects where there is a significant amount of feedback data; and

– an epistemic form in areas with human and organizational factors.

Focusing on the human factor of the IRA, a complementary aspect to MARATHON, the SOMAIR project (the role of the 'human–machine system' interface in integrated risk analysis) led to the collaboration between CRAN, EDF and LAMIH in GIS 3SGS who aimed to consider the individual above the collective. As a result, the MARATHON and SOMAIR projects aimed to contribute to an original integrated risk analysis approach that accounts for human (individual and/or organization), environmental, technical and organizational dimensions, such as those identified by EDF. Using the resulting model, from both a designer and end-user perspective, this enables us to evaluate the performance of maintenance protocols, safety barriers etc., and to assist with the elimination, reduction and limitation of risks.

14.2. State of the art

14.2.1. *Context of the study*

For a number of years, risk analysis was seen from a purely categorical perspective [ZIO 09], i.e. only one category of risk (e.g. technical, environmental, human or organizational) was examined at a time, and it mainly focused on technological risk analysis in systems. However, the occurrence of a number of serious disasters since the 1970s highlighted the strong links between these different categories. Among these, notable examples include Three Mile Island [KEM 79], Bhopal [UCB 85], AZF [BAR 01] and, more recently, Fukushima. This is because industrial systems are becoming increasingly complex [LEM 99] with an increasing number of views and interactions between the technical system and its environment (physical, regulatory, human, organizational, etc.). These are therefore referred to as "complex sociotechnical systems".

These different disasters have underlined the need to account for the strong connections between a technical system, its environment (physical and regulatory), the human actions to which it is subject and their underlying organizational aspect. This new view of risk analysis therefore allows the analyst to consider the interaction between different specialist views involved in a "global" risk analysis. Since these accidents, various pieces of research have been started (some of which are still underway) to develop a new approach to risk analysis that integrates the different specialist views mentioned previously (technical, environmental, human and organizational). Notable research includes:

– [PAT 96], with the development of the system-action-management (SAM) approach;

– [SVE 02], highlighting the importance of considering different actors in risk analysis in an industrial system through the graphic representation of causal flow of accidents (AcciMap);

– [PAP 03] developing the I-Risk method, accounting for both technical and organizational characteristics in system risk analysis in the chemical industry;

– [PLO 04] with MIRIAM[3]–ATHOS[4] to evaluate major risk management systems examining technical, human and organizational factors;

– [CHE 06], focusing on the representation of accidental scenarios via the bow-tie formalism (developed within the ARAMIS project framework[5]) to facilitate the organizational learning process; and

– [MOH 09], who has proposed a means of carrying out probabilistic safety studies taking organizational factors into account.

While this research has allowed us to develop new risk analysis methods, comparison between them shows that while the majority of them are multidisciplinary, they do not cover the four specialist views previously identified and often focus on a single area of application (e.g. chemistry with the I-Risk method). We therefore need to develop a new methodology that allows us to account for technical, human, organizational and environmental factors while ensuring the general applicability of the resulting methodology with regards to the different areas of application.

The aim of the MARATHON project is to contribute to developing an integrated risk analysis methodology responding to the need for a multidisciplinary method

---

3 *Maitrise Intégrée des Risques Industriels d'Accidents Majeurs*, Integrated Control of Industrial Risks for major accidents.

4 *Analyse Technique, Humaine et Organisationnelle de Sécurité*, Technical, Human and Organisational Safety Analysis.

5 *Accidental Risk Assessment Methodology for Industries in the framework of Seveso II directive.*

that is as generic as possible. This proposed methodology allows us to study technical systems in their environment that are subject to maintenance or control actions within an organizational context. MARATHON needs to fulfill this general applicability requirement, identify its limitations and difficulties and open perspectives for improvement, mainly in terms of issues of uncertainty.

14.2.2. *Towards an "integrated" approach to risk: combining several specialist disciplines*

As highlighted previously, the aim of IRA is to propose a cross-sector generic risk analysis methodology. MARATHON aims to reinforce this general applicability. This kind of implementation initially requires the identification and modeling of connections between different specialist views. The general problems with implementing this kind of methodology relate to obtaining a risk model coupling several of the four specialist views. The specific nature of a risk analysis approach such as this lies in the differences in information being examined in each of these specialist views. As such, one of the main objectives is to resolve inconsistencies (semantic, syntactic, etc.) between the information provided by these different disciplines by bringing together these semantics into a single approach and risk analysis model.

There are two possible solutions that can be used to combine these different views: *unifying* or *sharing* information. In the case of *unification* all of the information is directly represented in a single risk model by bringing together different information while ensuring coherence. With this type of information combination, each discipline must share a single semantic and formalism. This is because coding is unique and must represent the four views according to predefined patterns[6]. The creation of a risk model is therefore carried out according to a "global" perspective and not a "specialist" view and requires the semantics used to include all of the specialist views using the model. This solution relies on all the key concepts from each of the specialist views. In this type of configuration, one of the main problems involves ensuring that we use the tools and methods in each individual discipline to their maximum potential. Even when using the key concepts taken from each of these views, however, one of the risks of this unification approach is a loss of semantics caused by simplifications and modeling incapacities in relation to "sharp" specialist tools.

In the case of *sharing* information: information is shared between different views by identifying corresponding semantics, converting information, defining

6 Term used for a model, structure, feature, etc. It refers to phenomena that have been repeatedly observed, which gives them specific characteristics.

characteristics, etc. As such, the data from each specialist view are examined independently from each other using the tools and methods from each of these specialist views. Conversion rules or semantic correspondence allow us to exchange information between different specialist views in order to carry out global risk analysis. Each specialist view brings together specific and adapted tools and methods. An independent analysis of each view using its specific tools and methods therefore provides a more comprehensive representation of information used in the risk analysis being carried out. The main problem with sharing information is "interoperation" between models to ensure coherence.

The notions of unifying and sharing information can be approached in two ways; *integration* and *interoperability*. Bringing together different specialist knowledge in a single model is an integrated approach while examining them independently and comparing them for a global risk analysis is an "interoperable" approach. The notion of interoperability is defined by [IEE 90] as "the ability of two or more systems or views to exchange information and to use the information that has been exchanged". Other definitions are also given in [IEE 00, EIF 04] and [CHE 05]. While these definitions often relate to IT and commerce, they can also be applied to the general notion of interoperability, as highlighted by [VER 07] in a comparison of IT with complex sociotechnical systems. Comparing notions of integration and interoperation, the standard ISO 14258 [AFN 98] clarifies the difference between these two notions by identifying the main difference between an "integrated" and an "interoperated" system as the dependency of their view elements. Integration goes beyond interoperation in the sense that a certain level of functional dependence is introduced between the components of the system being studied. As stated previously, when these two approaches aim to bring together the different elements of a system and ensure proper overall operation, integration requires the interfaces and barriers between the different elements to be lifted for the formation of a coherent whole. With this in mind, [VER 07] distinguishes three types of integration: complete, strong and weak. He approaches the notion of interoperation from the scalar view of "weak integration", where the different elements exist separately and contribute to the system's final overall state via other tools that allow intercommunication (as a kind of portal).

There are other ways of approaching this issue of knowledge unification–sharing, such as the "system of systems" (SoS) concept in a systems engineering (SE) approach. [LUZ 10], who view SoS as "a collection of systems that can potentially be acquired and/or used independently, for which the designer, purchaser and/or user seeks to maximize the performance of the overall value stream". SoS are also linked to criteria similar to those encountered in our IRA methodology, such as the interdisciplinarity and variation of constitutive elements, "the emergence of a whole" from a lasting or punctual interaction, etc. In addition, "system integration" is one of the views of the SE approach. SE is defined [HAM 10] as "a collaborative, interdisciplinary methodological approach, encompassing all activities required in the design, development, evolution and verification of a set of processes". SE involves a logic that is broken down into subsystems or sub-problems that must be carried out using the IRA methodology, since the systems studied are complex (anticipating different types of risks, characterizing uncertainty, etc.)

In relation to this double "integrated–interoperated" vision, different methodologies have been developed over recent years in the field of multidisciplinary risk analysis that have often revolved around an "integrated" view of the different specialisms concerned. As such, when implementing our new IRA approach and following research led by [LEG 09a] into knowledge formalisms, we consistently apply an "integrated vision".

Recent applications of the IRA methodology have often raised issues, however, when an integrated approach has been chosen. The main advantage of this choice (including the use a single risk model with key concepts from each specialist view) can in some cases be a hindrance in the development of the methodology. For example, in terms of knowledge formalization and the choice of specific factors representing the human view – do they necessarily represent all the information available? If not, how can we compare this to a MERMOS study [BIE 98], for example? Equally, in terms of the issue of uncertainty, different solutions can be applied in terms of the data being used. The choice of an integrated model with a very specific formalism (that of Bayesian networks), however, introduces constraints in terms of the use of different solutions that cannot be directly implemented in this integrated formalism. These different issues raise a new object of study: taking the line opposite to the integrated approach, with knowledge formalization that involves interoperation (moving beyond the project's framework). As such, seeking to capitalize on research led in the development of the IRA. MARATHON involves adapting existing methods to the current demands and constraints of our models to capitalize on and strengthen work carried out to date.

14.3. Integrated risk analysis

14.3.1. *Concepts*

This new "integrated" approach between several disciplines, such as operational safety, risk analysis, human reliability and organizational analysis, aims to develop methods, aid the choice of appropriate tools and refine them for risk analysis in these technical systems:

– in close relation to the physical and regulatory environment;

– integrating human actions, such as maintenance, surveillance and management taken in their organizational context;

– guaranteeing key issues in the system – its availability as well as safety, maintaining property, its environmental impact;

– to prioritize these different kinds of risks in relation to different issues with the aim of setting aside appropriate resources; and

– thereby contributing to a better risk culture [AIE 91] and better communication about risks.

This therefore results in a "tiered" approach that can cover two or three or all of these areas, depending on the focus of the study and the anticipated risks.

In parallel with these methodological developments, EDF has implemented these advances to illustrate the significance of this new approach. In the framework of these demonstrations, representations of these risk models have been constructed for key systems in the nuclear industry in close relation to the surrounding environment, with the cooling source being retained. This system keeps the cooling water supply in the nuclear installation by drawing water from nearby rivers, the sea or oceans and pumping it towards the exchangers, thereby ensuring cooling via suction pumps passing through several filters. This system is technical in nature, is under the strong influence of its physical and regulatory environment and relies on a number of human actions, such as maintenance and management, which are themselves influenced by the organizational context.

The key principles of the approach can be summarized as follows:

– integrating risk models to obtain a unique model covering technical and human aspects (potentially also examined within their organizational context) which could impact on different aspects of the system such as safety, operation, etc.;

– describing the system using levels relative to these different aspects and relationships of influence between these levels;

– introducing the notion of a *barrier* into the technical model to account for technical means of reducing these risks, the human actions of maintenance and management, potentially taken within their organizational context. The *effectiveness of these barriers* is assessed using the aforementioned relationships between the different levels.

14.3.2. *A description of the approach*

An introduction to the approach developed featured was presented in the *Journal of Risk and Reliability* [LEG 09b], which will be summarized here.

The conceptual framework (see Figure 14.1) is based on those cited in the SAM approach developed by Paté-Cornell and Murphy [PAT 96], which proposes that organizations influence human actions, and as such the operation of the technical system (see Figure 14.2).

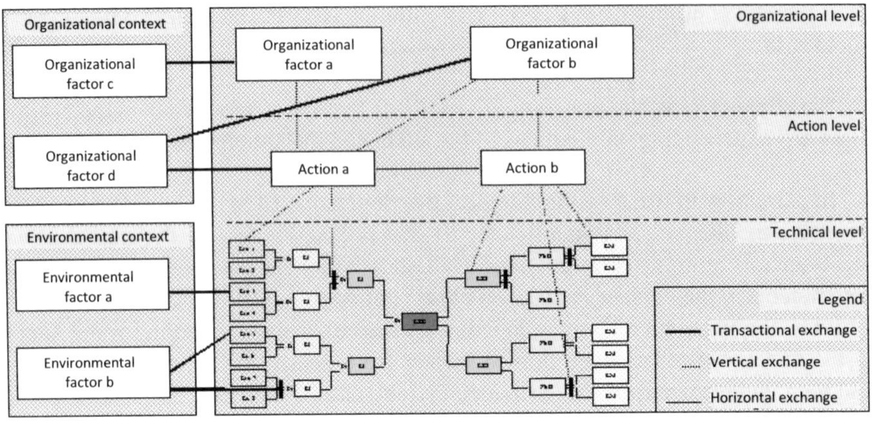

Figure 14.1. *Conceptual framework of the IRA*

The system is divided into three representative levels that interact through horizontal and vertical exchanges (see Figure 14.1): technical, human and organizational levels. This system is also influenced by transactional exchanges (see Figure 14.2) using external constraints; the context of the natural environment and organization [DUV 07].

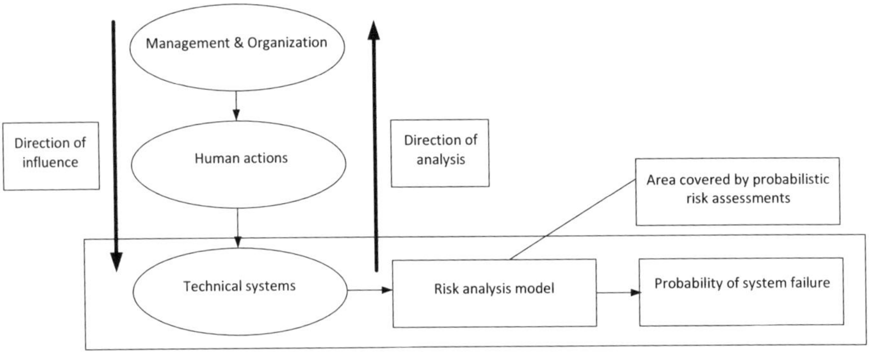

Figure 14.2. *Structure of the effects of humans and organizations on risk*

This distinction is made because the characteristics of the system's variables are different from those of contextual variables. The system's variables can be controlled while the latter influence the system but can be subject to them.

14.3.2.1. *Technical dimension*

Technical analysis is carried out using the "bow-tie" method (at the technical level shown in Figure 14.1) developed in the European ARAMIS project [AND 01].

A bowtie contains a fault tree and an event tree in which each path identifies an accident scenario. As a result, it allows us to describe an accident scenario from the trigger event to the final consequences while accounting for the influence of barriers.

14.3.2.2. *Human dimension*

This level characterizes the effect of specific human actions. This can be measured by comparing real and targeted performances. These actions are classified into two categories: "management actions" (supervision, diagnostics, etc.) that keep the system in its operational state; and "maintenance actions" that include all technical, administrative and management actions carried out during the lifecycle of a product that keep it (or direct it towards) in a state that allows it to fulfill its required function [PIE 07]. In terms of this dimension, the approach focuses on collective (and not individual) behavior.

Each action can be divided into three stages or general phases [GER 99] that structure the collection of information necessary to estimate the efficiency of actions:

– the preparation stage (which includes planning, defining and characterizing all of the conditions required for the correct execution of an intervention);

– the implementation stage (implementing this intervention into the system's operation); and

– the closing stage (reinforcement that guarantees the correct implementation of this intervention and confirms its continuation).

Each action is characterized by indicators which affect its efficacy. These indicators include:

– delegation (De);

– aids (Ai);

– training (Tr);

– experience (Ex);

– compliance with work specifications (Cws);

– contextual factors (Cf);

– collective management and group dynamic (Cmgd);

– real-time control (Rtc); and

– feedback experience (Fe).

Delegation (De) refers to giving responsibility for a task to another person who is generally subordinate. This delegation may (formal delegation) or may not be (spontaneous or informal) imposed by the organization.

Experience (Ex) can be defined as the knowledge acquired by the practice accompanied by a reflection on this practice, as expressed by Henry Fayol who wrote that experience is ‘the memory of the lessons we have ourselves drawn from facts’ [FAY 99].

Training (Tr) is defined as the series of activities (implemented by an organization) that aims to ensure the acquisition of practical skills, knowledge and attitude required to fulfill a position of employment (and thereby ensure the ability of operators to complete actions).

Collective management (Cdgm) of objectives and group dynamics identifies means implemented by the work collective to achieve specific objections (making decisions before acting, etc.).

Others are specific to tools that can be used by the collective to carry out these actions:

– (Ai) is the series of procedures (prescriptive documents) and tools (other documents and directions) used by operators to aid their activity;

– (Fe) is the relationship between the results of interventions in readapting the objectives that were initially fixed. To characterize these indicators, we will question whether the collective can always apply and it whether it is always applied;

– (Rtc) is an assessment of the conformity by observation and accompanying judgment, if necessary, of measures, tests or calibrating (ISO 9000) and the means implemented to judge the visibility of the results of an action;

– (Fe) feedback is a structured approach to capitalizing on and exploiting information from the analysis of positive and/or negative events. It uses a series of human and technological resources that must be managed to contribute to reducing repeated errors and encourage good practice.

The last variable (Cf), contextual factors, relates to external elements which influence the collective when carrying out an action (e.g. lighting, smoky environment etc.).

These indicators have been identified on the basis of work focused on human reliability based on fire probabilistic safety assessments [PES 08] and the ATOS (Analysis of Technical and Organizational Safety) method [LEC 01, LEC 02] developed by INERIS (French Institute for the environment and risk).

These indicators, established from our knowledge of the field, are characteristic of an action and must be evaluated using interviews. This list does not aim to be exhaustive, but can be added to according to specific contexts. These indicators can be regrouped and affected by the preparation/operation/closing stages/phases described above.

This regrouping process allows us to structure knowledge and can be useful during and after interviews to manage the collection and organization of information: preparation (delegation, aids and training), implementation (experience, respecting guidelines, environmental factors, collective management and group dynamics) and closing (real-time control and feedback).

14.3.2.3. *Organization dimension*

The modeling approach at this level is based on an overall pathogenic view: the approach [DIE 04, PIE 07] illustrates the organization as a whole and describes it using various pathogenic organizational factors (after an accident). These factors include the following:

– weakness in organizational culture of safety;

– failure in day-to-day management of safety;

– weakness of control mechanisms;

– poor handling of organizational complexity;

– difficulty in adapting to feedback;

– production pressures; and

– failure to re-examine design hypotheses.

These factors undermine safety and can provoke or trigger accidents. They are the result of a combination of indicators that are characteristic of a negative influence when an accident occurs. As a result, from an organizational perspective this is a common cause of failure (it includes a series of phenomena, processes and effects that have an impact on the organizational structure). These factors can also be used to provide a detailed description of events or to guide research into an organizational structure in order to establish a global (or partial) diagnostic for safety. As a result, they must be defined once for the system being studied using an organizational diagnostic.

Contextual dimensions (organizational context and the surrounding natural environment) are described, analyzed and modeled within this methodology but will not be examined within the scope of this chapter. The organizational context represents processes related to the situation in which the system develops (social situation, regulations and competition), while the context of the natural environment represents processes relating to the evolution of the physical and natural climate (meteorological data and geographic location).

This methodological description will be accompanied by a tooled platform to evaluate the efficacy of these barriers in a Bayesian network representation of a technical risk model.

14.4. Accounting for uncertainty in risk analysis

14.4.1. *Different kinds and sources of uncertainty*

Risk is by its very nature inherently uncertain and the kind of industrial processes with which EDF operates intrinsically cause risk. Furthermore, accounting for uncertainty in risk studies is now a legal obligation, since risk is increasingly socially unacceptable (there are requirements to be dependable, environmentally responsible, decision-making, etc.) [DER 08, ZIO 09]. As such, the growing complexity of systems, devices and the mixture of different study views in the analysis of sectorial risks (technical, human, organizational or environmental) reinforce the notion of uncertainty and the need to take it into account. Uncertainty is the key issue examined by MARATHON.

Uncertainty emerges as a difficulty encountered by analysts when attempting to propose a sufficiently detailed model with the correct parameter values. As such, two specific problems relating to uncertainty appear. The first problem concerns the type of uncertainty. If this consists of uncertainty related to a problem with knowledge about the source, then this can be said to be an epistemic uncertainty. The term "reducible" is also used because acquiring new information reduces uncertainty. If it is a repeated event where the values are precise but differ according to natural variations, however, uncertainty can be said to be random. The term irreducible is also employed [BAE 04]. The second problem concerns the place at which uncertainty occurs. For example, there may be uncertainty in the model or uncertainty in its parameters. With this in mind, and specifically for nuclear safety, in their technical report for the NRC, Drouin *et al.* [DRO 07] have observed that these two specific elements concern uncertainty.

An initial citation can be found in [NRC 07]:

> "An important aspect in understanding the base PRA results is knowing what are the sources of uncertainty and assumptions and understanding their potential impact. Uncertainties can be either parameter or model uncertainties, and assumptions can be related either to PRA scope and level of detail or to model uncertainties. The impact of parameter uncertainties is gained through the actual quantification process."

A second reference can be found in [BON 03]:

> "... a systematic treatment should include rigorous analyses for parametric uncertainties, sensitivity studies to identify the important epistemic uncertainties, and quantification of the latter. In a risk-informed environment, the proper role of sensitivity studies is to identify what is important to the results, not to replace uncertainty analyses."

Bouchon-Meunier has divided gaps in knowledge into three main types [BOU 95]:

– uncertainties that represent a doubt about the validity of knowledge;

– imprecisions that correspond to a difficulty in describing or obtaining information; and

– incompleteness due to a lack of knowledge or only partial knowledge about certain characteristics of the system.

Some researchers have added problems relating to imperfections, in terms of data or knowledge, to this list, viewed via these models. Smithson [SMI 89] has cited a lack of knowledge as a source of uncertainty, as shown in the tree in Figure 14.3.

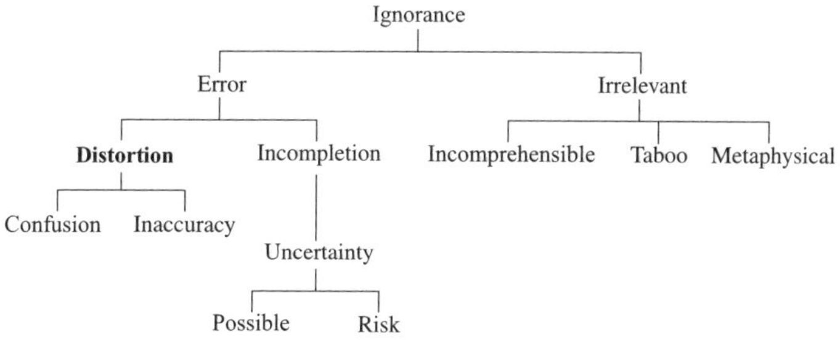

Figure 14.3. *Constituent elements of a lack of knowledge*

Incoherence should also be added to this list of imperfections. This occurs when knowledge is gathered from one or several sources that provide contradictory information [OSE 03].

14.4.2. *Frameworks for modeling uncertainty*

The presence of different types of uncertainty in risk analysis presupposes the use of one or several *ad hoc* modeling frameworks. Normally a probabilistic framework is used to model epistemic and random uncertainty. The scientific literature is abundant with articles showing that a probabilistic framework is not best able to handle the different aspects of epistemic uncertainty. This, however, is still the subject of debate. Further information on the different positions taken in relation to this problem can be found in the following references [OBE 04, FER 96, SHA 76] but does not fall within the scope of this chapter.

Several modeling frameworks are specifically used for modeling and handling uncertainty. First, the probabilistic framework is normally employed when modeling random phenomena (objective framework) and for modeling knowledge about a source (subjective framework), which is again a scalar of controversy. Second, the interval theory can be applied to probabilities or non-additive theories, such as the possibility theory proposed by Zadeh [ZAD 78] and further developed by Dubois and Prade [DUB 88], the framework of belief function theory [DEM 67, SHA 76], Dezert-Smarandach theory [SMA 09], and imprecise probability theory [WAL 91, FER 02], etc.

These different modeling frameworks are brought together in random phenomena modeling but offer different possibilities because they pose different initial hypotheses.

14.4.2.1. *Probability theory*

The probabilistic framework remains the reference framework for examining variables where the amount of information is sufficient to construct a non-erroneous distribution of probabilities and in presence of random uncertainty. This is the case, for example, in feedback, data readings, statistical measures, etc. It is found on a precise and well defined axiomatic [JAY 03]. In a frame of discernment $\Omega = \{H_1, \ldots, H_q\}$, a measure of probability p is associated with each hypothesis H_i such that $p : \Omega \to [0,1]$ under the additivity constraint $\sum_{H_i} p(H_i) = 1$. The hypotheses H_i are mutually exclusive and exhaustive. When the source of information is an expert, he/she can give his/her opinion by choosing a hypothesis with $p(H_i) = 1$, which marks his/her certainty with the occurrence of the hypothesis H_i. He/she can allocate a probability $p(H_i) < 1$ indicating his/her uncertainty with the occurrence of this hypothesis. The remaining probability quantity $1 - p(H_i)$ is uniformly distributed on $H_{i|j \neq i}$. Even if the expert does not then allocate a belief to H_j, a belief mass is assigned to these hypotheses. The expert can also choose several hypotheses by allocating $p(\{H_j \ldots H_k\}) = 1$, signifying that the occurrence is certain in one of these hypotheses. The probability is nevertheless uniformly distributed between the different hypotheses $H_j \ldots H_k$. It is a representation of partial ignorance.

Incompleteness cannot be analyzed since a basic hypothesis is that the state of affairs is exhaustive. Partial or total ignorance relies on the principle of maximum entropy [LEV 78] or even the principle of indifference (or sufficient reason) [KEY 21]. It is obtained by uniformly distributing the probabilities among hypotheses H_i. There is therefore no difference between objective equiprobability and ignorance in the epistemic sense. Every other form of distribution involves a supply of information (and not ignorance). Distribution should not be assumed in the absence of information. Lastly, imprecision cannot be modeled. An expert must allocate a measure to each hypothesis H_i.

14.4.2.2. *Interval theory*

The interval theory proposed by Moore [MOO 66] examines the problem of precision. This consists of a set approach postulating that an interval of values contains the right value. As such, when reexamining the probabilistic framework, it is possible that $p(H_i) \in [p_*(H_i), p^*(H_i)]$. This interval can be seen as a set of possible values or bounds. In the latter case, it will be manipulated by *ad hoc* arithmetic [MOO 66, BUC 05]. This arithmetic corresponds to the extension of conventional arithmetic operators. When the functions are monotonically increasing or decreasing, this type of arithmetic rarely poses a calculation problem. When the functions are not monotone or nonlinear, however, the difficulty of this calculation becomes considerable [JAU 01].

The interval approach allows us to focus on an imprecision problem alone by allocating probability measures. An expert can therefore allocate interval probability measures for one or more hypotheses by assuring that the real value is in the interval provided. The rest of the probabilistic framework is then applied.

14.4.2.3. *Possibility theory*

Possibility theory was proposed by Zadeh [ZAD 78] following the introduction of fuzzy sets to manipulate "non-probabilistic" uncertainty in order to represent incomplete, imprecise or vague information. Since then it has been further developed by Dubois and Prade [DUB 85, DUB 88]. This is a relatively recent method because while it was introduced in the 80s, its roots lie in research undertaken by Shackle [SHA 52]. This consists of reasoning around nested intervals by giving them degrees of confidence. A distribution of possibilities can therefore be seen as a set of nested confidence intervals of the values of the uncertain parameter to which degrees of confidence are assigned (confidence assigned to measuring devices or expert opinion).

Contrary to the probabilistic framework, where knowledge of an event indicates knowledge to the contrary, possibility theory does not impose this constraint. As such, an event may be possible as may its opposite counterpart. When measuring the possibility that defines the degree to which an event may occur without statistical knowledge, a necessity measure allows us to evaluate its credibility. It is possible to work with intervals to manage the problem of imprecision and, with fuzzy sets or linguistic forms, tackle the problem of vague information.

14.4.2.4. *Evidence theory*

Evidence theory was proposed by Dempster [DEM 67] and further developed by Shafer [SHA 76]. It has a number of interpretations, such as upper and lower probabilities, the hint model of Kholas and Monney's hint model [KHO 95] or Smets' transferable belief model [SME 94]. In its initial form, evidence theory uses two dual measures called a belief measure (*Bel*) and a measure of plausibility (*Pls*). $Bel(H_i)$ characterizes the degree of credibility for hypothesis H_i. $Pls(H_i)$ characterizes the amount of belief that does not contradict hypothesis H_i. As with probabilities, in evidence theory there is an assignment function m where: $m: 2^{\Omega} \rightarrow [0,1]$ and $\sum_{i=1}^{2^q} m(A_i)$. This function is applied to all subsets of hypotheses A_i in 2^{Ω}, which is called the power-set. The subsets A_i having a non-empty mass are called focal sets, i.e. the subsets on which the knowledge source focuses.

Knowledge of m is equivalent to that of Bel and Pls and there is a relationship between each of these measures. A belief mass allocated to a subset is not distributed across each of its elementary hypotheses.

A knowledge source can therefore express partial or total ignorance by allocating a belief mass to a subset of hypotheses or to Ω. If, in probability theory, the maximal entropy principle is the basis of modeling ignorance, in evidence theory it is the principle of minimal commitment that only allocates a belief mass to the subsets about which the source has information in the absence of being able to allocate more specific subsets. The remaining mass is therefore allocated to the set that requires the smallest amount of information, i.e. Ω. An allocation of mass on hypotheses H_i alone is the same as the probabilistic case.

In their initial form, the elementary hypotheses are exhaustive and exclusive so we have $m(\varnothing)=0$. This consists of a closed-world assumption where all the possible states are known. This is nevertheless not a constraint because it is possible to allocate a mass to $\varnothing$ characterizing an unknown situation. This is therefore an open-world assumption.

As is the case with possibility theory, knowledge about a hypothesis does not require knowledge of its opposite case based on the principle of minimal commitment. Evidence theory is a combination of set and probability theories. It converges with the latter when knowledge is Bayesian but can be used to better model partial or total ignorance. As such, evidence theory allows us to deal with imprecise information based on the *Bel* and *Pls* distributions. Exhaustivity of the hypotheses is not a constraint that allows us to analyze conflicts or inconsistencies in information, while exclusivity does. Furthermore, evidence theory does not restrict the modeling of consonant information and as such is therefore different from possibility theory.

14.4.2.5. *Dezert-Smarandache theory*

Evidence theory hypothesizes that a reasoning framework Ω is composed of exclusive and exhaustive hypotheses based on the excluded middle law. Dezert-Smarandache theory [SMA 09] challenges this framework in order to represent vague hypotheses with fuzzy limits between and model real problems by defining hybrid models authorizing partial intersections between hypotheses. The model is therefore less restrictive than evidence theory. The power-set in evidence theory is replaced by the hyper-power set D^{Ω}, which can be defined as the set of all the propositions possible for the elements 2^{Ω} constructed using conjunctive and disjunctive operators. The mass allocation principle is preserved but generalized as the hyper-power set $m: D^{\Omega} \rightarrow [0,1]$ and the belief and plausibility measures are also extended to this new framework.

This theory offers a much greater degree of freedom than the Dempster-Shafer model since this is only a specific case where intersections and unions of hypotheses cannot support belief masses. This freedom allows us to have much more realistic knowledge coding and enables us to cover other specific theories while it demands much more information and has a greater degree of complexity.

14.4.2.6. *Imprecise probability theory*

Imprecise probability theory is a highly general representation model proposed by Walley [WAL 91]. Its general nature allows us to cover the majority of models examined previously. It appears to be fairly natural to model knowledge that is both random and epistemic using a probability distribution family. The random view is represented by the distribution of probabilities and the epistemic view by introducing intervals. As such, imprecise probabilities can be interpreted as an extension of interval theory to a pair of distributions. The interval formed by this distribution pair is called a p-box [FER 02] and gives an incomplete picture of the information. The p-boxes can be constructed using known distributions and therefore appear to be a natural choice for parametric models with imprecise parameters, for example a Gaussian model where the average and/or variance are located in a prescribed interval. This can result in a thin closed p-box (even if the latter contains non-Gaussian distributions). It is also possible to directly formulate a p-box by defining the distributions of upper and lower bounds.

14.5. Modeling risk for a quantitative assessment of risk

14.5.1. *Bayesian networks*

Once a methodological framework has been identified, we need to implement a multidisciplinary risk analysis methodology such as IRA to propose a ‘tooled platform’ to enable evaluation of the risks being considered. The choice of platform focuses on the Bayesian network formalism. Léger [LEG 09a] has proposed a comparative study of several modeling tools that could respond to the problem of unifying knowledge to implement the IRA. Different tools, such as stochastic Petri nets, Markov chains and Bayesian networks [JEN 02], have been studied.

The final choice of a modeling framework based on Bayesian network formalism can be explained by both the advantages of this formalism and the limitations or lack of other modeling tools studied. Bayesian networks allow us to model multimodal, quantitative or qualitative variables that could have conditional dependences (notably via the notions of ‘nodes’ and ‘arcs’) and to carry out simulations and diagnostics on the systems studied. This formalism is increasingly being used to carry out maintenance, reliability and risk analysis studies. In a wide ranging literature review, [WEB 12] has examined the development of Bayesian networks in the period from 2001–2007 in which research on such networks has increased four-fold (of which 23% centers on risk analysis). The significance of Bayesian networks has also been highlighted in recent years in relation to other tools, such as fault trees, stochastic Petri nets or even Markov chains. Another benefit of Bayesian networks is the ease in modeling and analyzing complex systems such as those examined using IRA.

14.5.2. *Evaluating risk beyond a probabilistic framework*

Another important phase in the risk analysis process is the assessment phase. As noted in section 14.3 of this chapter, one of the original aims of the IRA was to handle both quantitative and qualitative data. Estimating technical or environmental risk is usually carried out using feedback analysis and human–organizational risk analysis is generally carried out by gathering expert opinions. During initial IRAs on different industrial systems, one of the main limitations that was highlighted was the difficulty in getting a qualitative value from expert opinion.

The probabilistic framework (used in Bayesian networks) therefore has certain limitations for 'consistently' analyzing and using expert advice. This is because experts often prefer to use expressions of preference, belief and order, etc., rather than unique probability values. This can be explained by the different types of knowledge that they can contain (see section 14.4.1). There are two possible solutions to this problem: restrict experts to expressing a single probabilistic value (at the risk of 'distorting' their knowledge) or proposing an alternative that allows them to fully represent the knowledge that they want to express. The different frameworks for modeling uncertainty examined in section 14.4.2 (interval theory, possibility theory and evidence theory) are the result of this.

Various pieces of research carried out in recent years examined in [FAL 10, FAL 11, SIM 09a] investigate other frameworks for modeling uncertainty with regards to the risk assessment examined in the IRA methodology. The main points of interest taken from this initial research include the ability of the different uncertainty modeling frameworks examined in this chapter to model different types of information in the IRA thanks to the adaptation of their basic concepts as well as the results obtained depending on the uncertainty modeling framework used. If, in the case of an expression of hard or soft evidence, the different modeling frameworks provide similar results ("scalar" values) in the case of epistemic uncertainty, the modeling frameworks either provide 'scalar' values or intervals whose size represents the uncertainty introduced by frameworks using "scalar" values. As such, this runs the risk of not "faithfully" representing expert knowledge, depending on the uncertainty modeling framework chosen. As such, expert knowledge should not be restricted (whether this concerns the modeling framework or the proposed elicitation grids) but should consistently represent the different types of knowledge that experts can provide.

Research is currently being led into the sensitivity of elicitation grids used in the IRA and the implementation of other uncertainty modeling frameworks in the IRA methodology in a real industrial case study.

In terms of validating the construction of the risk model itself, its representation and applicability to real-life situations is also an important scalar because current research focuses on handling uncertainty and the final aim of a model, such as the IRA, is to aid risk management, notably with regards to the safety and integrity of the systems being studied. It is for this reason that current research has focused on this aspect via an IMdR[7] project into the validation and representativity of a Bayesian network in risk analysis and reliability.

14.6. Conclusions and future perspectives

On the basis of the work carried out in [LEG 09a] seeking to develop the IRA methodology, the MARATHON project, which is scheduled to be completed at the end of 2012, principally aims to extend this IRA methodology to other areas of application and approach problems of epistemic and random uncertainty inherent to the model obtained using this methodology. In terms of results, this project has led to a bibliographic review of accounting for and characterizing uncertainty for different types of risk and, more specifically, those supported in the IRA (based on the industrial example used by EDF). Several tests of uncertainty modeling frameworks have been applied to parts of the integrated risk model in this industrial example. A subsequent study is being carried out into the problem of modeling uncertainty using Dempster-Shafer theory within evidential networks on noisy-OR and leaky-noisy-OR gates that are used in integrated models [SIM 09a, SIM 09b]. Lastly, research in this field aims to select a more integrated than interoperable modeling solution within the IRA framework.

In terms of ongoing research in the MARATHON project, attention is focusing on the issue of risk assessment (i.e. of different kinds, expert opinion, etc.), and on proposing tools to do this. These suggestions have subsequently been validated using the complete IRA model in the industrial example examined with the aim of evaluating the sensitivity of the results obtained to the input parameters.

In time, the MARATHON project will also integrate the results of the SOMAIR project, aiming to integrate the problem of the individual versus the collective into the part examining human factors as initially envisaged.

14.7. Bibliography

[AIE 91] AIEA, Collection sécurité no. 75-INSAG-4, AIEA, 1991.

[AND 01] ANDERSEN H. *et al.*, The ARAMIS User Guide, European Commission, 2001.

---

7 www.imdr.eu

[BAE 04] BAE H.-R., GRANDHI R.V., CANFIELD R.A., "Epistemic uncertainty quantification techniques including evidence theory for large-scale structures", *Computers & Structures*, vol. 82, no. 13-14, pp. 1101-1112, 2004.

[BAR 01] BARTHELEMY, F., HORNUS H., ROUSSOT J., HUFSCHMITT J.-P. and RAFFOUX J.-F., Report of the General Inspectorate for the Environment: Accident on 21st September 2001 at a factory belonging to the Grande Paroisse Company in Toulouse, 2001.

[BIE 98] BIEDER C. *et al.*, "MERMOS: EDF's new advanced HRA method", in: *Probabilistic Safety Assessment and Management, PSAM 2004*, New York, USA, Springer-Verlag, London, pp. 129-134, 1998.

[BON 03] BONACA M., Letter to Diaz L., Proposed Resolution of Public Comments on Draft Regulatory Guide (DG)-1122, An Approach for Determining the Technical Adequacy of Probabilistic Risk Results for Risk-Informed Decision Making, 2003.

[BOU 95] BOUCHON-MEUNIER B., "Advances in fuzzy systems: applications and theory", *Fuzzy Logic and Soft Computing*, Addison-Wesley, Paris, 1995.

[BUC 05] BUCKLEY J.J., *Fuzzy Probabilities: New Approach and Applications*, Springer, Berlin, 2005.

[CHE 05] CHEN D., DASSISTI M., TSALGATIDOU A., *Interoperability Knowledge Corpus*, European Commission, 2005.

[CHE 06] CHEVREAU F.R., WYBO J.L., CAUCHOIS D., "Organizing learning processes on risks by using the bow-tie representation", *Journal of Hazardous Materials*, vol. 130, no. 3, pp. 276-283, 2006.

[DEM 67] DEMPSTER A.P., "Upper and lower probabilities induced by a multivalued mapping", *Annals of Mathematical Statistics*, vol. 38, pp. 325-339, 1967.

[DER 08] DE ROCQUIGNY E. *et al.* (eds), *Uncertainty in Industrial Practice: a Guide to Quantitative Uncertainty Management*, John Wiley & Sons, 2008.

[DIE 04] DIEN Y., LLORY M., MONTMAYEUL R., "Organisational accidents investigation methodology and lessons learned", *Journal of Hazardous Materials*, vol. 111, no. 1-3, pp. 147-153, 2004.

[DRO 07] DROUIN M. *et al.*, *Guidance on the Treatment of Uncertainties Associated with PRAs in Risk-Informed Decision Making*, NRC, USA, 2007.

[DUB 85] DUBOIS D., PRADE H., "Evidence measure based on fuzzy information", *Automatica*, vol. 21, no. 5, pp. 547-562, 1985.

[DUB 88] DUBOIS D., PRADE H., *Possibility Theory. An Approach to Computerized Processing of Uncertainty*, Plenum Press, New York, 1988.

[DUT 97] DUTUIT Y. *et al.*, "Dependability modelling and evaluation by using stochastic Petri nets: application to two test cases", *Reliability Engineering and System Safety*, vol. 55, pp. 117-124, 1997.

[DUV 07] DUVAL C. *et al.*, "Choice of a risk analysis method for complex socio-technical systems", *Safety and Reliability Conference*, ESREL'07, Stavanger, 2007.

[EIF 04] European Interoperability Framework for pan-European eGovernment Services, Technical Report, European Communities, November 2004.

[FAL 10] FALLET G. *et al.*, "Characterization and propagation of uncertainties in complex socio-technical system risk analyses", *Workshop on Theory of Belief Functions Workshop on Theory of Belief Functions*, Brest, France, 2010.

[FAL 11] FALLET G. *et al.*, "Expert judgment collecting and modeling: Application to the Integrated Risks Analysis (IRA) methodology", *DCDS 2011 3rd International Workshop on Dependable Control of Discrete Systems*, Saarbrücken, Germany, 2011.

[FAY 99] FAYOL H., *Administration Industrielle et Générale*, 2nd edition, Dunod, Paris, 1999.

[FER 96] FERSON S., GINZBURG, L., "Different methods are needed to propaport ignorance and variability", *Reliability Engineering & System Safety*, vol. 54, pp. 133-144, 1996.

[FER 02] FERSON S. *et al.*, Constructing Probability Boxes and Dempster-Shafer Structures, Sandia National Laboratory, Technical Report SAND2002-4015, SNL Albuquerque, New Mexico, USA, 2002.

[GER 99] GERAM, GERAM: Generalised Enterprise Reference Architecture and Methodology, version 1.6.3, IFIP-IFAC Task Force on Architectures for Enterprise Integration , 1999.

[HAM 10] HAMELIN R., WALDEN D., KRUEGER M., INCOSE Systems Engineering Handbook v3.2: Improving the process for SE Practioners, INCOSE, 2010

[IEE 90] INSTITUTE OF ELECTRICAL AND ELECTRONICS ENGINEERS, *IEEE Standard Computer Dictionary: A Compilation of IEEE Standard Computer Glossaries*. IEEE, New York, NY, 1990.

[IEE 00] INSTITUTE OF ELECTRICAL AND ELECTRONICS ENGINEERS, *Standards Information Network, IEEE 100, The Authoritative Dictionary of IEEE Standard Terms*, IEEE, New York, 2000.

[JAU 01] JAULIN L. *et al.*, *Applied Interval Analysis with Examples in Parameter and State Assessment, Robust Control and Robotics*, Springer, London, 2001.

[JAY 03] JAYNES E., *Probability Theory: The Logic of Science*, Cambridge University Press, Cambridge, 2003.

[JEN 02] JENSEN F.V., *Bayesian Networks and Decision Graphs Corrected*, Springer, New York, 2002.

[KEM 79] KEMENY J.G., *Report of the President's Commission on the Accident at Three Mile Island*, Pergamon Press, New York, 1979.

[KEY 21] KEYNES J.M., "The principle of indifference", *A Treatise on Probability*, MacMillan and Co, London, 1921.

[KHO 95] KHOLAS J., MONNEY P., *A Mathematical Theory of Hints: An approach to Dempster-Shafer Theory of Evidence*, Springer, Berlin-New York, 1995.

[LEC 01] LE COZE J., Organizational Performance in High-risk Industries – State of the Art, Development of ATOS Concept, Master thesis, Ecole des Mines d'Alès, 2001.

[LEC 02] LE COZE J. *et al.*, "Development of the ATOS concept, analysis of technical and organizational safety", *ESREL 200,* Lyon, France, 2002.

[LÉG 09a] LÉGER A. *et al.*, "Methodological developments for probabilistic risk analyses of socio-technical systems", *Proceedings of the Institution of Mechanical Engineers, Part O: Journal of Risk and Reliability*, vol. 223, pp. 313-332, 2009.

[LÉG 09b] LÉGER A., Contribution à la formalisation unifiée des connaissances fonctionnelles et organisationnelles d'un système industriel en vue d'une évaluation quantitative des risques et de l'impact des barrières envisages, PhD thesis, Henri Poincaré University, Nancy 1, 2009.

[LEM 99] LE MOIGNE J.-L., *La Modélisation des Systèmes Complexes*, Dunod, Paris, 1999.

[LEV 78] LEVINE R.D., TRIBUS M., *The Maximum Entropy Formalism*, MIT Press, Cambridge, 1978.

[LUZ 10] LUZEAUX D., RUAULT J.-R., *Systems of Systems*, ISTE Ltd, London, John Wiley & Sons, New York, 2010.

[MOH 09] MOHAGHEGH Z., KAZEMI R., MOSLEH A., "Incorporating organizational factors into Probabilistic Risk Assessment (PRA) of complex socio-technical systems: A hybrid technique formalization", *Reliability Engineering & System Safety*, vol. 94, no. 5, pp. 1000-1018, 2009.

[MOO 66] MOORE R., *Interval Analysis*, Prentice-Hall, New York, 1966.

[NRC 07] U.S. NUCLEAR REGULATORY COMMISSION, *An Approach for Determining the Technical Adequacy of Probabilistic Risk Assessment Results for Risk-Informed Activities*, U.S. Nuclear Regulatory Commission 2007.

[OBE 03] OBERKAMPF W. *et al.*, "Challenge problems: Uncertainty in system response given uncertain parameters", *Reliability Engineering & System Safety*, vol. 85, no. 1-3, pp. 11-19, 2004.

[OSE 03] OSEI-BRYSON K.-M., "Supporting knowledge elicitation and consensus building for Dempster-Shafer decision models', *International Journal of Intelligent Systems*, vol. 18, no. 1, pp. 129-148, 2003.

[PAP 03] PAPAZOGLOU I.A. *et al.*, "I-Risk: development of an integrated technical and management risk methodology for chemical installations", *Journal of Loss Prevention in the Process Industries*, vol. 16, no. 6, pp. 575-591, 2003.

[PAT 96] PATÉ-CORNELL M.E., MURPHY D.M., "Human and management factors in probabilistic risk analysis: the SAM approach and observations from recent applications", *Reliability Engineering & System Safety*, vol. 53, no. 2, pp. 115-126, 1996.

[PES 08] PESME H., LE BOT P., MEYER P., "HRA insights from international empirical study in 2007: the EDF scalar of view", *PSAM 2009*, Hong Kong, China, 2008.

[PIE 07] PIERLOT S., DIEN Y., LLORY M., "From organisational factors for an organisational diagnosis of the safety", *Safety and Reliability Conference*, ESREL 2007, Stavanger, Springer, Berlin-Heidelberg, pp. 1329-1325, 2007.

[PLO 04] PLOT E., Guide de la prise en compte du facteur humain dans l'évaluation des systèmes de gestion des risques majeurs, Présentation du référentiel MIRIAM et des protocoles ATHOS, 2004.

[SHA 52] SHACKLE G.L.S., *Expectation in Economics*, Cambridge University Press, Cambridge, 1952.

[SHA 76] SHAFER G., *A Mathematical Theory of Evidence*, Princeton University Press, Princeton, 1976.

[SIM 09a] SIMON C., WEBER P., "Evidential networks for reliability analysis and performance evaluation of systems with imprecise knowledge", *IEEE Transactions on Reliability*, 58, 1, 69-87, 2009.

[SIM 09b] SIMON C., WEBER P., "Imprecise reliability by evidential networks", *Proceedings of the Institution of Mechanical Engineers, Part O: Journal of Risk and Reliability*, 223, 2, 119-131, 2009.

[SMA 09] SMARANDACH F., DEZERT J., *Advances and Applications of DSmT for Information Fusion*, America Research Press, Rehoboth, 2009.

[SME 94] SMETS P., KENNES R., "The transferable belief model", *Artificial Intelligence*, vol. 66, pp. 191-243, 1994.

[SMI 89] SMITHSON M., *Ignorance and Uncertainty: Emerging Paradigms*, Springer, New York, Berlin, 1989.

[SVE 02] SVEDUNG I., RASMUSSEN J., "Graphic representation of accident scenarios: mapping system structure and the causation of accidents", *Safety Science*, vol. 40, no. 5, pp. 397-417, 2002.

[UCC 85] UNION CARBIDE CORPORATION, Bhopal Methyl Isocyanate Incident Investigation Team Report, Danbury, *USA*, UCC, 1985.

[VER 07] VERDANAT F.B., "Interoperable enterprise systems: Principles, concepts, and methods", *Annual Reviews in Control*, vol. 31, no. 1, pp. 137-145, 2007.

[WAL 91] WALLEY P., *Statistical Reasoning with Imprecise Probabilities*, Chapman and Hall, London, 1991.

[WEB 12] WEBER P., MEDINA-OLIVA G., IUNG B., "Overview on Bayesian networks application for dependability, risk analysis and maintenance areas", *Engineering Applications of Articial Intelligence*, 25, 4, 671-682, 2012.

[ZAD 78] ZADEH L., "Fuzzy sets as a basis for a theory of possibility", *Fuzzy Sets and Systems*, vol. 1, pp. 3-28, 1978.

[ZIO 09] ZIO E., "Reliability engineering: old problems and new challenges", *Reliability Engineering & System Safety*, vol. 94, no. 2, pp. 125-141, 2009.

Chapter 15

Integrating Human and Organizational Factors into the BCD Risk Analysis Model: An Influence Diagram-based Approach

15.1. Introduction

With technological advances and the complexity of industrial systems, it is necessary to examine security in these systems also considering their human operators. As such, the study or analysis of risk in industrial systems has developed over recent years. Initial research into risk analysis has focused more on considering technical aspects, but this has resulted in an increase in serious incidents (such as the nuclear disaster in Chernobyl in 1986 and the 1984 Bhopal disaster). This has, therefore, highlighted the need to develop an integrated risk analysis approach considering technical, environmental, human, and organizational aspects together.

In this section, we examine research carried out within the SOMAIR Project[1] that aims to provide a detailed study of human factors in risk analysis based on research (undertaken by project partners EDF Energy and CRAN[2]) into integrated risk analysis [LÉG 09, LÉG 08] and research into analyzing human aspects in a human–machine system, particularly within the framework of the benefit-cost-deficit (BCD) model by another project partner, LAMIH[3] [POL 09, PAC 07, ROB 04, VAN 11]. The different

Chapter written by Karima SEDKI, Philippe POLET and Frédéric VANDERHAEGEN.

1 Human–machine system for integrated risk analysis.

2 Research Center for Automatics Control.

3 Automatics, mechanic, industrial and human technology laboratory.

partners in the project have a common scientific interest, notably in the evaluation of ways of limiting risks, i.e. barriers that are a means of avoiding the occurrence of a serious incident and limiting its consequences [HOL 04, POL 01].

The risk analysis model developed by EDF and CRAN [LÉG 09, LÉG 08] integrates the four technical, environmental, organizational, and human dimensions. A representation of this model is based on a Bayesian network to indicate information relating to each of these dimensions. In terms of the organizational dimension, several factors (known as pathogenic organizational factors (POF) related to the system's organization) have been examined. In terms of the human aspect, several factors (or items) are identified relating to the characteristics of human actions. In terms of the technical aspect, the operation of defense barriers and the causes, which may also include environmental factors, and consequences of serious incidents are studied. This dimension is based on the combination of an inductive and a deductive (node-butterfly) method. When organizational factors are pathogenic, they have a negative impact on the characteristics (items) of human actions, which, in turn, can have negative consequences for their performance. Human actions are ineffective when they fail to influence the operation of one or several barriers. When these barriers are not available (non-operational), this can cause one or several serious incidents. The original aim of the proposed Bayesian network-based representation consists of identifying the occurrence of a serious incident accounting for all the knowledge and factors from each dimension.

The other partner in the project, LAMIH [POL 09, ROB 04], has developed a model called BCD. This model uses the behavior and opinions of human operators when faced with barriers (voluntarily infringing barriers) in order to achieve objectives in productivity, workload, time, etc. Since the purpose of any barrier is to protect the system and the operator, the BCD model also allows us to evaluate costs and risks that may cause barriers to be broken. The BCD model allows us to compare action plans from a multicriteria approach. The benefit corresponds to the improvement between these two plans, while the cost corresponds to loss and the defect corresponds to losses likely to occur in the case of failure (undesirable consequences). The BCD model has also been studied for developing pattern recognition tools (case-based reasoning, neural networks) and allows us to predict the behavior of human operators in relation to barriers. These studies have also been carried out within the context of vehicle behavior as well as guided transport [POL 06]. The BCD model therefore allows us to evaluate the two (human operators and barriers) in terms of effectiveness.

Probabilistic graphic models such as Bayesian networks [JEN 96, NAI 07] and influence diagrams [OLI 90] are often used in risk analysis because they have several advantages: processing data taken from feedback and others provided by experiment opinions, handling simulations and diagnostics in the same model, modeling different types of information (discrete, continuous, symbolic, etc.), representing uncertain and incomplete information, and facilitating the design of a risk analysis model using existing tools such as Netica and Bayesia.

In this section, we propose an approach that allows us to analyze human actions. This approach is based on the BCD model, which integrates human and organizational factors that may affect the actions of human operators and the system. In the proposed approach, each action is evaluated in terms of its benefit and/or risk in the presence of multiple criteria (such as productivity, security, workload, and time). The approach also allows us to analyze the system's operational process (the measures taken in the case of high productivity, for example). To account for all these elements and to design the final model, we will use influence diagrams. This choice is motivated by the fact that influence diagrams contain chance nodes, as in the case of Bayesian networks, and decision and utility nodes. In effect, influence diagrams allow us to classify the different alternatives (i.e. actions or decisions) with the least to the most amount of risk.

The remainder of the chapter is organized as follows: section 15.2 examines the BCD model and section 15.3 presents our own analysis model for human behavior. In section 15.4, we illustrate the proposed model using an example of an industrial printing press. Finally, section 15.5 concludes the chapter.

15.2. Introduction of the BCD (benefit-cost-deficit) approach

Human operators are always present in the running of a system, whether at the design stage or at the operation phase [VIL 92]. Their roles are often challenged by new technologies and productivity and security targets [VAN 03]. Their actions can be considered not only as a source of performance and security [HOL 06], but also as a potential source of error [REA 90], and, paradoxically, the human operator is one of the main causes of the lack of security. For example, as cited in [TRU 08], statistics from the Transportation Safety Board of Canada show that in 74% of cases accidents at sea are due to human factors (such as errors in understanding between the pilot and the captain and the lack of communication).

Therefore, human operators work within a system with intentions, aims, and with their own abilities, intelligence, pressures, tiredness, lack of motivation, etc. All these factors and external pressures can cause deviations with regard to commands and the system's operation. These types of deviations are commonly observed events. For example, in 2003, there were 20 million infractions of the traffic law in France alone (right of way, traffic lights, speed limits, etc.) [DJE 07].

The Benefit-Cost-Deficit (BCD) model [POL 09, PAC 07, POL 01] has been developed to study deviating behavior in human operators when faced with barriers, i.e. ignoring them (*infringing barriers*). The implementation of barriers by a system's human-machine designer protects it from failures and harmful consequences and ensures the system's security and that of the human operator (see Figure 15.1).

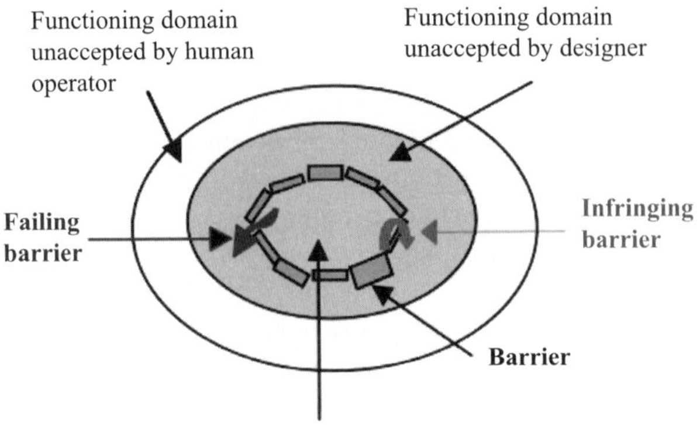

Figure 15.1. *Infringing barriers [POL 02]*

DEFINITION 15.1.– *A barrier is defined as an obstacle, obstruction, or difficulty, which may (1) indicate the execution of an action or the occurrence of a serious event or (2) predict or reduce the impact of its consequences [HOL 99].*

There are different types of barriers:

Material barriers: These are physical barriers that protect the system from hazardous actions by human operators. They do not need to be noticed or interpreted by the human operator to accomplish their purpose, for example metal grids that prevent access to high-tension zones.

Functional barriers: These are barriers that cause an action to be carried out establishing, for example, a logical or temporal dependence in order to activate them. These barriers do not need to be seen by the human operator but require the presence of preconditions that must be verified before obtaining a result, for example autopilot mode in trains.

Symbolic barriers: These are barriers that require interpretion in order for a human operator to react or respond to the messages they contain, for example signs and posters indicating the need to wear a helmet on construction sites.

Immaterial barriers: These are barriers that are not necessarily present or representable in the work environment but need to be recognized by the operator in order to be activated, e.g. operational procedures.

According to [POL 09, PAC 07, POL 01, POL 02], an infringement of a barrier is an intentional behavioral drift whose consequences can be analyzed following three parameters (benefit, cost, and deficit):

– *Immediate benefits (B)*: These represent associated gains (sought by the human operator) at the infringement of the barrier.

– *Immediate costs (C)*: These are acceptable losses for the operator in order to reach anticipated benefits.

– *Potential deficits (D)*: These are unacceptable losses due to a possible failure or due to the lack of respect for a barrier.

An action A of an infringement or respect for barriers when it occurs can lead to two kinds of situations (consequences):

– *Success*: This refers to a situation where the execution of an action is correct and the consequence is written as $CS(A)$. This is the consequence sought by the operator.

– *Failure*: This refers to a situation where the execution of an action is incorrect and does not cause a desired result. This consequence will be written as $CF(A)$.

The consequences of an action correspond to changes in the system's states where the actions are carried out. They can be evaluated according to several criteria: the task's duration, workload, quality, productivity, etc.

When the human operator has to decide whether to respect or ignore a barrier, two actions must be evaluated:

– The prescribed action that corresponds to the respect of the barrier. This action will be denoted by P.

– The action resulting from an infringement of the barrier. This will be written as FB.

To determine the benefits, costs, and deficits [POL 01, POL 02] has compared the consequences of prescribed behavior and deviating behavior. The comparison is as follows:

$B_i(FB)$ (respectively $C_i(FB)$ and $D_i(FB)$) represents the benefits (respectively the costs and deficits) related to the infringement of the barrier for each criterion i:

If $(CS_i(FB) - CS_i(P)) > 0$, then

$B_i(FB) = CS_i(FB) - CS_i(P)$ and $C_i(FB) = 0$

Otherwise, $C_i(FB) = CS_i(P) - CS_i(FB)$ and $B_i(FB) = 0$

If $(CF_i(FB) - CF_i(P)) > 0$, then

$D_i(FB) = CF_i(FB) - CF_i(P)$

Otherwise, $D_i(FB) = 0$.

Figure 15.2 illustrates the calculation of consequences for the infringement of barriers [POL 02].

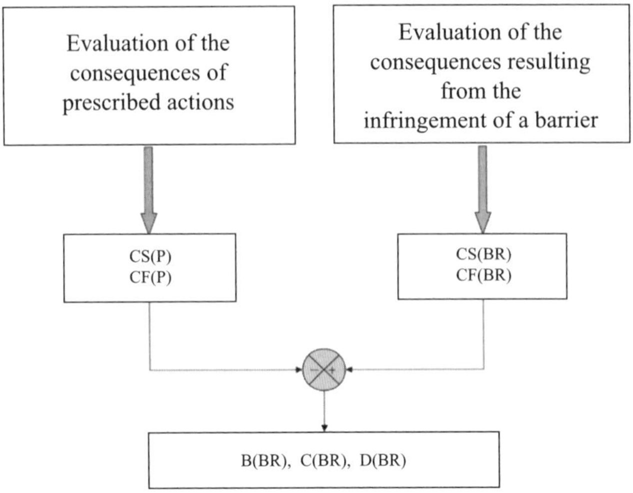

Figure 15.2. *Evaluation of benefits, costs and deficits*

The BCD model has also been used to solve several problems such as the control and analysis of human operator behavior within a system [VAN 10, PAC 07, ROB 04, VAN 03], biomechanical applications [ROB 06], prediction human reactions to barriers, and identification of errors using pattern recognition tools (neural networks and case-based reasoning) [ZHA 04, VAN 09].

EXAMPLE 15.1.– An example that explains the benefits, costs, and deficits likely to be obtained during deviant behavior is an egg [SAV 54]. A user has a bowl with an omelet made using five eggs with a sixth whose state (whether good or bad) is unknown and an empty cup. Suppose that the problem involves a restaurant with specific objective to be achieved: to satisfy clients by assuring a minimal waiting time and a good omelet (these last two criteria considered in this example are time and production). Since there is uncertainty concerning the state of one egg, the prescribed behavior concerning this problem is to first crack the egg into the empty cup (P) to verify whether it is good or bad. Directly cracking the egg into the bowl that already contains five eggs ($FB1$) or throwing it away ($FB2$) constitutes deviating behavior by the user (the user may not respect the prescribed behavior because he/she also has objectives to achieve (avoiding washing a cup, for example)) or due to other pressures. The decision by the user therefore concerns a choice among the three following actions:

1) Crack the sixth egg into the empty cup to check its state (good or bad).

2) Crack the sixth egg directly into the bowl.

3) Throw the egg away without cracking it.

	States of the egg	
Actions	Good egg	Bad egg
Crack the egg (*P*)	Six-egg omelet with a cup to wash	Five-egg omelet with a cup to wash
Crack the egg into the bowl (FB1)	Six-egg omelet	Spoilt omelet
Throw the egg away (FB2)	Five-egg omelet (one egg wasted)	Five egg omelet

Table 15.1. *Example representation of a problem using the BCD model*

Table 15.1 shows the consequences of each action according to its benefit, cost, and deficit.

Concerning this problem, when the prescribed behavior actions are successfully carried out, the consequences of the production concerning a six-egg omelet ($CS\_Prodution = 6$ eggs) and the time required to make the omelet is 10 min ($CS\_Temps = 10$ min). When the infringment (or deviation from the user's prescribed behavior) fails, the consequences of the production are a five-egg omelet ($CF\_Prodution = 5$ eggs) and we suppose that we will not have to remake the omelet ($CF\_Temps = 10$ min). This indicates that, concerning the prescribed behavior P, there are 0 benefits, 0 costs, and 0 deficits in the two cases. However, for the deviant component ($FB1$), when the egg is good, we have a benefit in terms of time (2 min gained if the user cracks the egg directly into the bowl and therefore there is no cup to wash). However, if the egg is bad, this action ($FB1$) means that the user loses the whole omelet (deficit $= -5$ eggs). Concerning the action ($FB2$), when the egg is good, the consequences are a production cost (one good egg lost) and a benefit in terms of time ($=$ 2 min by throwing away the egg). This indicates that deviant behaviors have benefits if their positive consequences outweigh those of prescribed behavior in the case of success (i.e. a benefit if the production involves more than six eggs and the time is less than 10 min) and provide costs and/or deficits if their negative consequences are worse than those of the behavior described in case of failure (i.e. cost and/or benefit if the production concerns fewer than five eggs and the time is less than 10 min).

15.3. Analysis model for human actions

The decision to infringe or respect one or several barriers generally relies on the human operator. His/ her decision is motivated by specific objectives that he/she wants to carry out such as an increase in productivity and time gains. These objectives are related to his/her motivations or personal intentions as well as the organization (where, depending on the system, there are priorities concerning the attained objectives such as security, productivity, and quality). Factors such as these relating to the organization and others relating to human operators must be taken into account in the analysis process for human actions.

	States of the egg	
Actions	Good egg (CS)	Bad egg (CF)
P	$CS\_Prodution = 6$ eggs $CS\_Temps = 10$ min	$CF\_Production = 5$ eggs
$FB1$	$CS\_Prodution = 6$ eggs $CS\_Temps = 8$ min (time benefit)	$CF\_Prodution = -5$ eggs (deficit production)
$FB2$	$CS\_Prodution = 5$ eggs (production) $CS\_Temps = 8$ min (time benefit)	$CF\_Prodution = 5$ eggs

Table 15.2. *Problem of an omelet (consequences in terms of BCD)*

As such, our proposition consists of extending the BCD model by introducing several organizational and human factors influencing operators' actions. We have proposed a model based on the structure of an influence diagram, an extension of a Bayesian network, to represent and analyze the actions of a human operator in a system. The factors considered can be defined according to the probability being studied and the objectives of the system in question. The approach also accounts for the presence of multiple criteria such as productivity, security, and workload. In the proposed model, organizational and human factors that have been defined within the integrated risk analysis developed by the partners of the SOMAIR project (EDF Energy and CRAN) are considered [DUV 08, LÉG 08, LÉG 09]. The final model allows us to calculate the benefit of an action. The benefit value of a prescribed action must be zero (which signifies that if all the barriers are respected by human operators, the consequences in relation to each criterion correspond to zero benefits, costs, or deficits). If the benefit of the deviated action (or infringed barrier) is greater than that of the prescribed action, there is a benefit on one or several criteria; otherwise, there is a risk (deficit). The objective of developing this kind of model lies in the ability to analyze human actions, notably those that concern the influence of different factors affecting the efficacy of actions.

15.3.1. *Accounting for organizational and human factors*

The aim is to consider organizational factors in risk analysis and identify factors likely to affect the actions of human operators and cause negative consequences. These factors can include production, security, etc. In [LÉG 08, LÉG 09], seven organizational factors are defined in order to study the risk analysis in industrial systems. These factors include weakness in the organization's safety culture

(OSC), failure in day-to-day safety management (DSM), weakness of monitoring organizations (MO) (or watchdogs), poor treatment of organizational complexity (OC), difficulty in implementing feedback (FB), production pressures (PP), and failure to reexamine design hypotheses (DH).

Human factors can influence the efficacy of operators' actions. They can include experience, training, lack of respect for regulations, feedback, etc. When an organizational factor has a pathogenic (or dangerous) effect, e.g. when there are production pressures, this can impact the actions of human operators (lack of concentration by users, for example). In the proposed model, we consider the human factors proposed by [LÉG 08, LÉG 09], which are as follows: delegation (De), aids (Ai), training (Tr), experience (Ex), the possibility of respecting guidelines (RG), contextual factors (CF), dynamic management and group collective (DMGC), management and achievement of objectives (MAO), and feedback (Fee).

Before examining the proposed model, we will provide a brief analysis of influence diagrams.

15.3.2. *Influence diagrams*

An influence diagram (ID) [OLI 90] is a directed acyclical graph (DAC) commonly used to model decision problems in the presence of uncertain information. An ID contains three types of nodes. The first is decision nodes that represent the decisions of the problem studied. The second is chance nodes that represent the random variables. If the variables do not rely on any variable, then we introduce *a priori* information (i.e. the *a priori* probability of the state "raining" is less than 0.7). In the opposite case, information is shown as conditional probability tables that are introduced manually or automatically using mathematical, logical, or other functions. Finally, the utility nodes (represented by diamonds) represent the utility value for a decision. Note that the domain of each variable is not always binary (several states can be associated with a variable). Also note that an influence diagram is an extension of a Bayesian network that only contains chance nodes.

The functioning of the influence diagram consists of propagating a set of known information toward a set of desired variables in the designed model. The calculation of *a posteriori* probabilities and utilities (or inference) can be carried out on any set of variables in the model for both observations and important values. For the evaluation (inference) of an ID, there are several algorithms [COO 88, SHA 92].

Figure 15.3 shows an example of a simple ID illustrating the situation of taking or leaving behind an umbrella when going out [JEN 96].

The ID in this example contains two chance nodes (forecasts and weather), with probabilistic information about them. The rain decision is a decision node with two states (take an umbrella and leave the umbrella at home). Satisfaction is a utility node that represents the utility of each decision. The aim is to calculate the desired utility for each decision. The one with the greatest amount of use is considered to be the optimal decision.

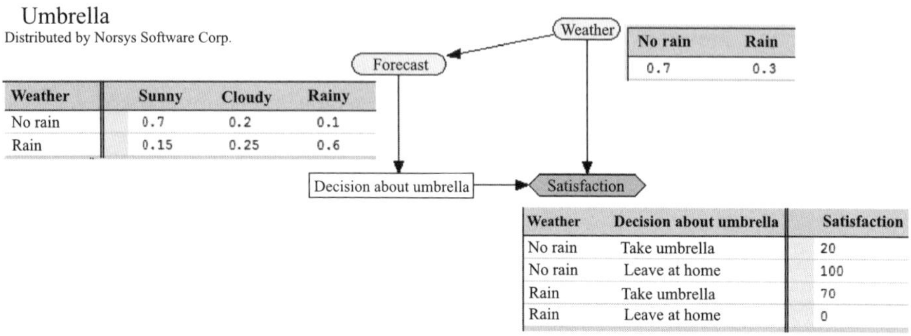

Figure 15.3. *Example of an influence diagram*

15.3.3. *Structure and parameters associated with the risk analysis model*

To model a given problem with a probabilistic graphic model such as an influence diagram or a Bayesian network, we need to define its components: (1) a qualitative component that defines the model's different variables and the relationships between these variables (the graph or model's structure) and (2) a quantitative component that defines the local probabilities or benefits associated with the model's variables (the model's parameters).

Figure 15.4 illustrates the proposed risk analysis model's structure.

The model is composed of several variables that can be described as follows:

– variables representing the decisions to infringe or respect barriers (two states associated with each variable: respected and infringe);

– variables concerning organizational factors;

– variables representing human factors;

– variables indicating the probability of success or failure for human actions (infringement or respect of one or several barriers);

– variables representing different criteria (e.g. quality, productivity, time, and workload that are associated with a weight that shows the significance of each criterion for the organization);

– several variables representing the consequences of infringing barriers in terms of benefits, costs, deficits, and utility.

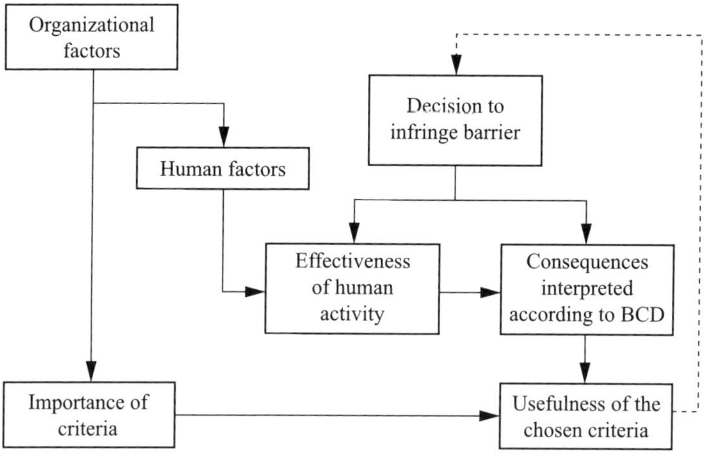

Figure 15.4. *Structure of the proposed risk analysis model*

When designing the risk analysis model, each of the variables examined above corresponds to a node in the influence diagram. The variables representing organizational and human factors, representing the importance of criteria, and associating with the consequences of actions (benefit, cost, deficit) are represented as chance nodes in the influence diagram. Human actions concerning the infringement or respect of barriers are represented by decision nodes. The utility node (U) quantifies the value of each action (infringement of one or several barriers). When the utility value of an action is positive, it signifies that the action carries a benefit; otherwise, the action can be said to be hazardous.

Each organizational factor takes two potential states (present and absent) with a probability for each state, and this is the same for each human factor (present and damaged). The significance of each criterion (symbolized by Imp_i) is evaluated in an interval [0.1] (see Table 15.3).

The variables allowing us to evaluate the consequences for each action are defined as follows:

– *Benefit* (B_i)*, cost* (C_i)*, and deficit* (D_i): These variables are defined in section 15.2 (BCD model). Each decision to infringe or respect a barrier is evaluated in relation to each criterion, i.e. each criterion is allocated with three probability tables (benefit, cost, and deficit). The values associated with benefit, cost, and deficit are in an interval [0.1] (see Table 15.4);

– *Total benefit* (B_t): This variable represents the overall benefit in relation to all the criteria. B_t is calculated as follows:

$$B_t = \sum_{i=1,\ldots,n} (Imp_i * B_i)$$

where Imp_i is the value of importance associated with each criterion i.

– *Total cost* (C_t): this variable represents the overall cost in relation to all the criteria. C_t is calculated as follows:

$$C_t = \sum_{i=1,\ldots,n} (Imp_i * C_i)$$

where Imp_i is the value of importance associated with each criterion i;

– *Total deficit* (D_t): This variable represents the overall deficit by considering all the criteria. D_t is calculated as follows:

$$D_t = \sum_{i=1,\ldots,n} (Imp_i * D_i)$$

– *Utility* (U): This variable represents the risk or benefit of the decision taken (infringement or respect for barriers). U is calculated as follows: $U = B_t - C_t - D_t$.

The values associated with the overall benefit, cost, and deficit are located in one interval [0.1] (see Table 15.4).

Evaluation	Importance criterion (Imp_i)
0–0.2	Not important
0.2–0.4	Not very important
0.4–0.6	Average
0.6–0.8	Important
0.8–1	Very important

Table 15.3. *Values associated with the importance criterion*

Evaluation	Benefit (cost and deficit, respectively)	Overall benefit (cost and deficit, respectively)
0–0.2	Zero	Zero
0.2–0.4	Weak	Weak
0.4–0.6	Average	Average
0.6–1	Heightened	Heightened

Table 15.4. *Values associated with the consequences of human actions*

15.4. Example application

15.4.1. *Description of the case study: industrial printing presses*

In this chapter, we will examine an example application in an industrial printing press [POL 02, VAN 03]. An industrial printing press is generally composed of operational blocks distributed over three levels. At the beginning of each line, a roller feeds the line with paper. It is followed by four printing blocks (black, blue, magenta, and yellow), which create the offset print. The roll of paper is then passed through a drier and a cooler. It is then fed into the folding machine to be cut and folded depending on the type of product or book. Books are then sent to the mock-up room and, if the sample copies are acceptable (from the point of view of the operators), they are sent to the receiver; otherwise, they are discarded. The minimum number of people required to run this dual line is four: two machine operators, a winder, and a receiver. The role of the paper feeder is carried out at the front of the lines and he/she feeds the rollers with paper. In addition, he/she ensures that the process of sticking between the rollers occurs correctly. The receivers stay at the end of the line to stack the paper, ensuring a consistent supply of palettes, etc. With regard to the machine operators, they work in several control posts from where they can manage lines (color level and superposition, speed, etc.)

The procedures set out by the manufacturer are the following:

1) to use appropriate protective equipment (gloves, safety glasses);

2) press the *Print button*;

3) press the *Emergency stop* button;

4) clean all visible surfaces with a sponge and an appropriate cleaning product;

5) dry the surface with a cloth;

6) release the *Emergency stop* button;

7) press the *Maintenance* button;

8) repeat steps 3–7 as required.

After preliminary observations [POL 02, VAN 03] the actions carried out by human operators are as follows:

1) press the *Print* button;

2) continue slow rotation;

3) clean the surface of the blanket using a sponge and a appropriate cleaning product;

4) dry the surface of the blanket;

5) press the *Stop* button.

After these actions are carried out by the human machine operators, the blanket is washed and dried while in rotation. The barriers that are unobserved by the human operators are as follows: not using protective equipment and interfering with the machine while operating at high speed. This reduces the amount of time required to clean the rollers but exposes the human operators to three hazards: crushing of hands, skin irritation, and getting solvent in the eyes.

15.4.2. *Presentation of the model for the test case*

In this section, we examine a modelization of this problem using the influence diagram formalism. We will consider the two barriers (means of protecting and intervening in the machine when stopped) and three evaluation criteria (workload, safety, and time). Figure 15.5 shows the model's structure representing the problem of an industrial printing press.

We consider three organizational factors: production pressures (PP), weakness in organizational safety culture (OSC), and weakness in management (WM). We also introduce the five human factors introduced in section 15.3.1. As the model shows (see Figure 15.5), each human action is split into three phases (preparation, implementation, and closing) and each phase has two states (effective and ineffective). Concerning the human factors influencing the efficacy of human actions, for example, when the factors delegation, training, and aids are impaired, there is a higher probability that the action preparation will be ineffective. Taking into account the two barriers (safety measures and intervention in the machine when stopped), decisions or human actions are represented by the decision node. We have four actions:

1) $B1\_B2$: This action represents the respect for the two barriers by the human operator, i.e. using protective equipment and not interfering with the machine while operating.

2) $NB1\_B2$: This action concerns the infringement of the first barrier (not using protective equipment) and respecting the second (not interfering with the machine while operating).

3) $B1\_NB2$: This action concerns respect for the first barrier and infringement of the second.

4) $NB1\_NB2$: This action concerns the infringement of the two barriers.

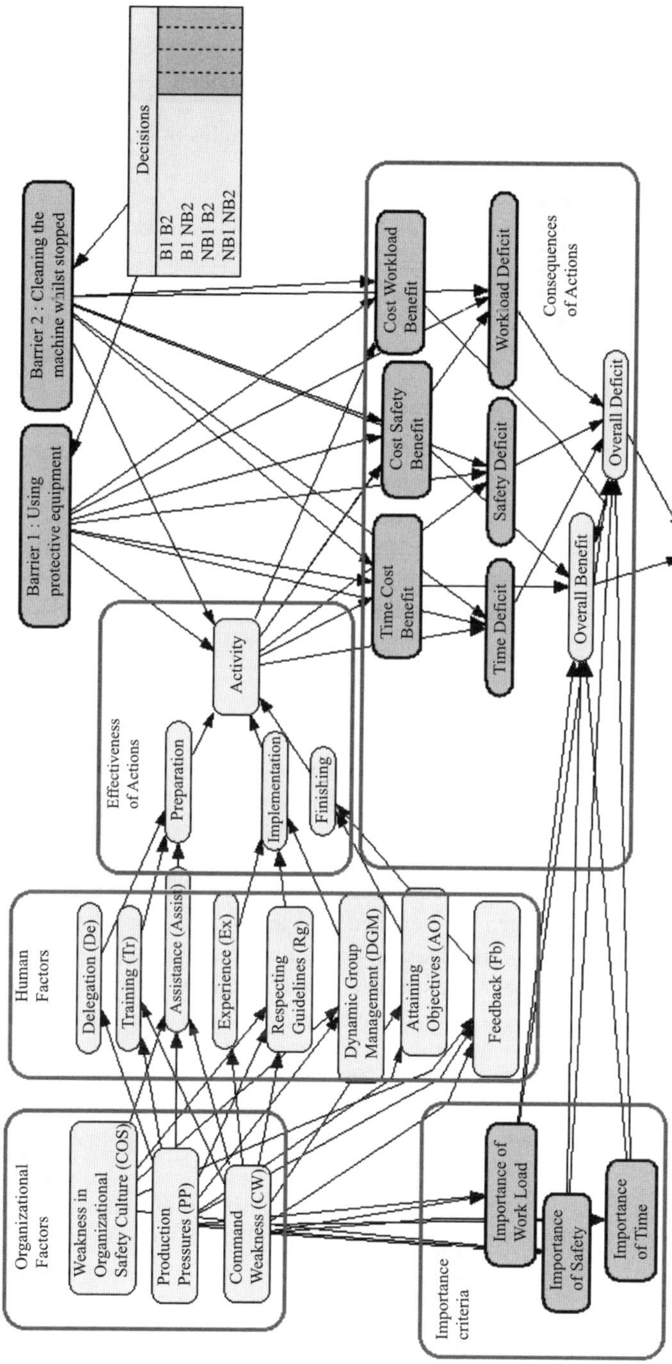

Figure 15.5. *Analysis model for human actions corresponding to the issue of an industrial printing press*

The aim of our model is to calculate the usefulness of each of these actions and determine the most hazardous or the most beneficial action. The advantage is that this allows us to identify factors causing the deficit or benefit.

In the remainder of this chapter, we present some examples of our model in terms of observations and analysis of the results.

Example illustration 1: In this case, we have introduced observations about three organizational factors that we have considered in our example: $COS = Present$, $OC = Present$, and $PP = Present$. This indicates that there is a weakness in the organizational safety culture and in the management of production pressures. The inference application in the designed model allows us to update the probabilities of other variables and calculate the usefulness of each action. In Figure 15.6, we can see that the three actions ($NB1\_B2$, $B1\_NB2$, and $NB1\_NB2$) correspond to infringement of the first barrier, and the second indicating the two barriers that present a risk (negative utility). The actions are ordered according to their degree of risk as follows (action $NB1\_NB2$ with a risk equal to -0.3073, action $B1\_NB2$ with a risk equal to -0.1211, and the action $NB1\_B2$ with a degree of risk equal to -0.0799). This result shows that the infringement of the two barriers is the most hazardous action. This is due to the fact that the majority of human factors are impaired and each of the preparation, implementation, and closing actions are ineffective at high probabilities.

Figure 15.7 shows the consequences of the action $NB1\_NB2$ with the same observations introduced for the first example illustration. The model in this figure shows that this action has a higher impairment of security (with a probability equal to 0.567) and an average deficit for the criteria of time and workload.

Example illustration 2: In this case, we have introduced observations concerning organizational factors where each is associated with the state absent: $COS = Absent$, $OC = Absent$, and $PP = Absent$. This indicates that there is no weakness in the organizational safety culture or in the management of production pressures. We can see in Figure 15.8 that the activity is carried out successfully (with a probability equal to 0.868). As such, the activity where the barriers are successfully infringed, the three actions ($B1\_NB2$, $NB1\_B2$, and $NB1\_NB2$) present a slight benefit, in particular with regard to workload. This is shown in Figure 15.9 concerning the consequences of the action $NB1\_NB2$ with the same actions introduced for this example illustration.

The model in Figure 15.9 shows that the action $NB1\_NB2$ presents a higher benefit for the two criteria, workload and time, but zero benefit and cost for the safety criterion.

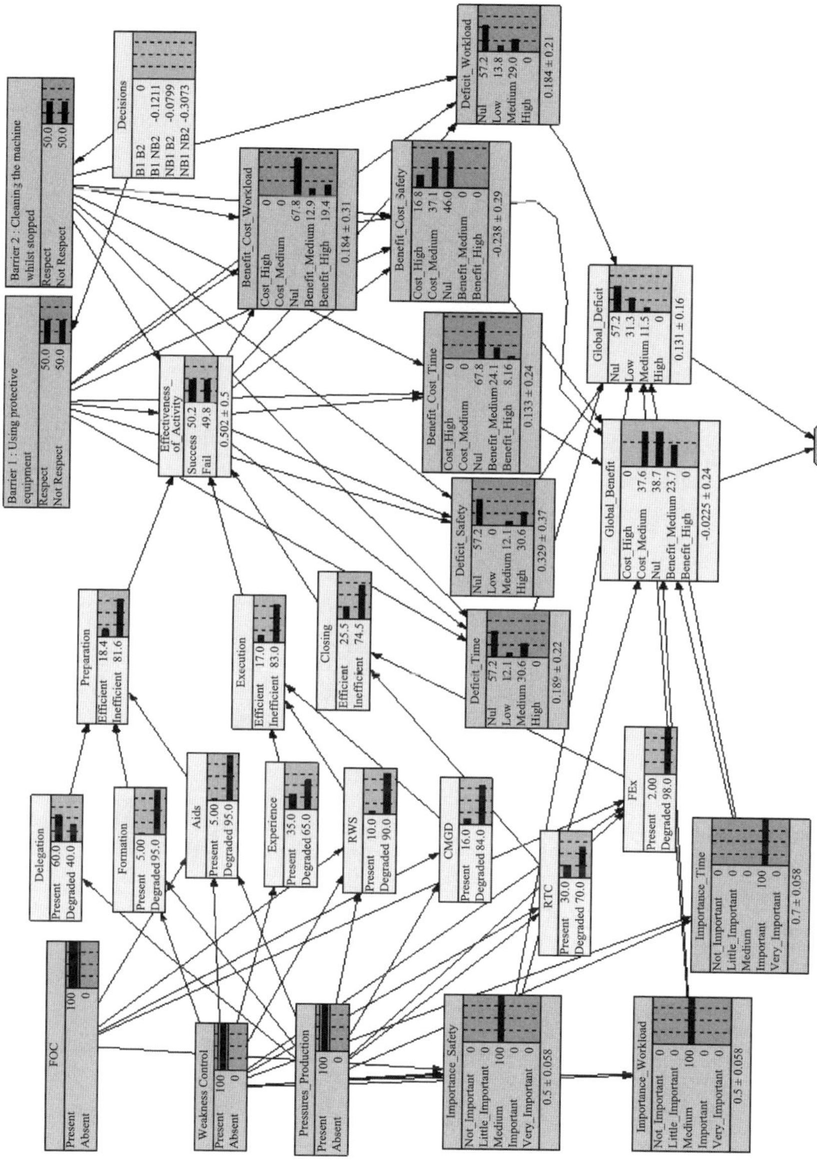

Figure 15.6. *Model corresponding to example illustration 1*

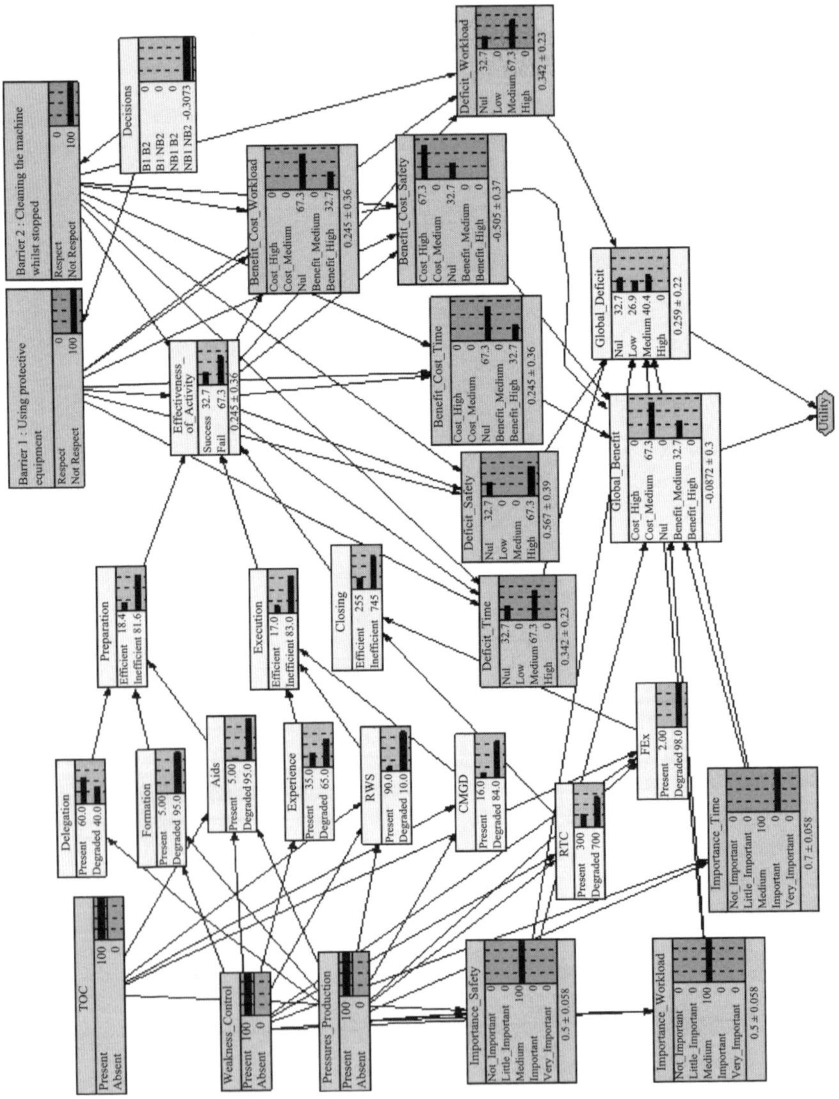

Figure 15.7. *Consequences of the action $NB1\_NB2$ concerning example illustration 1*

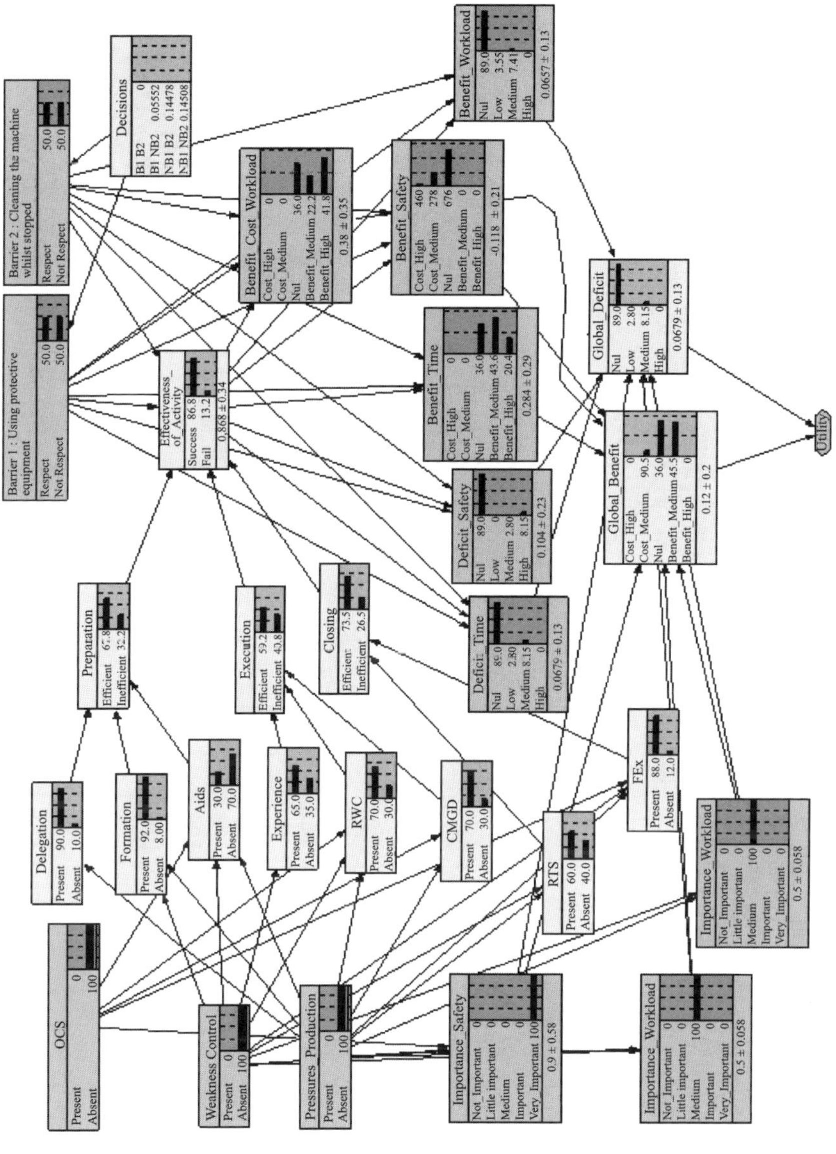

Figure 15.8. *Model corresponding to example illustration 2*

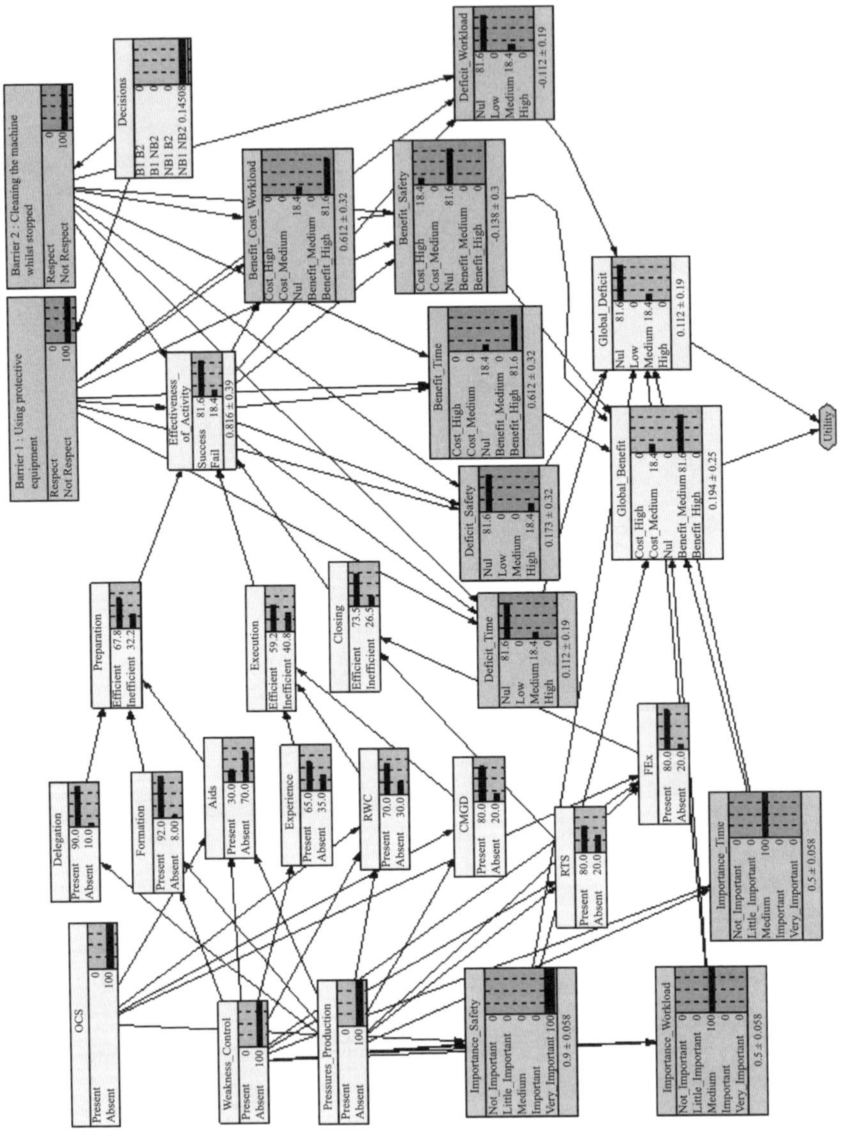

Figure 15.9. *Consequences of the action $NB1\_NB2$ concerning example illustration 2*

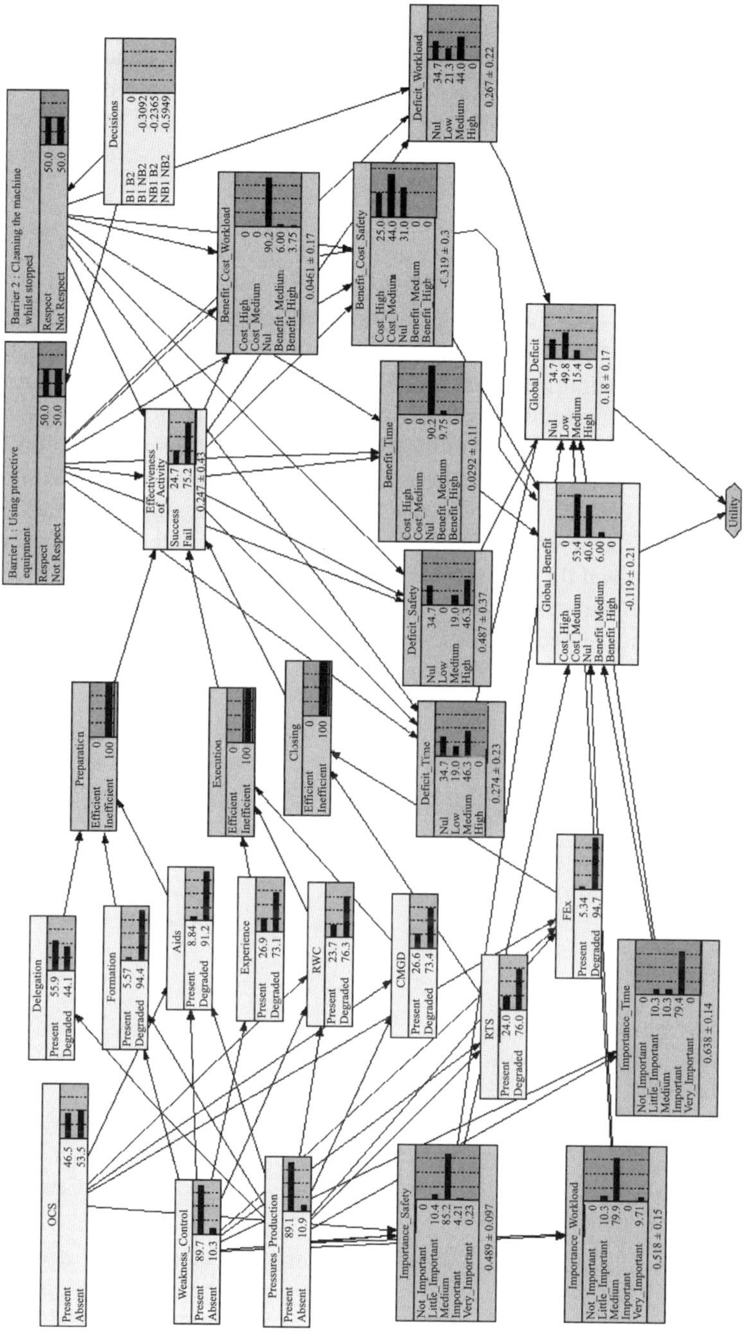

Figure 15.10. *Model corresponding to example illustration 3*

Example illustration 3: In this case, we have introduced observations about the three different stages: preparation, implementation, closing, that correspond to the state "ineffective". We can see in Figure 15.10 that nearly all human factors are impaired and the activity corresponds to the state of failure with a probability equal to 0.752 and, therefore, all the infringements of one or several of the barriers are hazardous.

15.5. Conclusion

In this section, we have proposed an approach for studying and analyzing human actions on the basis of BCD model. The model allows us to study and evaluate the consequences of deviation from prescribed behavior in the system (infringement of barriers). To design the proposed model, we have used a graphic representation using an influence diagram that is a powerful and effective model for modeling decision problems and accounting for uncertain information. We have also examined human and organizational factors. The advantage of this approach is that it allows us to analyze and predict the behavior of human operators and also to revise the prescribed behavior regulations and the barriers implemented in the system examined. We have illustrated our model with a real example of an industrial printing press where human operators often do not respect the prescribed regulations for cleaning blankets.

15.6. Acknowledgments

This chapter is written within the context of the SOMAIR Project that is funded by the Scientific Interest Group "Supervision, Safety and Security of Complex Systems (SIG 3SG)".

15.7. Bibliography

[COO 88] COOPER G., "A method for using belief networks as influence diagrams", *Procceedings of the 12th Conference on Uncertainty in Artificial Intelligence*, Misovalko, pp. 55–63, 1988.

[DJE 07] DJELASSI A.C., Modélisation et prédiction des franchissements de barrières basée sur l'utilité espérée et le renforcement de l'apprentissage. Application à la conduite automobile, Doctoral Thesis, University of Valenciennes and Hainaut-Cambresis of Valenciennes, Valenciennes, 2007.

[DUV 08] DUVAL C., LÉGER A., FARRET R., WEBER P., "Méthodologie d'analyse de risques pour les systèmes socio-techniques complexes et application à un cas industriel", *16e Congrès de Maîtrise des Risques et de Sûreté de Fonctionnement, Lambda Mu 16*, Avignon, France, CDROM, 2008.

[HOL 99] HOLLNAGEL E., "Accident and barriers", *Proceedings of European Conference on Cognitive Science Approaches to Process Control*, Villeneuve d'Ascq, France, pp. 175–180, 1999.

[HOL 04] HOLLNAGEL E., *Barriers and Accident Prevention*, Ashgate, Aldershot, UK, 2004.

[HOL 06] HOLLNAGEL E., WOODS D.D., LEVESON N., *Resilience Engineering: Concepts and Precepts*, Ashgate, Aldershot, UK, New York, 2006.

[JEN 96] JENSEN F.V., *An Introduction to Bayesian Networks*, UCL Press, London, 1996.

[LÉG 08] LÉGER A., FARRET R., DUVAL C., LEVRAT E., WEBER P., LUNG B., "A safety barriers-based approach for the risk analysis of socio-technical systems", *17th IFAC Word Congress*, Seoul, South Korea, 2008.

[LÉG 09] LÉGER A., Contribution à la formalisation unifiée des connaissances fonctionnelles et organisationnelles d'un système industriel en vue d'une évaluation quantitative des risques et de l'impact des barrières envisagées, Doctoral Thesis, University Henri Poincaré, Nancy 1, 2009.

[NAI 07] NAIM P., WUILLEMIN P.-H., LERAY P., POURRET O., BECKER A., *Réseaux Bayésiens*, Eyrolles, Paris, 3rd ed., 2007.

[OLI 90] OLIVER R.M., SMITH J.Q., *Influence Diagrams, Belief Nets and Decision Analysis*, John Wiley, New York, 1990.

[PAC 07] PACAUX-LEMOINE M.-P., VANDERHAEGEN F., "BCD model for human state identification", *Proceedings of the 10th IFAC/IFIP/IFORS/IEA Symposium on Analysis, Design, and Evaluation of Human-Machine Systems*, Seoul, Korea, September 2007.

[POL 01] POLET P., VANDERHAEGEN F., MILLOT P., WIERINGA P., "Barriers and risk analysis", *Proceedings of IFAC/IFIP/IFORS/IEA Symposium on Analysis Design and Evaluation of Human Machine Systems*, Kassel Germany, September 2001.

[POL 02] POLET P., Modélisation des franchissements de barrières pour l'analyse des risques des systèmes homme-machine, Doctoral Thesis, University of Valenciennes and Hainaut-Cambrésis, Valenciennes, 2002.

[POL 06] POLET P., CHAALI-DJELASSI A., VANDERHAEGEN F., "Comparaison de deux méthodes de prédiction des erreurs humaines en conduite automobile", *Actes de ErgoIA 2006 "L'humain comme facteur de performance des systèmes complexes"*, France, pp. 193–200, 2006.

[POL 09] POLET P., VANDERHAEGEN F., MILLOT P., WIERINGA P.A., "Human behaviour analysis of barrier deviation using the benefit-cost-deficit model", *Journal of Advances in Human-Computer Interaction*, pp. 10–19, 2009.

[REA 90] REASON J., *Human Error*, Cambridge University Press, New York, 1990.

[ROB 04] ROBACHE F., MORVAN H., POLET P., PACAUX-LEMOINE M.-P., VANDERHAEGEN F., "The benefit-cost-deficit (BCD) model for human analysis and control", *Proceedings of the 9th IFAC/IFORS/IFIP/IEA symposium on Analysis, Design, and Evaluation of Human-Machine Systems*, Atlanta, GA, September 2004.

[ROB 06] ROBACHE F., MORVAN H., POLET P., PACAUX-LEMOINE M.-P., VANDERHAEGEN F., "The BCD model for biomechanical application", *Proceedings of the 25th European Annual Conference on Human Decision Making and Manual Control (EAM'06)*, ValenSciences, PUV, Valenciennes, September 2006.

[SAV 54] SAVAGE L.J., *The Foundations of Statistics*, John Wiley & Sons, New York, 1954.

[SHA 92] SHACHTER R., PEOT M., "Decision making using probabilistic inference methods", *Procceedings of the 8th Conference on Uncertainty in Artificial Intelligence*, pp. 276–283, 1992.

[TRU 08] TRUCCO P., CAGNO E., GRANDE O., RUGGERI F., GRANDE O.A., "A Bayesian belief network approach modelling of organisational factors in risk analysis: a case study for the maritime transportation", *Reliability Engineering & System Safety*, vol. 93, no. 6, pp. 845–856, 2008.

[VAN 03] VANDERHAEGEN F., *Analyse et contrôle de l'erreur humaine*, Hermès, Paris, France, 2003.

[VAN 09] VANDERHAEGEN F., ZIEBA S., POLET P., "A reinforced iterative formalism to learn from human errors and uncertainty", *Journal of Engineering Applications of Artificial Intelligence*, vol. 22, nos. 4–5, pp. 654–659, 2009.

[VAN 10] VANDERHAEGEN F., "Human-error-based design of barriers and analysis of their uses", *Cognition, Technology and Work* (Special issue in Honor of E. Hollnagel), vol. 12, no. 2, pp. 133–142, 2010.

[VAN 11] VANDERHAEGEN F., ZIEBA S., ENJALBERT S., POLET P., "A benefit/cost/deficit (BCD) model for learning from human errors", *Reliability Engineering and System Safety*, vol. 96, pp. 757–766, 2011.

[VIL 92] VILLEMEUR A., *Reliability, Availability, Maintainability and Safety Assessment, Volume 1, Methods and Techniques*, Wiley, 4 February 1992.

[ZHA 04] ZHANG Z., POLET P., VANDERHAEGEN F., MILLOT P., "Artificial neural network for violation analysis", *Reliability Engineering and System Safety*, vol. 84, pp. 3–18, 2004.

Conclusion

Managing risks during ongoing developments and system safety have become major issues in a number of industrial sectors. Depending on the needs of applications, different attributes of safety – reliability, availability and security – must be considered [AVI 04]. These considerations are both major challenges for industry and interesting challenges for research.

This book covers the main research carried out by academic and industrial partners involved in the Scientific Interest Group: Supervision, Safety and Security of Complex Systems (GIS 3SGS). This project highlights the synergies established through partnership between both companies and research laboratories. In practice, this results in a number of projects aiming to respond to the needs of industrial partners. Contributions and recommendations from members of the scientific community have been decisive in selecting, accompanying and evaluating these projects. Table C.1 lists the members of scientific community involved.

The main needs expressed are the subject of part 1 of this book, which focuses on *industrial issues*.

In the context of new generations of nuclear plants, the CEA has identified improving structural inspectability as an essential factor in equipment safety. In the field of instrumentation and supervision, acoustic detection, while still under development, appears to be a promising approach.

Conclusion written by Jean ARLAT and Nada MATTA.

CS Member	Organization
Jean ARLAT	LAAS–CNRS (president of CS)
Christine BAMIERE	SGDSN
Michèle BASSEVILLE	IRISA
Guy BOY	Institute of Human–Machine Cognition
Stéphane COLLAS	TOTAL
Philippe CORNU	UTT (Director of GIS 3SGS 2007–2009)
Bruno DEBRAY	INERIS
Raymond FOURNIER	CEA
Hubert GARAVEL	INRIA
Sylviane GENTIL	INPG (ex. President of CS)
Hubert GUILLERMAIN	AREVA–TA
Erik HOLLNAGEL	Mines ParisTech
Guy JOIGNAUX	INRETS (IFFSTAR)
Pierre-Etienne LABEAU	The Free University of Brussels
André LANNOY	IMDR
Nadine MARTIN	INPG
Michel ROUSSIGNOL	Univ. Paris-Est Marne La Vallée (Vice president of CS)
Yves VANDENBOOMGAERDE	Scientific director of GIS 3SGS

Table C.1. *List of members in the scientific board for GIS 3SGS*

Within this framework, the long-term management of radioactive waste is another issue that has been identified by ANDRA (French national agency for managing radioactive waste); notably regarding the observation–supervision of research into a deep geological repository.

This clearly highlights that, in addition to aspects of *safety* (as regards accidental faults both externally and internally, including faults in the design of a device), *security* considerations (relating to attacks – intrusions into confidentiality or information integrity, attack on services, inclusion and exploitation of logical malignancies, etc.) are increasingly important for designers, installers and users of critical systems.

Ten of the projects carried out through GIS 3SGS have been the subject of discussion in this book. Table C.2 provides a summary for each of these different projects, which indicates the main activities carried out and the main results as well as the participating partners.

Three main themes are examined in these projects:

– supervising and modeling complex systems (seven projects);

– supervising nuclear reactors by characterizing background noise and identifying characteristic signatures; and

– accounting for human, organizational and environmental factors in risk analysis (two projects).

Within the framework of supervising and modeling complex systems, the FD2S project (Chapter 4) has integrated the concept of fault tolerance into controls in the case of failure or imprecisions in sensors. The aim is therefore to integrate the detection, diagnosis and treatment of anomalies in sensor data. The proposed model is based on a probabilistic approach: Bayesian networks. These techniques have been used to model and locate a vehicle convoy. The main significance of Bayesian networks shown by this research is that this formalism integrates the different dimensions of the problem: fusing data, diagnosis and controlling the learning. The results obtained are refined within the framework of a new project entitled FDADI. The aim is to ensure the exact location of each vehicle in a convoy using open GPS. This consists of approaching the problem with a "tight pair" in comparison to loose pairing that was practiced previously.

Projects	Activities and results	**Partners** *Chapter*
Part 2 – Supervising and modeling complex systems		
FD2S/FDADI: fusing data for the supervision of a complex application system in guided transport systems	Development of fusion approaches for multi-sensor data resistant to sensor failure for control supervision and reconfiguration. Development of a dynamic functional simulator in a systems system. Application to a vehicle convoy.	LAGIS, LORIA, Tecnomade *See Chapter 4*
DEPRADEM: modeling processes assisting maintenance decisions	Quantification, optimization and rationalization of maintenance strategies. Identification of weaknesses and limitations of different approaches.	ICD, CRAN, LAGIS, EDF *See Chapter 5*
COSMOS: design and observation of systems with multiple safety operational modes **DICOP**: diagnostics of power converters, and their application to rail transport	Developing tools and methods for observations and diagnostics (detection, location and identification of faults) in systems with several operational modes.	CresTIC, CRAN, LAGIS, *See Chapter 6*
PARDI: using the "park" effect in diagnosing system operation	A new approach to multitask learning using single class support vector machines (1-SVM). Application to several areas, such as health, environment, etc.	CRAN, ICD, EDF, ANDRA *See Chapter 7*
APPRODYN: dynamic reliability approaches to modeling critical systems	Fault-resistant command: compensating for the influence of faults by synthesizing an adapted control law. Study into its application to industrial-sized systems, particularly risk systems (aviation, chemical reactors, etc.).	EDF, ICD, INRIA, CRAN *See Chapter 8*

Table C.2. *Overview of projects examined in the book*

Projects	**Activities and results**	**Partners** ***Chapters***
Part 2 – Characterizing background noise and identifying characteristic signatures in test cases *Detecting noise inside reactors*		
PFR: research program into the supervision of *prototype fast reactors*. Proposed approach to analyzing acoustic data from steam generators in a PFR	Detailed analysis of spectra and spectrograms highlighting the existence of parasitic rays and frequencies indicating the stable nature of injections. This analysis allows us to propose potential frequency bands for detection and validating mean DSP-based approaches	CEA, ICD, CRAN, CresTIC, LAGIS *See Chapters 9–13*
Part 4 – Human, organizational and environmental factors in risk analysis		
MARATHON: monitoring and analyzing risk in systems with technical, human, organizational and environmental aspects	Characterizing and modeling the uncertain and developing an approach for predicting risks	CRAN, EDF *See Chapter 14*
SOMAIR: "human–machine" system for integrated risk analysis	Analysis and prediction of human behavior as regards security barriers	CRAN, LAMIH, EDF *See Chapter 15*

Table C.2. *(continued) Overview of projects examined in the book*

The DEPRADEM project (Chapter 5) focuses on optimization strategies for aiding maintenance decisions. Two categories of approaches – probabilistic and non-probabilistic – have been examined and combined. The results have allowed us to identify two important aspects of the proposed approach:

– the approach that consists of constructing a failure indicator clearly makes sense; and

– the use of a stochastic approach can be beneficial for the quality of prognosis and prediction. In this case the stochastic process is used to model what can be attributed to a damage phenomenon (a change from the operational state to failure).

The COSMOS and DICOP projects (Chapter 6) have concentrated on developing tools and methods to observe and diagnose systems with several operational modes. Specifically, this chapter focuses on studying dynamic systems whose temporal evolution of state variables are described, either by discrete communications between subsystems or the accumulation of a set of systems sharing all or part of their state variables. The main results examine numerous algorithmic and methodological developments as well as their practical implementation. Simulations are carried out in order to validate the proposed solutions and experimental applications are underway to further refine this validation.

The PARDI project (Chapter 7) studies the role of the "park" effect in the diagnostic process. In effect, the redundancy inherent to a park machine or similar systems may contain much more information and measures relating to the operation or malfunction modes in the systems examined. The proposed approach is based on methods optimizing a decision function within a function class, particularly those based on support vector machines (SVMs). More precisely, this approach introduces the single-class SVM into the multitask learning framework. On the basis of an experimental validation carried out using an artificially generated mono-class data set, the results show that learning several tasks simultaneously increases performance with regards independent learning in each class.

The APPRODYN project (Chapter 8) experiments with different approaches to dynamic reliability within the framework of a probabilistic study into safety in critical control/command systems, notably those used in the energy industry and process manufacturing. This consists of exploring the possibilities created by innovative approaches, still underused in industrial contexts, notably hybrid stochastic automatons (HSAs), by comparing them with more widely used and better equipped approaches, such as stochastic Petri nets, on the basis of their application to a representative case study. The case study consists of part of a secondary circuit in a pressurized water reactor. The main indications concern the fact that the significant expansion of the model's size can be relativized by accounting for the capacity of modern technical tools and applicable simplification methods. Another area of progress involves the development of formal HSA composition tools. When only specific automatons are available, the rest of the model can be automatically obtained without requiring a graphic visualization of the complete automaton.

The project has demonstrated the potential of modeling fairly large hybrid systems fusing a combinatory part with discrete events and a hybrid part across several models. Initial indications show the significant expansion in the model's size can be relativized by accounting for the capacity of modern technical tools and simplification methods that are applicable. Another area of progress concerns the development of formal composition HSA composition tools.

In the section examining the characterization of background noise and identification of characteristic signatures in test cases, the PFR (*prototype fast reactor*) project (Chapter 9) presents a detailed study into acoustic methods for detecting noise inside reactors. With an intrinsically very poor response time, these methods have a strong diagnostic potential. The research has reexamined analyses previously conducted on the basis of data established during the 1990s in test recordings of injections of argon, hydrogen and water into sodium. Several current methods of analyzing signals are used: semi-supervised dynamic fuzzy K nearest neighbors, vast margin supervised classification, power spectral density-based approaches and wavelet decomposition. Using these methods and their accompanying tools, the studies conducted have highlighted progress in signal recognition. The results show that acoustic emissions from injections of water and argon are significantly different, at least around some frequencies. This is conveyed notably by the fact that argon is not an ideal substitute for water when testing and calibrating a 'water–sodium' detection system. They also show that, for the data considered, the power spectral density approach gives better results than wavelet decomposition.

In terms of the section examining the human, organizational and environmental factors in risk analysis, the MARATHON project (Chapter 14) aims to validate the integrated risk analysis approach (IRA) methodology based on developing a unified knowledge model using Bayesian networks. The application of IRA to new fields has highlighted its limitations and potential improvements. It tackles the problems of both epistemic and random uncertainty inherent to the model obtained using this methodology. This study has quantified the influence of uncertainty in the quantitative evaluation of risk, knowing that it takes:

– a random (probabilistic) form for technical or environmental aspects where feedback is quantitatively significant; and

– an epistemic form in the area with a human and organizational factor.

The SOMAIR project (Chapter 15) centers on further examining information taken from the MARATHON project with regards the human factor, emphasizing the role of the individual rather than the collective. The proposed approach for the study and analysis of human actions relies on an original method based on the BCD (benefit, cost, deficit) model, which allows us to study and evaluate the consequences of deviations in the prescribed behavior within a system (infringement of barriers). To describe behavior and construct this type of model, a graphic representation using an influence diagram (which is particularly powerful and effective for modeling decision problems and accounting for uncertain information) has been proposed. This approach is illustrated using a real example of an industrial printing press, showing that human operators do not necessarily always respect prescribed regulations.

The projects that have been supported by GIS 3SGS and the research carried out have allowed us to cover the majority of the various dimensions surrounding the issue of supervision, safety and security in large installations, infrastructures or organizations. These dimensions include rigorous development techniques (verification and tests), mechanisms for online detection, diagnosis, modeling and evaluating aspects of safety and architecture resistant to faults [ARL 06].

The projects examined in this book have explicitly focused on risks that are accidental in nature: failure affecting materials and/or devices and faults in interaction (misunderstandings by operators or users). Research carried out by GIS 3SGS has also examined intentional faults (intrusions and malicious faults). These are threats with which information-processing systems are increasingly faced, including in critical applications, due to their "openness" (the existence of communication channels, whether explicit or implicit) towards the environment (users, strong interactions between systems, etc.), see [VER 06] for example. Another major challenge for such strongly-dependent systems relates to monitoring their continuing evolutions (improving functionality, adapting interfaces, accounting for the mobility of devices, adapting to new threats, etc.). With this in mind, the notion of *resilience* [LAP 08], introduced and developed within the framework of the excellence network in the 6th PCRD ReSIST[1], provides an overview that coherently integrates this dimension related to the development in the concept of safety.

As such, it seems evident that the boundaries between safety, supervision and overall security are gradually becoming less clear-cut. Approaches to supervision in technical systems are a subject of increasing interest in order to identify security barriers in sociotechnical systems. Equally, studies into group and organizational behavior are considered in the supervision of technical systems. This overlap between the 21 projects in the GIS 3SGS framework, of which only 12 are examined in this book, is evident.

---

1 *Resilience for Survivability in IST*. For more information, see: www.resist-noe.org.

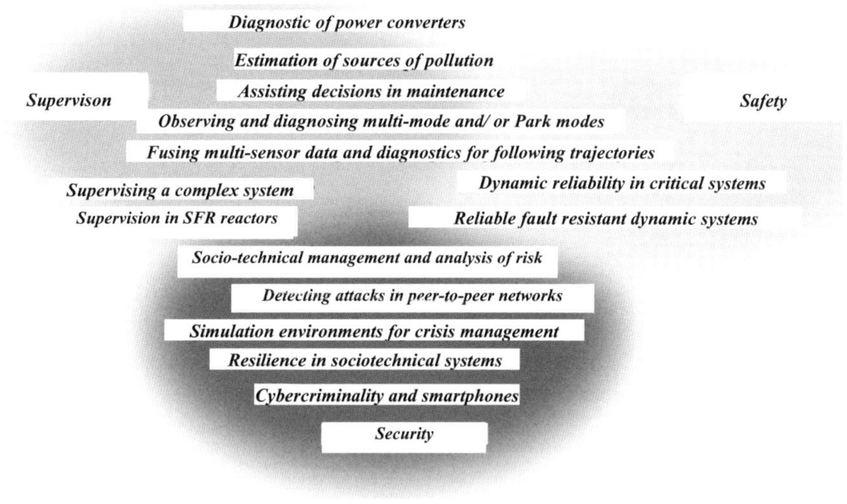

Figure C.1. *The projects carried out within the GIS 3SGS framework*

Bibliography

[AVI 04] AVIŽIENIS A., LAPRIE J.-C., RANDELL B., LANDWEHR C., "Basic concepts and taxonomy of dependable and secure computing", *IEEE Transactions on Dependable and Secure Computing*, vol. 1, no. 1, pp. 11-33, 2004.

[ARL 06] ARLAT J., CROUZET Y., DESWARTE Y., FABRE J.-C., LAPRIE J.-C., POWELL D., " Tolérance aux fautes", in: BANATRE M., (section ed.), AKOKA J. and COMYN-WATTIAU I. (eds), *Encyclopédie de l'informatique et des systèmes d'information*, Part 1: La dimension technologique des systèmes d'information – Section 2: L'architecture et les systèmes, pp. 241-270, Vuibert, Paris, 2006.

[LAP 08] LAPRIE J.-C., "From Dependability to Resilience", *Proceedings of the38th Annual IEEE/IFIP International Conference on Dependable Systems and Networks (DSN-2008) – Fast Abstracts*, Anchorage, AK, USA, 2008.

[VER 06] VERÍSSIMO P., NEVES N., CACHIN C., PORITZ J., DESWARTE Y., POWELL D., STROUD R., WELCH I., "Intrusion-tolerant middleware: The road to automatic security", *IEEE Security & Privacy*, vol. 4, no. 4, pp. 54-62, 2006.

List of Authors

Jean ARLAT
LAAS-CNRS
University of Toulouse
France

Jean-François AUBRY
CRAN
INLP University of Lorraine
Nancy
France

Genia BABYKINA
CRAN
INLP University of Lorraine
Nancy
France

François BAQUÉ
CEA
Cadarache
France

Anne BARROS
STMR
University of Technology of Troyes
France

Mireille BAYART
LAGIS
University of Lille1
Polytech-Lille
France

Pierre BEAUSEROY
STMR
University of Technology of Troyes
France

Djamel Eddine Chouaib BELKHIAT
CReSTIC
University of Reims Champagne-Ardenne
France

Ludovic BÉNÉTRIX
MRI
EDF
Chatou
France

Christophe BERENGUER
STMR
University of Technology of Troyes
France

Eric BESSON
Former French Minister of Industry, Energy and the Digital Economy
France

Taha BOUKHOBZA
CRAN
University of Lorraine
Nancy
France

David BRIE
CRAN
University of Lorraine
Nancy
France

Nicolae BRINZEI
CRAN
INLP University of Lorraine
Nancy
France

Stéphane BUSCHAERT
ANDRA
Châtenay-Malabry
France

Jean-Philippe CASSAR
LAGIS
University of Lille 1
France

François CHARPILLET
INRIA-LORIA
Nancy
France

Vincent COCQUEMPOT
LAGIS
University of Lille 1
France

Michael DEFOORT
LAMIH
University of Valenciennes and Hainaut Cambrésis
France

Gilles DELEUZE
MRI
EDF
Clamart
France

Olivier DESCOMBIN
CEA
Cadarache
France

Antoine DESPUJOLS
MRI
EDF
Chatou
France

Mohammed DJEMAI
LAMIH
University of Valenciennes and Hainaut Cambrésis
France

El-Hadi DJERMOUNE
CRAN
University of Lorraine
Nancy
France

François DUFOUR
INRIA-CQFD
Bordeaux
France

Carole DUVAL
MRI
EDF
Clamart
France

Maan EL BADAOUI EL NAJJAR
LAGIS
University of Lille 1
France

Geoffrey FALLET-FIDRY
MRI
EDF
Clamart
France

Mitra FOULADIRAD
CRAN
University of Lorraine
Nancy
France

Olivier GASTALDI
CEA
Cadarache
France

Antoine GRALL
STMR
University of Technology of Troyes
France

Edith GRALL-MAËS
STMR
University of Technology of Troyes
France

Kevin GUELTON
CReSTIC
University of Reims Champagne-Ardenne
France

Assia HAKEM
LAGIS
University of Lille 1
France

Frédéric HAMELIN
CRAN
University of Lorraine
Nancy
France

Laurent HARTERT
CReSTIC
University of Reims Champagne-Ardenne
France

Xiyan HE
CRAN
University of Lorraine
Nancy
France

Simon HENROT
CRAN
University of Lorraine
Nancy
France

Benoît IUNG
CRAN
University of Lorraine
Nancy
France

Dalel JABRI
CReSTIC
University of Reims Champagne-Ardenne
France

Jean-Philippe JEANNOT
CEA
Cadarache
France

Philippe KLEIN
MRI
EDF
Clamart
France

Yves LANGERON
STMR
University of Technology of Troyes
France

Christian LERMINIAUX
STMR
University of Technology of Troyes
France

Sylvie LESOILLE
ANDRA
Châtenay-Malabry
France

Khanh LE SON
STMR
University of Technology of Troyes
France

Eric LEVRAT
CRAN
University of Lorraine
Nancy
France

Noureddine MANAMANNI
CReSTIC
University of Reims Champagne-Ardenne
France

Didier MAQUIN
CRAN
University of Lorraine
Nancy
France

Sinuhé MARTINEZ
CReSTIC
University of Reims Champagne-Ardenne
France

Benoît MARX
CRAN
University of Lorraine
Nancy
France

Nada MATTA
STMR
University of Technology of Troyes
France

Slimane MEDJAHER
CRAN
INLP University of Lorraine
Nancy
France

Nadhir MESSAI
CReSTIC
University of Reims Champagne-Ardenne
France

Maxime MONNIN
CRAN
University of Lorraine
Nancy
France

Gilles MOUROT
CRAN
University of Lorraine
Nancy
France

Jean-Philippe NABOT
CEA
Saclay
France

Anca Maria NAGY
CRAN
University of Lorraine
Nancy
France

Danh NGOC NGUYEN
STMR
University of Technology of Troyes
France

Igor NIKIFOROV
STMR
University of Technology of Troyes
France

Danielle NUZILLARD
CReSTIC
University of Reims Champagne-Ardenne
France

Kévin PAUMEL
CEA
Cadarache
France

Komi Midzodzi PEKPE
LAGIS
University of Lille 1
France

Philippe POLET
LAMIH
University of Valenciennes and Hainaut Cambrésis
France

Denis POMORSKI
LAGIS
University of Lille 1
France

Jean PRIMET
MRI
EDF
Clamart
France

José RAGOT
CRAN
University of Lorraine
Nancy
France

Emmanuel RÉMY
CRAN
University of Lorraine
Nancy
France

Thomas RUIN
CRAN
University of Lorraine
Nancy
France

Benoîte DE SAPORTA
INRIA-CQFD
Bordeaux
France

Moamar SAYED-MOUCHAWEH
IA
University of Lille 1
Mines Douai
France

Karima SEDKI
LAMIH
University of Valenciennes and Hainaut Cambrésis
France

Christophe SIMON
CRAN
University of Lorraine
Nancy
France

Cherif SMAILI
LAGIS
University of Lille 1
France

André SMOLARZ
STMR
University of Technology of Troyes
France

Frédéric VANDERHAEGEN
LAMIH
University of Valenciennes and Hainaut Cambrésis
France

Yves VANDENBOOMGAERDE
STMR
University of Technology of Troyes
France

Jérémy VAN GORP
LAMIH
University of Valenciennes and Hainaut Cambrésis
France

Alexandre VOISIN
CRAN
University of Lorraine
Nancy
France

Philippe WEBER
CRAN
University of Lorraine
Nancy
France

Huilong ZHANG
INRIA-CQFD
Bordeaux
France

Talel ZOUARI
LAGIS
University of Lille 1
France

Index

H, I

K, L

M

N, O

P

R

S

U, V, W, Y